**Heidelberg
Science
Library**

Heidelberg
Science
Library

I. M. Yaglom

A Simple Non-Euclidean Geometry and Its Physical Basis

Springer-Verlag
New York
Heidelberg
Berlin

An Elementary Account of
Galilean Geometry and the
Galilean Principle of Relativity

Translated from the Russian
by Abe Shenitzer
With the Editorial Assistance
of Basil Gordon

With 227 Figures

I. M. Yaglom
Department of Mathematics
University of Yaroslav
Yaroslav
USSR

Abe Shenitzer
Department of Mathematics
York University
4700 Keele Street
Downsview, Ontario M3J 1P3
Canada

AMS Subject Classifications: 50.01, 50A10

Library of Congress Cataloging in Publication Data
Iaglom, Isaak Moiseevich, 1921–
 A simple non-Euclidean geometry and its physical basis.

 (Heidelberg science library)
 Translation of Printsip otnositel'nosti Galileia i neevklidova geometriia.
 Bibliography: p.
 Includes index.
 1. Geometry, Non-Euclidean. 2. Relativity (Physics)
I. Title. II. Series.
 QA685.I2413   516'.9   78-27788

Title of the Russian original edition: *Printsipi otnositelnosti Galileya i Neevklidova Geometriya*. Nauka, Moscow, 1969.

All rights reserved.

No part of this book may be translated or reproduced in any form without written permission from Springer-Verlag.

© 1979 by Springer-Verlag New York Inc.

Printed in the United States of America.
9 8 7 6 5 4 3 2 1

ISBN 0-387-90332-1 New York Heidelberg Berlin
ISBN 3-540-90332-1 Berlin Heidelberg New York

Preface

There are many technical and popular accounts, both in Russian and in other languages, of the non-Euclidean geometry of Lobachevsky and Bolyai, a few of which are listed in the Bibliography. This geometry, also called hyperbolic geometry, is part of the required subject matter of many mathematics departments in universities and teachers' colleges—a reflection of the view that familiarity with the elements of hyperbolic geometry is a useful part of the background of future high school teachers. Much attention is paid to hyperbolic geometry by school mathematics clubs. Some mathematicians and educators concerned with reform of the high school curriculum believe that the required part of the curriculum should include elements of hyperbolic geometry, and that the optional part of the curriculum should include a topic related to hyperbolic geometry.[1]

The broad interest in hyperbolic geometry is not surprising. This interest has little to do with mathematical and scientific applications of hyperbolic geometry, since the applications (for instance, in the theory of automorphic functions) are rather specialized, and are likely to be encountered by very few of the many students who conscientiously study (and then present to examiners) the definition of parallels in hyperbolic geometry and the special features of configurations of lines in the hyperbolic plane. The principal reason for the interest in hyperbolic geometry is the important fact of "non-uniqueness" of geometry; of the existence of many geometric systems. The non-uniqueness of geometry sheds new light on basic features of mathematics; on the role of idealization in science; on deductive knowledge (Aristotle's "inferential knowledge"), i.e., knowledge deduced from a definite system of axioms; on the role of axiom systems in mathematics, and on the requirements that must be satisfied by such systems; and on the relation between abstract "mathematical geometry" and the "physical geometry" concerned with certain properties of physical

---

[1]See, for example, Chapter 16 in [6]. (The numbers in brackets refer to the Bibliography.)

space. The non-uniqueness of geometry already justifies the effort to dislodge from the minds of prospective high school teachers the notions that Euclidean geometry is "innate," "unique," "natural," or "god-given."

While bearing in mind the non-uniqueness of Euclidean geometry, we should not lose sight of the fact that hyperbolic geometry is not the only possible non-Euclidean geometry. In fact, in addition to Euclidean and hyperbolic geometry, there are countless other geometric systems.

Historically, the view of what constitutes geometry changed radically on a number of occasions. For centuries it was thought that the sole aim of geometry is the thorough exploration of the properties of ordinary three-dimensional Euclidean space. Despite the development of other points of view (spherical geometry, the earliest non-Euclidean geometric system, was well known in antiquity[2]), there was not a shadow of doubt about the universality of the concept of Euclidean space until the discovery of hyperbolic geometry. The revolutionary findings of C. F. GAUSS (1777–1855), N. I. LOBACHEVSKY (1793–1856), and J. BOLYAI (1802–1860) in the first third of the 19th century[3] were unique in the history of mathematics because they shattered views which had lasted for millennia.

Each of the three discoverers of hyperbolic geometry has his own special merits. Formal priority belongs to Lobachevsky, who first published findings on this topic (1829) and who devoted his life to hyperbolic geometry and developed it more extensively than either Gauss or Bolyai. It appears that it was Gauss who, about 1816, first arrived at the idea of the existence of a new geometric system just as valid as ordinary Euclidean geometry, and that Lobachevsky and Bolyai arrived at this idea almost simultaneously but later than Gauss (Bolyai approximately in 1824; Lobachevsky not later than 1826). Finally, it is likely that Bolyai appreciated more deeply the significance of the discovery than did Gauss or Lobachevsky. Bolyai was very much troubled by the thought that he had no proof of the consistency of his geometry, while to Gauss and Lobachevsky this issue did not seem a matter of great concern. In fact, Lobachevsky was very close to the discovery of what is now known as the Klein–Beltrami model of hyperbolic geometry, but he saw no need for such a model and did not pursue this line of thought. That Bolyai, unlike Gauss and Lobachevsky, made no attempt to decide experimentally whether the geometry of the universe is hyperbolic or Euclidean might, in view of present knowledge, be considered an argument for rather than against him.

---

[2]The fact that spherical geometry did not fit the Euclidean scheme was first stressed by B. Riemann, whose geometric ideas were strongly influenced by the discovery of hyperbolic geometry. See text below.

[3]In the Russian literature there is a detailed account of the dramatic story of the discovery of hyperbolic geometry in V. F. Kagan's book [63]. [Gauss, J. Bolyai, and Lobachevsky arrived at the ideas of hyperbolic geometry entirely independently and at approximately the same time. This prompted F. Bolyai, the father of J. Bolyai, to remark that "there seems to be a right time for certain ideas and they are then discovered simultaneously in different places; they are like violets which bloom in the spring wherever the sun shines."]

Preface

After the discovery of hyperbolic geometry there arose the notion that there are just *two* admissible geometric systems, namely, Euclidean and hyperbolic geometry. This was the firm conviction of the discoverers of hyperbolic geometry. However, this point of view did not last very long. The 19th century was a period of rapid development in geometry. In 1854 the eminent German mathematician G. F. B. RIEMANN (1826–1866) formulated, in a famous memoir [74], an extremely general view of geometry which greatly widened its scope.[4] Riemann also noted that there are *three* (rather than two) related but distinct geometric systems, namely, the usual Euclidean geometry studied in high school, hyperbolic geometry and so-called elliptic geometry, which is very close to spherical geometry. This list of geometries was extended in 1870 by the German mathematician F. KLEIN (1849–1925), [73] (see also [56]). According to Klein, there are *nine* related plane geometries[5] including Euclidean geometry, hyperbolic geometry and elliptic geometry (in this connection, see Supplement A). Klein's views, which were in a way a synthesis of the geometric views of his predecessors and of the work of the English algebraist A. CAYLEY (1821–1891), appeared in 1872 in his *Erlanger Programm* (see Klein [9]). Klein's broad view of geometry has a universality comparable to that of Riemann.

Thus, just as the fundamental discoveries of Lobachevsky (published in 1829), Bolyai (published in 1832), and Gauss destroyed the exclusive position of Euclidean geometry, so, too, the classical investigations of Riemann and Klein (1854–1872) destroyed the exclusive position of hyperbolic geometry. Nevertheless, even today the term "non-Euclidean geometry" frequently stands for just hyperbolic geometry (less frequently, the plural "non-Euclidean geometries" is used to denote just hyperbolic geometry) and elliptic geometry, and the existence of other geometric systems is known only to specialists. It seems that this is largely due to the influence of dated discussions about the nature of physical space. The views presented in those discussions have long ago lost all scientific significance.

---

[4]This paper [74] was an inaugural lecture presented by Riemann to the faculty of Göttingen University. The presentation of a probationary inaugural lecture was required of all prospective professors. Riemann's lecture was so far ahead of its time that in all likelihood only Gauss understood and appreciated it. (The fact that Riemann was granted a teaching position was probably due to Gauss's high opinion of the lecture.) Riemann's lecture was published posthumously in 1864 by his student R. Dedekind, who could hardly have understood it completely. The full range of Riemann's ideas entered mathematics only with A. Einstein's 1916 memoir "On the Foundations of the General Theory of Relativity," which opens with a remarkably clear presentation of "Riemannian geometry," and with H. Weyl's 1919 book [39] containing a new edition of Riemann's paper and a penetrating discussion of Riemann's ideas.
   The second universal view of geometry which will be discussed in this book is due to F. Klein. It is of interest to note that Klein's view was also contained in an inaugural lecture—in this case, a lecture presented to the faculty of the University of Erlangen in 1872.

[5]Klein distinguished *seven* plane geometries (cf. for example, his book [56]). The subdivision of "Kleinian geometries" carried out in 1910 by the English geometer D. M. Y. Sommerville increased the number of plane geometries from seven to *nine*.

Thus, for example, even in Klein's "Non-Euclidean Geometry," [56] originally published in German in 1928, we find the assertion that the geometry of our universe must be either Euclidean, hyperbolic, or elliptic[6]; this in spite of the fact that the scientific unsoundness of this viewpoint, at least in its original formulation, followed from Einstein's special theory of relativity of 1905 and, even more decisively, from his general theory of relativity of 1916. The hypnotic effect of these antiquated cosmological discussions[7] has resulted in an unfortunate imbalance: too much attention is paid in scientific and popular-scientific literature to hyperbolic geometry (for example, a number of Russian books and countless papers are concerned with as special a problem as the theory of geometric constructions in the hyperbolic plane; this in spite of the fact that the problem clearly does not merit so much attention) and too little to the remaining "non-Euclidean geometries of Klein." The aim of this book is to help redress this imbalance by presenting a widely accessible account of one of these geometries, namely the geometry of the two-dimensional manifold of "events" (x,t) (x is the coordinate of a point on a line and t is time) whose "motions" are the Galilean transformations of classical kinematics.

After this unavoidably long historical introduction I should like to return to the question of the pedagogical soundness of the stress on hyperbolic geometry. It is undoubtedly important to familiarize future teachers of mathematics (and also, to some extent, senior high school students interested in mathematics) with a geometry different from the Euclidean geometry which they know so well. What is worth debating, I think, is the question of the choice of a non-Euclidean geometry. I am not inclined either to disregard completely or to regard as decisive the fact that hyperbolic geometry was the first non-Euclidean geometry to be discovered. The fact that hyperbolic geometry is linked to the issue of the independence of the parallel axiom and clarifies the role of that axiom in Euclidean geometry, is a strong argument in favor of its pedagogical value.[8] On the other hand, hyperbolic geometry is rather complex—it is definitely more complex than Euclidean geometry—and yet the non-Euclidean nature of a geometry need not imply complexity. In fact, the geometry presented in this book—Galilean geometry—is the simplest of

---

[6]Klein's book, published posthumously, is an edited version of mimeographed lecture notes published by Klein in 1892 and again in 1893 as a textbook for students of the University of Göttingen. I regard the efforts at updating undertaken by students of Klein, who prepared these lecture notes for publication, as entirely inadequate.

[7]We shall not concern ourselves with modern cosmological theories which discuss (in different formulations) the question whether the geometry of the universe is, in its essential features, elliptic, Euclidean, or hyperbolic.

[8]This point is of major importance only in developments of a geometry such as those based on Euclid [1]–[3] or Hilbert [4] (see, for example, the textbook [7]). If high school texts on geometry used the vector approach (an approach supported by many mathematicians and teachers; cf. the fervently written book of Dieudonné [8]), then it would be better for present and future teachers of mathematics to know Minkowskian and Galilean geometry rather than hyperbolic geometry (cf. Supplement B).

all Kleinian geometries; in many respects it is simpler than Euclidean geometry. The main distinction of this geometry is its relative simplicity, for it enables the student to study it in relative detail without losing a great deal of time and intellectual energy. Put differently, the simplicity of Galilean geometry makes its extensive development an easy matter, and extensive development of a new geometric system is a precondition for an effective comparison of it with Euclidean geometry. Also, extensive development is likely to give the student the psychological assurance of the consistency of the investigated structure. Another distinction of Galilean geometry is the fact that it exemplifies the fruitful geometric idea of duality. And last, but certainly not least, a major merit of the geometry presented in this book is that it illustrates the important connection between Klein's *Erlanger Programm* and the principles of relativity, and sheds additional light on Klein's conception as well as on the role of the principles of relativity in physics.[9] These reasons make me think that one should give serious thought to a mathematics program for teachers' colleges which would include a comparative study of three simple geometries, namely, Euclidean geometry, the geometry associated with the Galilean principle of relativity, and the geometry associated with Einstein's principle of relativity (Minkowskian geometry; cf. Section 12 of this book), as well as an introduction to the special theory of relativity.

The present book, with its many possibly interesting but nonessential details,[10] is intended for high school seniors, mathematics teachers, and students and lecturers in universities and teachers' colleges, but is not meant to be a blueprint for the reform of the curriculum of teachers' colleges. At the moment, we are certainly not ready for such a reform. The evaluation of the questions raised above requires extensive knowledge of the geometries associated with the principles of relativity of Galileo and Einstein, whereas apparently this is the first popular scientific book to analyze in detail the geometry associated with the Galilean principle of relativity.

Serious scientific accounts dealing with this geometric system are relatively recent. References to it are found in the works of Klein (cf. [56]) and in textbooks dealing with Klein's ideas. However, these references are brief and insubstantial. The first detailed investigations of this geometry appeared in the years 1913–1915, and are contained in papers of the German geometers H. Beck, F. Boehm, and

---

[9]It is frequently claimed that hyperbolic geometry can be used in judging the merits of various cosmological hypotheses. However, to imply that only Euclidean or hyperbolic geometry can fit the universe seems to me to hinder the acceptance of the ideas of the theory of relativity. From this point of view there is, I think, more merit in the geometry associated with the Galilean principle of relativity, since it is likely to pave the way for an understanding of just such ideas.

[10]For example, in spite of the lack of scientific applications of the theorem of K. W. Feuerbach on triangles (*the nine-point circle of a triangle touches its inscribed as well as its three escribed circles*; cf. [33]), we give in this book three proofs, counting exercises, of its Galilean analogue.

L. Berwald.[11] In 1925–1928 this geometry reappeared in the papers of the English mathematician L. Silberstein, the Pole S. Glass, the Russian A. P. Kotel'nikov, and the Dane D. Fog[12]; these authors seem to have been the first to note the connection between the geometry they investigated and the Galilean principle of relativity. However, it was not until the 1950s that Galilean geometry was analyzed in some sense even more extensively than Euclidean geometry. This thorough study involved the eminent Dutch geometer N. Kuiper, the German geometer K. Strubecker, and my student N. M. Makarova (see the Bibliography).

The question of the name of the geometry studied in this book requires some comment. In the literature it is variously referred to as "semi-Euclidean," "flag," "parabolic," "isotropic," and "Galilean," but none of these terms is particularly apt. The terms "flag geometry," "parabolic geometry" (or "doubly parabolic geometry"), and "isotropic geometry" are justified by considerations which cannot be entered into in an elementary account aimed at readers without substantial mathematical background. A merit of the term "semi-Euclidean geometry" is its closeness to the term "pseudo-Euclidean geometry", which is used in connection with the geometry associated with the Einsteinian principle of relativity. However, the true merit of the term "semi-Euclidean geometry" can be appreciated only by those familiar with the "pseudo-Euclidean geometry of Minkowski." Finally, the now popular name "Galilean geometry" is historically inaccurate: Galileo, whose works date from the beginning of the 17th century, did not in fact know this geometry, whose discovery was necessarily preceded by one of the greatest intellectual triumphs of the 19th century—the emergence of the idea that many legitimate geometric systems exist. A more accurate name would be "the geometry associated with the Galilean principle of relativity." This name is too long for repeated use and that is why we have decided, somewhat reluctantly, to use the name "Galilean geometry." This name is partially justified by the brilliant clarity and completeness with which Galileo formulated his principle of relativity, which leads directly to the (non-Euclidean!) geometry considered in the present book.

Finally, a few remarks about the plan of this somewhat unusual book. The long Introduction consisting of two sections, the Conclusion, consisting of three sections, the three Supplements, the extensive Bibliography, the unusually (and perhaps excessively) long Preface are features not frequently found in books aimed, in part, at beginners. The complicated

---

[11]Cf. H. BECK: "Zur Geometrie in der Minimalebene," *Sitzungsber. Leipziger Berliner Math. Ges.* **12**: 14–30, 1913; F. BOEHM: "Beiträge zum Äquivalenzproblem der Raumkurven." *Sitzungsber. Akad. München* **2**: 257–280, 1915; L. BERWALD: "Über Bewegungsinvarianten und elementare Geometrie in der Minimalebene." *Monatsh. Math. Phys.* **26**: 211–228, 1915.
[12]Cf. L. SILBERSTEIN: "Projective geometry of Galilean space–time." *Philos. Mag.* **10**: 681–696, 1925; S. GLASS: "Sur les géométries de Galilée et sur une géométrie plane particulière." *Ann. Soc. Polonaise Math.* **5**: 20–36, 1926; A. P. KOTEL'NIKOV: "The principle of relativity and the geometry of Lobachevsky." A paper in the collection *In memoriam Lobatschevskii*, 2nd ed. Kazan', Glavnauka, 1927, pp. 37–66; D. FOG: "Den isotrope Plans elementare Geometri." *Math. Tidskrift, Ser. B* **xx**: 21–33, 1928.

plan of the book reflects the unusual nature of the topic (I was forced to begin the book with a long preface when I discovered that its very title[13] was bewildering) and, more important, the fact that the book is intended for different categories of readers. While the Bibliography includes many items which can be read by beginners, it is intended primarily for teachers and can be ignored by high school students. Study of the material in fine print (partly intended for high school teachers of mathematics) is not necessary for the understanding of the rest of the book. Familiarity with the introduction is indispensable for the understanding of the rest of the book. In a first reading, the Conclusion can be ignored. However it seems to me that Sections 11 and 12 of that chapter could be read with profit at a later time. Incidentally, the style of the Conclusion is concise; it was not the author's intention to add another item to the extensive list of popular accounts of Einstein's special theory of relativity, and readers of the very brief Section 11 will undoubtedly find it helpful to acquaint themselves first with one of the expositions [38]–[55] (we highly recommend the remarkable book [40] of M. Born). Section 12, devoted to the "non-Euclidean geometry of Minkowski," is also quite concise and is not meant for rapid reading. This section contains almost no proofs of the quoted results; the enterprising reader may try to supply his own proofs (here the bibliographical references may turn out to be helpful). I wish to note that I made a consistent effort to bring together phenomena of Galilean geometry and related phenomena of Euclidean geometry, and this parallelism of presentation accounts in part for the relatively large size of the book.

The few problems and exercises at the end of each section are intended to stimulate the reader's initiative and to encourage reflection, and so are often formulated somewhat vaguely. The reader will undoubtedly pose other problems himself. The Exercises (numbered with Arabic numerals) are meant to help the reader check his grasp of the subject matter. The Problems (numbered with Roman numerals) are more difficult than the Exercises, but beyond that the distinction is rather subjective. The solution of the Problems may involve considerable effort as well as a measure of research, and may require knowledge in excess of that needed for the reading of the book itself. However, even these problems are of methodological rather than scientific interest. The selected answers and hints at the end of the book are not uniformly detailed and are intended to give the reader an idea of the nature of the required solutions.

A word about Supplements A, B, and C: Just as there are no prerequisites for the reading of the main part of the text, so there are none for the reading of the Supplements; they can be read, in principle, by a persistent high school student. But persistence is required. The Supplements, written in a very concise manner (and containing practically no proofs), form the most complex part of the book and are meant for

---

[13]The Russian title of the book is *The Galilean Principle of Relativity and Non-Euclidean Geometry* (translator's note).

thoughtful readers. In spite of the fact that the Supplements do not explicitly rely on the knowledge of concepts and theorems unknown to readers with minimal background, they are probably accessible only to students who are familiar with analytic geometry and the elements of elliptic or hyperbolic geometry. The material of the Supplements is not needed for the understanding of the rest of the book. Nevertheless, the Supplements play a rather important role: while the rest of the book contains a detailed exposition of one simple non-Euclidean geometry, the Supplements afford a glimpse of a universe of geometries of which Galilean geometry is the simplest member.

Formulas (numbered with Arabic numerals), exercises, and problems are numbered independently in the Introduction, in Chapter I, in Chapter II, in the Conclusion, and in the Supplements. Hence if a formula, exercise, or problem is referred to in some part of the book by number alone, then it is found in that part. In other cases, the reader is referred to an appropriate section or page.

The book is based on lectures which I delivered in 1956–57 to high school juniors and seniors who were members of the high school mathematics club at the Moscow State University. In 1963–64 an expanded version of these lectures was presented to high school juniors at the Evening Mathematics School of the Moscow State University. The book was influenced to some extent by the contents of a (more advanced and extensive) special course, entitled "Principles of Relativity and Non-Euclidean Geometries," which I taught at the Lenin Moscow State Teachers' College.

It is a pleasant task for an author to thank those who have helped him. I wish to express my gratitude to G. B. Gurevich for friendly and helpful criticism. I am also most grateful to F. I. Kizner, the editor of my books, for her valuable assistance in the preparation of the final version of this book. Finally, I wish to acknowledge the fine work of M. S. Koroleva, who prepared the drawings for the book, and whose help and enthusiasm made it possible for it to see the light of day.

I. M. Yaglom

Translator's Preface

This book is remarkable in that it relies only on precalculus mathematics and yet has an "idea density" exceeding that of many advanced texts. It is a fascinating story which flows from one geometry to another, from one model to another, from geometry to algebra, and from geometry to kinematics, and in so doing crosses artificial boundaries separating one area of mathematics from another and mathematics from physics.

The book abounds in helpful comments, discussions, examples, and applications. Some of the exercises and problems following each section are what the terms indicate. Others sketch paths to be followed and explored by more experienced readers.

Enough has been said to indicate that this is an unusual book. However, only by reading it can one gain an appreciation of the skill with which the author develops its many themes. It is safe to say that Professor Yaglom has created a royal road to a large part of geometry and to parts of algebra and physics for readers ranging from high school students to college teachers.

The following is a somewhat technical description of the content of the book.

The Introduction discusses Klein's concept of a geometry, mechanics, and the geometrization of mechanics. Chapters I and II (about half the book) are an elementary but nontrivial introduction to plane Galilean geometry and to Galilean inversive geometry, with plane and inversive Euclidean geometry providing both background and contrast.

Plane Galilean geometry can be thought of as the geometry of either classical kinematics on a line or of shears and translations. In many ways, Galilean geometry is simpler than Euclidean geometry; for example, we have for sing and cosg, the Galilean analogs of sin and cos, sing $A \equiv A$ and cosg $A \equiv 1$ for all angles A. Like projective geometry, Galilean geometry has a principle of duality.

The characteristic property of the transformations of EIG (Euclidean inversive geometry) is that they preserve (lines and) circles. Similarly, the characteristic property of the transformations of GIG (Galilean inversive geometry) is that they preserve (lines and) Galilean circles and cycles, the Galilean analogs of Euclidean circles. A more surprising characterization of the transformations of EIG and GIG is that they are the restrictions of the collineations of space to a sphere and to a cylinder, respectively. In Supplement C, EIG is shown to be the geometry of the group of fractional linear transformations

$$w = (az+b)/(cz+d) \quad \text{and} \quad w = (a\bar{z}+b)/(c\bar{z}+d), \quad z = a+bi, \quad i^2 = -1$$

and GIG the geometry of the group of transformations

$$w = (az+b)/(cz+d) \quad \text{and} \quad w = (a\bar{z}+b)/(c\bar{z}+d), \quad z = a+b\varepsilon, \quad \varepsilon^2 = 0.$$

Chapters I and II are followed by a chapter called "Conclusion." Here classical kinematics, the physical basis of Galilean geometry, is replaced by relativistic kinematics. The geometrization of relativistic kinematics leads to Minkowskian geometry. The earlier themes are discussed in the context of the new geometry. In particular, MIG (Minkowskian inversive geometry) is characterized by the study of collineations of space restricted to a hyperboloid of one sheet. In Supplement C, MIG is shown to be the geometry of the group of fractional linear transformations

$$w = (az+b)/(cz+d) \quad \text{and} \quad w = (a\bar{z}+b)/(c\bar{z}+d), \quad z = a+be, e^2 = 1.$$

Supplement A introduces a new range of ideas. The author points out that in each of the three geometries studied earlier length is measured "the same way" but angles are not. By measuring angles as well as lengths in three different ways we are led to *nine* plane geometries. In Supplement A some of these geometries are realized as the intrinsic geometries of unit spheres in appropriate spaces. In Supplement B they are characterized by suitable sets of axioms. In Supplement C some of them are realized in terms of analytic models involving the algebras of complex, dual, and double numbers.

I wish to thank the author for bibliographical suggestions (the German, French, and even Russian items may not be used by high school students who read the book, but may be used by other readers). The staff at Springer-Verlag New York was most helpful. I wish to thank Hardy Grant for reading an earlier version of the translation. My special thanks go to Basil Gordon for his untiring efforts to improve the translation.

Remarks on notation

The book poses a notational dilemma in the sense that an absolutely precise and consistent notation would be very cumbersome. The author's notation is sufficiently precise and yet simple, and the translator has followed the author's example.

Translator's Preface

The reader may find useful the following list of symbols and comments on notation.

(1) (p. 2). $d = AA_1 = \sqrt{(x_1-x)^2 + (y_1-y)^2}$ is the positive Euclidean length of the segment AA_1.
(2) (p. 16). In the case of motion along a fixed line the velocity can be given by a number rather than a vector.
(3) (p. 38). $\tan \delta_{l l_1} = (k_1 - k)/(1 + kk_1)$ is the signed magnitude of the tangent of the directed Euclidean angle from l to l_1.
(4) (p. 38). $d_{l l_1} = |s_1 - s|/\sqrt{1+k^2}$ is the positive Euclidean distance between the parallel lines l and l_1 with common slope k and y-intercepts s and s_1, respectively.
(5) (p. 38). $d_{AA_1} = x_1 - x$ is the signed Galilean length of the directed segment AA_1.
(6) (p. 39). $\delta_{AA_1} = y_1 - y$ is the signed special length of the directed special segment AA_1.
(7) (p. 41). $\delta_{l l_1}$ denotes the signed Galilean magnitude of the directed Galilean angle from l to l_1. If the slopes of l and l_1 are k and k_1, respectively, then $\delta_{l l_1} = k_1 - k$.
(8) (p. 42). $d_{l l_1}$ denotes the signed Galilean distance between the parallel lines l and l_1. If l and l_1 have y-intercepts s and s_1, respectively, then $d_{l l_1} = s_1 - s$.
(9) (p. 42). d_{Ml} denotes the signed Galilean distance from the point M to the line l.
(10) (pp. 41 and 51). $\overline{NN_1}$ is the signed length of the directed segment NN_1. $\bar{a}$ is the signed length of the directed segment a. $\bar{A}$ is the signed magnitude of the directed angle A.
(11) (p. 44). The mechanical significance of $\delta_{l l_1}$ is that of $v_{l,l}$, the velocity of the uniform motion represented by l_1 relative to the uniform motion represented by l. If the slopes of l and l_1 are k and k_1, respectively, then $v_{l,l} = \delta_{l l_1} = k_1 - k$. Since $\tan g \delta_{l l_1} = \delta_{l l_1}$ (see p. 47, Ex. 3), we have $v_{l,l} = \tan g \delta_{l l_1}$ (constrast this with the relation $v_{l,l} = \tanh_{l l_1}$ in relativistic mechanics; see p. 188, footnote 16).
(12) (p. 179). $d_{AB} = \sqrt{2S(AKBL)}$ is the positive Minkowskian length of AB.
(13) (p. 180). $d_{AB} = \sqrt{|(x_2-x_1)^2 - (y_2-y_1)^2|}$ is the positive Minkowskian length of AB.
(14) (p. 180). $d_{AB} = \sqrt{(x_2-x_1)^2 - (y_2-y_1)^2}$ is the positive real, or complex Minkowskian length of AB.
(15) (p. 258). The modulus $|z|$ of a complex number $z = x + iy$ is given by $z = \sqrt{x^2 + y^2}$.
(16) (p. 267). The modulus $|z|$ of a dual number $z = x + \varepsilon y$ is given by $|z| = x$.
(17) (p. 267). The modulus $|z|$ of a double number $z = x + ey$ with $|x| > |y|$ is given by $\pm \sqrt{x^2 - y^2}$, and the sign of $|z|$ is that of x. The modulus

of a double number $z = x + ey$ with $|y| > |x|$ is given by $\pm\sqrt{y^2 - x^2}$, and the sign of $|z|$ is that of y.

(18) (pp. 259 and 272). $d_{z,z_1} = |z_1 - z|$ is the signed distance from z to z_1. Here z and z_1 can be complex, dual, or double numbers.

(19). In Supplement C, lower-case Latin characters denote complex, dual, or double numbers. If a is such a number, then $\bar{a}$ is its conjugate.

Contents

Introduction   **1**

   1. What is geometry?   1
   2. What is mechanics?   15

**Chapter I. Distance and Angle;
Triangles and Quadrilaterals**   **33**

   3. Distance between points and angle between lines   33
   4. The triangle   47
   5. Principle of duality; coparallelograms and cotrapezoids   54
   6. Proofs of the principle of duality   67

Chapter II. Circles and Cycles   **77**

   7. Definition of a cycle; radius and curvature   77
   8. Cyclic rotation; diameters of a cycle   91
   9. The circumcycle and incycle of a triangle   104
   10. Power of a point with respect to a circle or cycle; inversion   117

Conclusion   **158**

   11. Einstein's principle of relativity and Lorentz transformations   158
   12. Minkowskian geometry   174
   13. Galilean geometry as a limiting case of Euclidean
       and Minkowskian geometry   201

Supplement A. Nine plane geometries   **214**

**Supplement B. Axiomatic characterization
of the nine plane geometries**   **242**

**Supplement C. Analytic models of the
nine plane geometries**   **258**

Bibliography   **289**

Answers and Hints to Problems and Exercises   **294**

Index of Names   **303**

Index of Subjects   **305**

Introduction

1. What is geometry?

Geometry, a subject with which we are all familiar from high school, investigates properties of figures, i.e., sets of points in the plane or in space.[1] The question arises, *which* properties of figures are of interest to the geometer? To answer this question, we can use two different approaches. Both lead to the same conclusions. Both will be of use to us in what follows, and so deserve our attention.

The first approach is bound up with the concept of congruence of figures. Figures are said to be *congruent* if they have the same geometric properties. It is perhaps more suggestive to say that congruent figures differ only in position but not in form or size, or that figures are congruent if and only if they can be obtained from each other by a motion. For example, triangles ABC and MNP in Figure 1 differ only in position, and so have the same (geometric) properties. Thus if the angle between the median AD and the angle bisector AE of $\triangle ABC$ is $15°$, then the same is true of the angle between the median MQ and the angle bisector MR of $\triangle MNP$. Congruent figures are frequently regarded as essentially the same. That is why, in geometry, triangles such as ABC and MNP are frequently regarded as one rather than two triangles. Similarly, we say that two sides $AB = c$ and $AC = b$ and the angle $\angle A = \alpha$ between them determine a *unique* triangle (Fig. 2a)[2] meaning that triangles ABC and $A'B'C'$, (Fig. 2b) with $AB = A'B'$ ($=c$), $AC = A'C'$ ($=b$), and $\angle A = \angle A'$ ($=\alpha$), are congruent.

---

[1] In this book we will deal mainly with *plane geometry*, which studies properties of figures in the plane.
[2] Note that in general, given two sides $AB = c$, $BC = a$ and the angle $\angle A = \alpha$ opposite one of the given sides, we can construct *two* triangles (Fig. 2b).

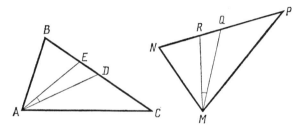

Figure 1

We wish to stress that, far from having given a mathematically acceptable definition of congruence, we have merely suggested its sense and equated statements such as "F and F' are congruent figures," "F and F' have the same geometric properties," and "F and F' can be obtained from each other by a motion." There are various ways out of this difficulty. One of them is to say that *two figures F and F' are congruent if it is possible to establish a one-to-one correspondence between the points of F and those of F' such that the distance between any two points A and B of F is the same as the distance between the corresponding points A' and B' of F'* (Fig. 3). While this definition may be unwieldy, it does not tie the concept of congruence of figures to that of motion. True, our definition introduces the new concept of "distance between two points," and a critical reader might well ask what is meant by the distance between two points. This question, however, is easy to answer: If, in some (arbitrary) rectangular coordinate system in the plane, the coordinates of A are x and y and those of A_1 are x_1 and y_1 (Fig. 4a), then we define the distance between A and A_1 by the formula[3]

$$d = AA_1 = \sqrt{(x_1 - x)^2 + (y_1 - y)^2} \ . \tag{1}$$

In this way the notion of distance is expressed by a simple algebraic

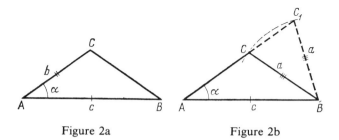

Figure 2a　　　　　Figure 2b

---

[3]Similarly, the distance between the points $A(x,y,z)$ and $A_1(x_1,y_1,z_1)$ in space is given by the formula

$$d = AA_1 = \sqrt{(x_1 - x)^2 + (y_1 - y)^2 + (z_1 - z)^2} \tag{1a}$$

(cf. Fig. 4b).

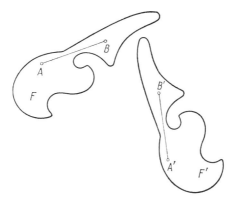

Figure 3

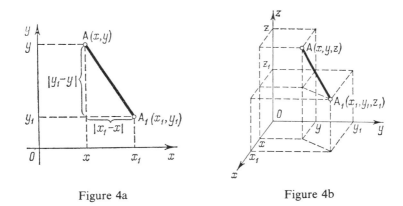

Figure 4a   Figure 4b

formula. That d is independent of the choice of a (rectangular) coordinate system is proved on p. 9.

Now that we have a definition of congruence, we also have a definition of the term "geometric property" and, incidentally, of "motion." Thus *a geometric property is one shared by all congruent figures*, and *a motion is a transformation* (of the plane) *which maps every figure to a congruent figure*, or, equivalently, *a transformation which preserves distances*.

At this point it is appropriate to give the definition of geometry due to Klein. According to Klein, *geometry is the study of invariant properties of figures*, i.e., properties unchanged under all motions.[4] This view of geometry makes it clear why an angle between two sides of a triangle drawn on a blackboard is of interest to a geometer, whereas the angle between a side of the traingle in question and an edge of the blackboard is not. The first angle is unaffected by motions and is therefore a geometric property of the

---

[4]Actually, Klein's definition of geometry is more general than this (see Klein's original paper [9], the introductions to Yaglom's books [10]–[12], or Section 6 of [14]). However, the above definition suffices for our purposes.

triangle. The second angle changes under motions and therefore is not a geometric property of the triangle; it depends on the location of the triangle on the blackboard.

We now return to our main theme. The use of coordinates is intimately connected with the second approach to the question of which properties of figures are geometric, that is, belong to the domain of geometry. As is well known, already in the 17th century the famous French mathematicians R. DÉSCARTES (1596–1650) and P. FERMAT (1601–1665) showed that Euclidean geometry can be based on the concept of number and reduced to analysis by the introduction of a coordinate system (of the kind shown in Fig. 4a) in the plane. Then to each point in the plane there corresponds a pair of numbers x and y, its coordinates, and to each (plane) figure F, i.e., to each set of points, there corresponds a set of number pairs, the coordinates of the points of F.

For example, to the set of points M at a fixed distance r from a given point $Q(a,b)$, i.e., the *circle S* with center Q and radius r (Fig. 5), there corresponds the set of number pairs (x,y) such that

$$\sqrt{(x-a)^2+(y-b)^2} = r \tag{1'}$$

or

$$(x-a)^2+(y-b)^2 = r^2, \tag{2}$$

that is,

$$x^2+y^2+2px+2qy+f=0, \tag{2'}$$

where

$$p=-a, \quad q=-b, \quad f=a^2+b^2-r^2. \tag{2a}$$

Similarly, to the *line l* which intersects the y-axis at the point $S(0,s)$ and forms with the x-axis an angle $\angle xQM = \varphi$ with $\tan\varphi = MP/QP = k$ (Fig. 6), there corresponds the set of number pairs (x,y) such that

$$y = kx+s, \tag{3}$$

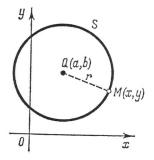

Figure 5

1. What is geometry?

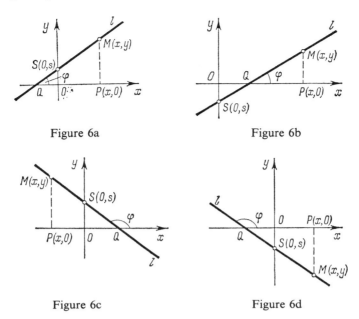

Figure 6a Figure 6b

Figure 6c Figure 6d

where the intercept s and the slope k may be positive, negative, or zero. (In this book angles are always measured counterclockwise, so that a negative value of $k = \tan\varphi$ means that $\angle xQM = \varphi$ is obtuse; cf. Figs. 6a–d). Again, to the line m parallel to the y-axis and intersecting the x-axis at $A(a,0)$ there corresponds the set of number pairs (x,y) with

$$x = a \tag{4}$$

(Fig. 7a). To the line n through $B(b,0)$ parallel to the x-axis (Fig. 7b) there corresponds the set of number pairs (x,y) with

$$y = b$$

[the latter equation is a special case of Eq. (3) with $k=0$.] In addition to the terminology employed so far, we also use the expressions "Equation (2) corresponds to the circle S," and "Equation (3) corresponds to the line l," or, briefly, "the circle S is given by Eq. (2)," and "the line l is given by Eq. (3)."

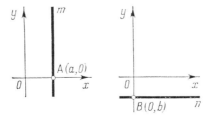

Figure 7a Figure 7b

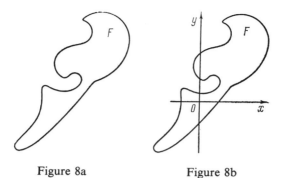

Figure 8a   Figure 8b

Now it is natural to ask what relations between the coordinates of points have geometric significance and what relations are accidental, i.e., depend on the choice of a coordinate system. For example, r in Eq. (2) expresses a fundamental characteristic of the circle. On the other hand, p in Eq. (2′) has no geometric significance; it is quite easy to find two congruent circles given by equations of the form (2′) with different values of p as well as two noncongruent circles with the same values of p.[5]

By now the reader may well have guessed the answer to the question just raised. The coordinates (x,y) of a point A depend not only on A but also on something not directly related to it, namely, the coordinate system. We can choose the (rectangular) coordinate axes Ox and Oy in various ways, and depending on these choices, the coordinates of the point A will vary. In much the same way, a figure, say a line, will be given in different coordinate systems by different sets of number pairs. If a quantity determined by the set of number pairs corresponding to a figure F does not depend on the coordinate system, i.e., remains the same upon transition from one coordinate system to another, then this quantity may be said to have an *invariant* character, to depend on the figure F alone, to reflect a geometric property of F. On the other hand, if the quantity in question changes upon transition from one coordinate system to another, then it depends on the position of the figure F in the chosen coordinate system; it is characteristic not of the figure F with which we are concerned (Fig. 8a), but of the more complex entity made up of F and the pair of coordinate axes Ox and Oy (Fig. 8b). In general, *a property of a figure is geometric if and only if it does not depend on the choice of a coordinate system*. This is our second answer to the question of which properties of a figure are geometric.

Our remarks can be reformulated so as to help us distinguish effectively between geometric quantities and quantities which depend on the choice of a coordinate system. Thus, let A be a point with coordinates x and y in a

---

[5]Clearly, the value $p = -a$ depends on the abscissa a of the center Q of the circle. This quantity is determined by the choice of the coordinate system, not by the geometric properties of the circle.

1. What is geometry? 7

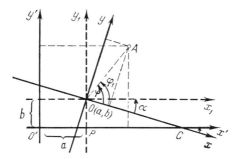

Figure 9

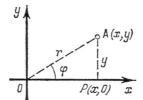

Figure 10

coordinate system xOy (or, as we shall say in what follows, in the coordinate system $\{x,y\}$), and with coordinates x' and y' in a coordinate system $x'O'y'$ (or $\{x',y'\}$). Suppose that the coordinates of the "old" origin O in the new coordinate system $x'O'y'$ are a and b, and the angle xCx' between the axes Ox and $O'x'$ is α (Fig. 9).[6] Our problem is to express the "new" coordinates (x',y') of a point A in terms of its old coordinates (x,y) and the quantities a, b, and α.

To solve this problem it is convenient to introduce the angle $\angle xOA = \varphi$ (as always, angles are laid off in the positive, i.e., counterclockwise direction from Ox) and the distance $OA = r$.[7] Then the coordinates x and y of A are expressed in terms of r and φ by means of the relations

$$x = r\cos\varphi, \qquad y = r\sin\varphi \tag{5}$$

(Fig. 10). We now introduce an auxiliary coordinate system $x_1 O y_1$, or $\{x_1, y_1\}$, whose origin coincides with that of $\{x,y\}$ and whose axes are parallel to those of $\{x',y'\}$ (see Fig. 9). Clearly, the angle from the x_1-axis to the ray OA is $\varphi_1 = \varphi - \alpha$ (see Fig. 9). Hence,

$$x_1 = r\cos\varphi_1 = r\cos(\varphi - \alpha) = r(\cos\varphi\cos\alpha + \sin\varphi\sin\alpha)$$
$$= r\cos\varphi \cdot \cos\alpha + r\sin\varphi \cdot \sin\alpha = x\cos\alpha + y\sin\alpha,$$

---

[6] In analytic geometry one usually employs "right-handed" coordinate systems. In such a coordinate system the 90° rotation which takes the ray Ox to the ray Oy is *counterclockwise*. Here we follow this convention.

[7] The quantities r and φ are called the *polar coordinates* of A.

and
$$y_1 = r\sin\varphi_1 = r\sin(\varphi - \alpha) = r(\sin\varphi\cos\alpha - \cos\varphi\sin\alpha)$$
$$= r\sin\varphi \cdot \cos\alpha - r\cos\varphi \cdot \sin\alpha = -x\sin\alpha + y\cos\alpha.$$

On the other hand, Figure 9 shows that
$$x' = x_1 + a,$$
$$y' = y_1 + b,$$
where $a = O'P$ and $b = PO$.

It thus follows that
$$\boxed{\begin{aligned} x' &= x\cos\alpha + y\sin\alpha + a, \\ y' &= -x\sin\alpha + y\cos\alpha + b. \end{aligned}} \tag{6}$$

Given x' and y', we can solve (6) for x and y, and so express the old coordinates of a point in terms of its new coordinates. The relevant formulas are:
$$\begin{aligned} x &= (x'-a)\cos\alpha - (y'-b)\sin\alpha, \\ y &= (x'-a)\sin\alpha + (y'-b)\cos\alpha. \end{aligned} \tag{6a}$$

Formulas (6a) can also be written as:
$$\begin{aligned} x &= x'\cos\bar{\alpha} + y'\sin\bar{\alpha} + \bar{a}, \\ y &= -x'\sin\bar{\alpha} + y'\cos\bar{\alpha} + \bar{b}, \end{aligned} \tag{6b}$$

where $\bar{a}$ and $\bar{b}$ are the coordinates of O' in the system $\{x,y\}$, and $\bar{\alpha} = \angle x'Cx = -\alpha$ is the angle between the axes $O'x'$ and Ox.[8] To justify this step we need only bear in mind the fact that we may regard $\{x',y'\}$ as the old coordinate system and $\{x,y\}$ as the new one and use formulas (6) (with a, b, α replaced by their analogues $\bar{a}, \bar{b}, \bar{\alpha}$).

We boxed formulas (6) to emphasize their importance. These formulas play a fundamental role in geometry, since they enable us to decide whether a quantity or property (defined analytically, i.e., by means of coordinates) has geometric significance. For example, the distance
$$d = \sqrt{(x_1 - x)^2 + (y_1 - y)^2} \tag{1}$$
between $A(x,y)$ and $A_1(x_1,y_1)$ has geometric significance. In fact, if we choose a coordinate system $\{x',y'\}$, then the new coordinates of A are given by formulas (6), and the new coordinates of A_1 by analogous formulas
$$\begin{aligned} x_1' &= x_1\cos\alpha + y_1\sin\alpha + a, \\ y_1' &= -x_1\sin\alpha + y_1\cos\alpha + b. \end{aligned}$$

---

[8] Strictly speaking we should put $\bar{\alpha} = 360° - \alpha$ because we agreed to lay off angles in the positive direction. However, this does not affect the values of the trigonometric functions in formulas (6b), and so may be overlooked.

Hence
$$(x_1'-x')^2+(y_1'-y')^2 = [(x_1-x)\cos\alpha+(y_1-y)\sin\alpha]^2$$
$$+[-(x_1-x)\sin\alpha+(y_1-y)\cos\alpha]^2$$
$$=(x_1-x)^2(\cos^2\alpha+\sin^2\alpha)+(y_1-y)^2(\cos^2\alpha+\sin^2\alpha)$$
$$=(x_1-x)^2+(y_1-y)^2,$$
which shows that d is indeed independent of the choice of a coordinate system.

Similarly, if the equation of a circle S in the old coordinate system $\{x,y\}$ is given by (2'), then in order to obtain its equation in the new system $\{x',y'\}$ we must replace x and y in (2') by their expressions in terms of the new coordinates [see (6b)]. This gives
$$[x'\cos\bar{\alpha}+y'\sin\bar{\alpha}+\bar{a}]^2+[-x'\sin\bar{\alpha}+y'\cos\bar{\alpha}+\bar{b}]^2$$
$$+2p(x'\cos\bar{\alpha}+y'\sin\bar{\alpha}+\bar{a})+2q(-x'\sin\bar{\alpha}+y'\cos\bar{\alpha}+\bar{b})+f=0,$$
or, after reductions,
$$x'^2+y'^2+2p'x'+2q'y'+f'=0,$$
where
$$p'=\bar{a}\cos\bar{\alpha}-\bar{b}\sin\bar{\alpha}+p\cos\bar{\alpha}-q\sin\bar{\alpha},$$
$$q'=\bar{a}\sin\bar{\alpha}+\bar{b}\cos\bar{\alpha}+p\sin\bar{\alpha}+q\cos\bar{\alpha},$$
$$f'=\bar{a}^2+\bar{b}^2+2p\bar{a}+2q\bar{b}+f.$$

This shows that p has no geometric significance (since, in general, $p'=\bar{a}\cos\bar{\alpha}-\bar{b}\sin\bar{\alpha}+p\cos\bar{\alpha}-q\sin\bar{\alpha}\neq p$). On the other hand,
$$p^2+q^2-f$$
has geometric significance, for it is easy to see that
$$p'^2+q'^2=[(\bar{a}\cos\bar{\alpha}-\bar{b}\sin\bar{\alpha})+(p\cos\bar{\alpha}-q\sin\bar{\alpha})]^2$$
$$+[(\bar{a}\sin\bar{\alpha}+\bar{b}\cos\bar{\alpha})+(p\sin\bar{\alpha}+q\cos\bar{\alpha})]^2$$
$$=\bar{a}^2+\bar{b}^2+p^2+q^2+2p\bar{a}+2q\bar{b},$$
and therefore
$$p'^2+q'^2-f'=p^2+q^2-f.$$
In fact, formulas (2a) show that
$$p^2+q^2-f=r^2,$$
where r is the *radius* of the circle S.

If a line l in the coordinate system $\{x,y\}$ is given by Equation (3), then in the system $\{x',y'\}$ it is given by the equation
$$-x'\sin\bar{\alpha}+y'\cos\bar{\alpha}+\bar{b}=k(x'\cos\bar{\alpha}+y'\sin\bar{\alpha}+\bar{a})+s.$$

If $\cos\bar\alpha - k\sin\bar\alpha \neq 0$, then this equation can be rewritten as
$$y' = k'x' + s',$$
where
$$k' = \frac{k\cos\bar\alpha + \sin\bar\alpha}{\cos\bar\alpha - k\sin\bar\alpha}, \qquad s' = \frac{k\bar a - \bar b + s}{\cos\bar\alpha - k\sin\bar\alpha},$$
and otherwise (i.e., if $\cos\bar\alpha - k\sin\bar\alpha = 0$), as
$$x' = a',$$
where
$$a' = \frac{\bar b - k\bar a - s}{k\cos\bar\alpha + \sin\bar\alpha}.$$

It follows that neither k nor s has geometric significance; both numbers reflect the position of the line l relative to the coordinate system. On the other hand, if the coefficient k of the line in (3) is equal to the coefficient k_1 of a line l_1 with equation
$$y = k_1 x + s_1, \tag{3a}$$
then this equality does have geometric significance; in fact, if $k = k_1$, transition to a new coordinate system yields the equality
$$k' = \frac{k\cos\bar\alpha + \sin\bar\alpha}{\cos\bar\alpha - k\sin\bar\alpha} = \frac{k_1\cos\bar\alpha + \sin\bar\alpha}{\cos\bar\alpha - k_1\cos\bar\alpha} = k'_1.$$

In geometric terms, $k = k_1$ signifies *parallelism* of the lines l and l_1.

If l and l_1 are parallel, i.e., if $k = k_1$, then the quantity
$$\frac{|s_1 - s|}{\sqrt{k^2 + 1}}$$
has geometric significance. In fact,
$$s'_1 - s' = \frac{k\bar a - \bar b + s_1}{\cos\bar\alpha - k\sin\bar\alpha} - \frac{k\bar a + \bar b + s}{\cos\bar\alpha - k\sin\bar\alpha}$$
$$= \frac{s_1 - s}{\cos\bar\alpha - k\sin\bar\alpha}$$
and
$$k'^2 + 1 = \left(\frac{k\cos\bar\alpha + \sin\bar\alpha}{\cos\bar\alpha - k\sin\bar\alpha}\right)^2 + 1$$
$$= \frac{(k\cos\bar\alpha + \sin\bar\alpha)^2 + (\cos\bar\alpha - k\sin\bar\alpha)^2}{(\cos\bar\alpha - k\sin\bar\alpha)^2}$$
$$= \frac{k^2 + 1}{(\cos\bar\alpha - k\sin\bar\alpha)^2},$$
so that
$$\frac{|s'_1 - s'|}{\sqrt{k'^2 + 1}} = \frac{|s_1 - s|/|\cos\bar\alpha - k\sin\bar\alpha|}{\sqrt{k^2 + 1}/|\cos\bar\alpha - k\sin\bar\alpha|} = \frac{|s_1 - s|}{\sqrt{k^2 + 1}}.$$

[Consideration of $\triangle SS_1 P$ in Fig. 11 shows that
$$\frac{|s_1 - s|}{\sqrt{k^2 + 1}} = \frac{SS_1}{\sqrt{\tan^2\varphi + 1}} = SS_1 \cos\varphi = S_1 P = d$$
is the *distance between the lines* l and l_1.]

1. What is geometry?

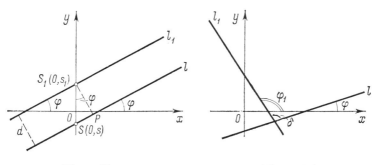

Figure 11a Figure 11b

Even if the lines l and l_1 given by Equations (3) and (3a) are not parallel, i.e., if $k \ne k_1$, then the rather complicated expression

$$\frac{k_1 - k}{k_1 k + 1}$$

has geometric significance.[9] To see this, note that

$$k_1' - k' = \frac{k_1 \cos \bar{\alpha} + \sin \bar{\alpha}}{\cos \bar{\alpha} - k_1 \sin \bar{\alpha}} - \frac{k \cos \bar{\alpha} + \sin \bar{\alpha}}{\cos \bar{\alpha} - k \sin \bar{\alpha}}$$

$$= \frac{(k_1 \cos \bar{\alpha} + \sin \bar{\alpha})(\cos \bar{\alpha} - k \sin \bar{\alpha}) - (k \cos \bar{\alpha} + \sin \bar{\alpha})(\cos \bar{\alpha} - k_1 \sin \bar{\alpha})}{(\cos \bar{\alpha} - k_1 \sin \bar{\alpha})(\cos \bar{\alpha} - k \sin \bar{\alpha})}$$

$$= \frac{k_1 - k}{(\cos \bar{\alpha} - k_1 \sin \bar{\alpha})(\cos \bar{\alpha} - k \sin \bar{\alpha})},$$

and

$$k_1' k' + 1 = \frac{k_1 \cos \bar{\alpha} + \sin \bar{\alpha}}{\cos \bar{\alpha} - k_1 \sin \bar{\alpha}} \cdot \frac{k \cos \bar{\alpha} + \sin \bar{\alpha}}{\cos \bar{\alpha} - k \sin \bar{\alpha}} + 1$$

$$= \frac{(k_1 \cos \bar{\alpha} + \sin \bar{\alpha})(k \cos \bar{\alpha} + \sin \bar{\alpha}) + (\cos \bar{\alpha} - k_1 \sin \bar{\alpha})(\cos \bar{\alpha} - k \sin \bar{\alpha})}{(\cos \bar{\alpha} - k_1 \sin \bar{\alpha})(\cos \bar{\alpha} - k \sin \bar{\alpha})}$$

$$= \frac{k_1 k + 1}{(\cos \bar{\alpha} - k_1 \sin \bar{\alpha})(\cos \bar{\alpha} - k \sin \bar{\alpha})},$$

so that[10]

$$\frac{k_1' - k'}{k_1' k' + 1} = \frac{k_1 - k}{k_1 k + 1}.$$

---

[9]If the denominator $k_1 k + 1 = 0$, we put $(k_1 - k)/(k_1 k + 1) = \infty$. (Translator's note.)

[10]If $\cos \bar{\alpha} - k \sin \bar{\alpha} = 0$ or $\cos \bar{\alpha} - k_1 \sin \bar{\alpha} = 0$, then in the expression $(k_1' - k')/(k_1' k' + 1)$ we put ∞ in place of k' or of k_1', respectively. For example, if $\cos \bar{\alpha} - k \sin \bar{\alpha} = 0$ (i.e., if the new equation of l is $x' = a'$), then we put

$$\frac{k_1' - k'}{k_1' k' + 1} = \lim_{k' \to \infty} \frac{k_1' - k'}{k_1' k' + 1} = \lim_{k' \to \infty} \frac{(k_1'/k') - 1}{k_1' + (1/k')} = -\frac{1}{k_1'}.$$

Then in this case we also have

$$\frac{k_1' - k'}{k_1' k' + 1} = -\frac{1}{k_1'} = \frac{-\cos \bar{\alpha} + k_1 \sin \bar{\alpha}}{k_1 \cos \bar{\alpha} + \sin \bar{\alpha}} = \frac{-\cot \bar{\alpha} + k_1}{k_1 \cot \bar{\alpha} + 1} = \frac{k_1 - k}{k_1 k + 1},$$

for clearly, $k = \cot \bar{\alpha}$.

[It is easy to see the geometric meaning of $(k_1-k)/(k_1k+1)$. If $k=\tan\varphi$ and $k_1=\tan\varphi_1$, where φ and φ_1 are the angles formed by the lines l, l_1 and the ray Ox (Fig. 11b), then

$$\frac{k_1-k}{k_1k+1} = \frac{\tan\varphi_1-\tan\varphi}{\tan\varphi_1\tan\varphi+1} = \tan(\varphi_1-\varphi) = \tan\delta,$$

where $\delta = \varphi_1 - \varphi$ is the *angle between the lines* l and l_1.]

The term "geometric property" has by now been given two different meanings. It remains to show that the two meanings are actually the same. This we do next.

We note that formulas (6) admit an "alibi" interpretation.[11] We can think of these formulas as defining a transition from the point $A(x,y)$ to the point $A'(x',y')$, where (x,y) and (x',y') are the coordinates of A and A', respectively, in the same (rectangular) coordinate system $\{x,y\}$. Thus in this interpretation formulas (6) define a *transformation of the plane*. Since any set of points A and its corresponding set of points A' have the same geometric properties [this follows from the original interpretation of formulas (6) as giving the connection between the coordinates of the same point in different coordinate systems], it follows that the transformation (6) must take a figure F into a congruent figure F', and so represents a **motion**. In fact, the transformation (6) may be thought of as a *rotation*

$$\begin{aligned} x_1 &= x\cos\alpha + y\sin\alpha, \\ y_1 &= -x\sin\alpha + y\cos\alpha \end{aligned} \tag{7}$$

through the angle α about the origin O (Fig. 12a) followed by the *translation*

$$\begin{aligned} x' &= x_1 + a, \\ y' &= y_1 + b \end{aligned} \tag{8}$$

(Fig. 12b). we thus return to the assertion that geometry studies properties of figures preserved by the motions (6) (cf. p. 3, lines 9–10).[12]

A similar treatment can be applied to three-dimensional geometry. In a given (rectangular) coordinate system $\{x,y,z\}$ each point is assigned three coordinates (x,y,z). Transition from a coordinate system $\{x,y,z\}$ to another system $\{x',y',z'\}$ is given by the complicated formulas

$$\begin{aligned} x &= (\cos\beta\cos\alpha - \cos\gamma\sin\beta\sin\alpha)\cdot x + (\sin\beta\cos\alpha + \cos\gamma\cos\beta\sin\alpha)\cdot y \\ &\quad + \sin\gamma\sin\alpha\cdot z + a, \\ y' &= -(\cos\beta\sin\alpha + \cos\gamma\sin\beta\cos\alpha)\cdot x + (-\sin\beta\sin\alpha + \cos\gamma\cos\beta\cos\alpha)\cdot y \\ &\quad + \sin\gamma\cos\alpha\cdot z + b, \\ z' &= \sin\gamma\sin\beta\cdot x - \sin\gamma\cos\beta\cdot y + \cos\gamma\cdot z + c, \end{aligned} \tag{9}$$

---

[11] By contrast, the original interpretation, where (x,y) and (x',y') are regarded as coordinates of the same point in two different coordinate systems, is called the "alias" interpretation (alibi = another place; alias = another name). (Translator's note.)

[12] Since we restricted ourselves to right-handed coordinate systems (cf. footnote 6), formulas (6) do not give all the motions of the plane but only the so-called "direct" motions of the plane which carry a right-handed coordinate system into a right-handed coordinate system (see Chap. II, Sec. 2 of [10]; Chap. 3 of [19] and Chap. 1, Sec. 18 of [32]).

1. What is geometry?

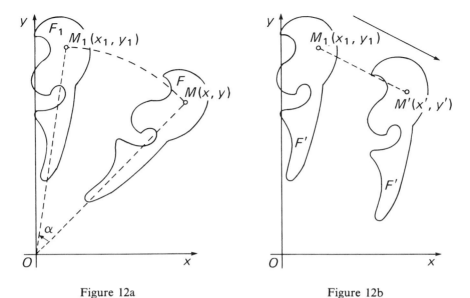

Figure 12a   Figure 12b

where a,b,c are the coordinates of the old origin O in the new system $\{x',y',z'\}$, and α,β,γ are the so-called *Euler angles*. These angles are shown in Figure 13, in which we use an auxiliary coordinate system $\{x_1,y_1,z_1\}$ whose origin is O, and whose axes are parallel to those of $\{x',y',z'\}$. The line of intersection of the planes Oxy and Ox_1y_1 is denoted in the figure by ON. We may say that *three dimensional geometry is concerned with those relations among the coordinates of points which are unchanged by the transformations* (9). Alternatively, we may regard the coordinate system as fixed and view (9) as defining a transition from a point $A(x,y,z)$ to a new

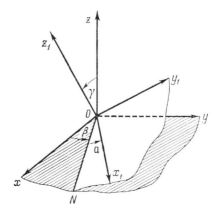

Figure 13

point $A'(x',y',z')$, that is, we may regard (9) as defining a *transformation of space*. Since the transformations (9) are **motions**,[13] we again come to Klein's point of view, which asserts that *geometry studies the properties of figures preserved by all motions*.

PROBLEMS AND EXERCISES

1   Prove formulas (9).

2   (a) How would formulas (6) change if one did not employ right-handed coordinate systems only (cf. footnote 6)? [It is clear that the "alibi" interpretation of the restriction to right-handed coordinate systems amounts to consideration of direct motions only (cf. Footnote 12)]. (b) How would formulas (9) change if one did not employ direct motions only (cf. Footnote 13)?

3   (a) Show that every direct motion of the plane is a *rotation* or a *translation*. (b) Show that every opposite motion of the plane is a *glide reflection*, i.e., a reflection in a line followed possibly by a translation in the direction of this line.

I   (a) Classify direct motions in space [cf. Exercise 3 (a)]. (b) Classify opposite motions in space.

II Interpret "motion" as translation [cf. Equations (8)]. Define the *geometry of translations* to be the study of properties of plane figures invariant under all translations. (a) Which of the following Euclidean concepts are also concepts of the geometry of translations: triangle, quadrilateral, trapezoid, rhombus, circle, parallelism, perpendicularity, median, angle bisector, altitude, length of a segment, magnitude of an angle, area? [*Hint*: In Euclidean geometry, a trapezoid is a quadrilateral with a pair of parallel opposite sides, and a circle is the locus of points equidistant from a fixed point Q. Could these concepts be defined in terms of our geometry of translations?] (b) Which of the following theorems of Euclidean geometry carry over in a natural way to the geometry of translations: the theorems about the midline of a triangle[14] and a trapezoid; the theorems about the concurrence of the medians and altitudes of a triangle; the theorem on the sum of the angles in a triangle; the theorems about properties of isosceles triangles; the theorem which asserts that triangles with equal bases and altitudes have equal areas; the theorem which asserts that triangles (polygons) with equal areas are equidecomposable[15]; Pythagoras' theorem? (c) Show that, in the geometry of translations: two triangles ABC and $A'B'C'$ are congruent if and only if the segments AA', BB', and CC' are congruent; if two opposite sides of a quadrilateral are congruent, then the remaining two sides are congruent; if the opposite sides of a $2n$-gon are congruent, then the diagonals joining opposite vertices intersect in a point which is their common midpoint.

III Define *central Euclidean geometry* to be the study of properties of the plane which are invariant under the rotations (7) about a fixed point O (which should be excluded from the plane since it plays a different role from the

---

[13] More precisely, "direct" motions; cf. Chap. 7 in [19], or Chap. 4, Sec. 14 of [32].

[14] This theorem states that the line joining the midpoints of two sides of a triangle is parallel to the remaining side. (Translator's note.)

other points). Develop elements of plane central Euclidean geometry using Problem **II** as a model.

IV Define *geometry of parallelism* to be the study of properties of plane figures invariant under translations and central dilatations (i.e., under motions which carry each line to a parallel line). **(a)** Write down equations [analogous to (6)] that yield all the "motions" of this geometry. **(b)** Using Problem **II** as a model, develop elements of the geometry of parallelism; in particular, you might try to decide the status of the theorem about the equidecomposability[15] of two triangles (polygons) with equal areas.

V Develop three-dimensional geometries corresponding to various "exotic" [i.e., different from (9)] sets of "motions" of space.

2. What is mechanics?

One of the basic aims of this book is to establish a definite connection between mechanics and geometry. This connection rests on a deep analogy between the role of motions in geometry (considered as distance-preserving transformations of the plane or of space, unrelated to such purely mechanical concepts as velocity or the path of a moving point) and the role of uniform motions in mechanics. (Here velocity is obviously a crucial concept, as the word "uniform" in the description indicates.) This connection can also be said to rest on the analogy between the respective roles of rectangular coordinate systems in geometry and so-called inertial reference frames in mechanics. The present section is devoted to a discussion of these analogies.

The three basic areas of mechanics are kinematics, statics, and dynamics. We shall be primarily concerned with *kinematics*, which studies the motions of material points and bodies. What we mean by "motion" in mechanics is so different from what we mean by "motion" in geometry that, were it not a matter of firmly rooted tradition, there would be every reason to give these concepts different names.

In geometry a "motion" is a certain type of point transformation which associates to each point A a definite point A'. The geometer regards the question of how A reaches A' as meaningless. He identifies the motion with the correspondence $A \mapsto A'$ and regards all else as irrelevant. In mechanics, on the other hand, motion is a definite *process* which takes A to a new point A', and what concerns us are the paths of individual points as well as their velocities or accelerations at various times. A principal difference between mechanical and geometric motions is that *time* is a factor in mechanical motions but not in geometric motions. In mechanics, a description of a motion which takes a figure F onto a figure F' is a rule which tells us how the position of each point A of F changes in time; in mathematical terms, we speak of the *functional dependence* $A = A(t)$ of the

---

[15]For a discussion of equidecomposable Euclidean figures see, for example, [5b].

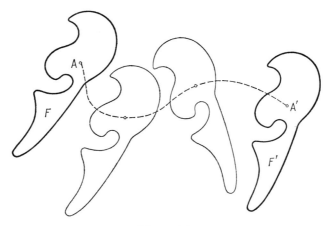

Figure 14

position of A in F on time t. This functional dependence is given in some time interval $t_0 \leq t \leq t_1$, where the moving figure occupies at time t_0 the initial position F [so that $A(t_0) = A$ is a point of F], and at time t_1 the position F' [so that $A(t_1) = A'$ is a point of F'; see Fig. 14].

In mechanics, with its technical applications, we are mainly concerned with motions of figures in space. Nevertheless, we shall also study plane motions of the kind illustrated in Figure 14. The concept of plane motion can be made physically real as follows. Consider a *plane-parallel* motion of an object Φ, that is, a motion in which the points of Φ move parallel to a fixed plane π (Fig. 15a). It is clear that the motion is completely characterized by the process of displacement of the points of a section of Φ parallel to the plane π, for example, by the motion in the plane π of the plane figure F in which π intersects Φ.[16]

A special case of this type of motion is *rectilinear* motion, in which the points of Φ move with the same velocity in a direction determined by some line o (Fig. 15b). It is clear that such a motion is fully characterized by the motion of an arbitrary point A of Φ. Since the direction of a rectilinear motion is along a fixed line, its velocity can be given by a number rather than a vector (the number being positive or negative depending on the direction of the motion along the line). The position of the point A on the line o is also given by a (positive or negative) number x, which measures the (positive or negative) distance of A from the origin O on the directed line o. We say that x is the magnitude of the directed segment $\overline{OA}$; this means that x is the length of this segment if its direction (from O to A) agrees with the positive direction on o, and the negative of that length otherwise. [We use the notations OA and $\overline{OA}$ for segments and directed segments, respectively.] The fact that in rectilinear motion velocity and

---

[16] Assuming, of course, that the intersection $F = \Phi \cap \pi$ is not the empty set. This is really no loss of generality, since π can always be replaced by any plane parallel to π throughout the discussion. (Translator's note.)

2. What is mechanics?

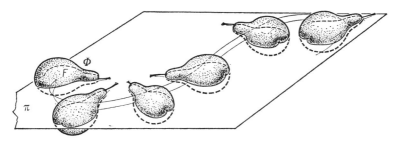

Figure 15a

Figure 15b

position of a point can each be given by a number is an obvious advantage. Rectilinear motion is also simpler than plane-parallel or general motion in space in other respects. The relative simplicity of rectilinear motion is the main reason why, in this book, we are mainly concerned with rectilinear motion of a (material) point A on line o.

We shall now explain which properties of moving objects are significant in mechanics (i.e., are mechanical phenomena), and which are not. At first it might seem that the fundamental notions of mechanics (more precisely, of kinematics) are "path," "velocity," and "acceleration." [By the *path* of a point A we mean the curve traced by the point in its process of motion (see, for example, Fig. 14). By the *mean velocity* of A in the time interval from t_0 to t_1 we mean the change in position $A(t_1) - A(t_0)$ divided by the difference $t_1 - t_0$; and by the *instantaneous velocity* at time t we mean the limit as $\Delta t \to 0$ of the mean velocity in the time interval from t to $t + \Delta t$ (derivative of displacement with respect to time). By the *mean acceleration* in the time interval from t_0 to t_1 we mean the change in velocity $v(t_1) - v(t_0)$ divided by the difference $t_1 - t_0$; and by the *instantaneous acceleration* at time t we mean the limit as $\Delta t \to 0$ of the mean acceleration in the time interval from t to $t + \Delta t$ (derivative of velocity with respect to time).] This is not the case, however. In and of themselves, paths and velocities of points have no mechanical significance and cannot be the objects of study in mechanics.

This assertion may surprise many readers. We leave the necessary explanation to the great GALILEO GALILEI (1564–1642)[17]:

> Shut yourself up with some friend in the main cabin below decks on some large ship, and have with you there some flies, butterflies, and other small

---

[17]Galileo Galilei, *Dialogue Concerning the Two Chief World Systems*. [15], pp. 186–187.

flying animals. Have a large bowl of water with some fish in it; hang up a bottle that empties drop by drop into a wide vessel beneath it. With the ship standing still, observe carefully how the little animals fly with equal speed to all sides of the cabin. The fish swim indifferently in all directions; the drops fall into the vessel beneath; and, in throwing something to your friend, you need throw it no more strongly in one direction than another, the distances being equal; jumping with your feet together, you pass equal spaces in every direction. When you have observed all these things carefully (though there is no doubt that when the ship is standing still everything must happen in this way), have the ship proceed with any speed you like, so long as the motion is uniform and not fluctuating this way and that. You will discover not the least change in all the effects named, nor could you tell from any of them whether the ship was moving or standing still. In jumping, you will pass on the floor the same spaces as before, nor will you make larger jumps toward the stern than toward the prow even though the ship is moving quite rapidly, despite the fact that during the time that you are in the air the floor under you will be going in a direction opposite to your jump. In throwing something to your companion, you will need no more force to get it to him whether he is in the direction of the bow or the stern, with yourself situated opposite. The droplets will fall as before into the vessel beneath without dropping toward the stern, although while the drops are in the air the ship runs many spans. The fish in their water will swim toward the front of their bowl with no more effort than toward the back, and will go with equal ease to bait placed anywhere around the edges of the bowl. Finally the butterflies and flies will continue their flights indifferently toward every side, nor will it ever happen that they are concentrated toward the stern, as if tired out from keeping up with the course of the ship, from which they will have been separated during long intervals by keeping themselves in the air. And if smoke is made by burning some incense, it will be seen going up in the form of a little cloud, remaining still and moving no more toward one side than the other.

This deservedly famous and frequently quoted passage contains a beautiful and most accomplished description of one of the fundamental principles of mechanics—**Galileo's principle of relativity**. This principle can be briefly formulated as follows. *No mechanical experiment conducted within a physical system can disclose the uniform motion of this system.* Thus mechanical phenomena taking place in two laboratories, one of which is in uniform motion with respect to the other (as, for example, in the case of the moving and stationary boats in Galileo's Dialogue), are indistinguishable from the point of view of observers in the two laboratories. Galileo's principle of relativity implies that all properties studied in mechanics are preserved under transformations of the physical system obtained by imparting to it a velocity which is constant in magnitude and direction (such transformations are called **Galilean transformations**). In other words, *mechanical properties of (moving) objects do not change under Galilean transformations* (cf. the definition of geometric properties given in Sec. 1 as properties invariant under motions).

The Galilean principle of relativity can be stated in a "geometric" form which links it directly to Klein's concept of geometry. For the sake of

2. What is mechanics?

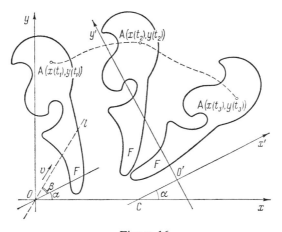

Figure 16

simplicity, we restrict ourselves to mechanical phenomena which may be regarded as taking place in a plane, such as, for example, motions of physical objects restricted to a portion of the earth small enough to be supposed flat. As usual, we introduce in the plane a rectangular coordinate system $\{x,y\}$. Then the mechanical motion of a point A is given by the formulas

$$
\begin{aligned}
x &= x(t), \\
y &= y(t),
\end{aligned}
\tag{10}
$$

which tell us how the coordinates (x,y) of the point vary with time t. [The motion of a figure F is described by similar formulas which tell us how all the points $A = A(x,y)$ in F move.[18]] It is quite clear that transition to a new coordinate system $\{x',y'\}$ obtained from $\{x,y\}$ by rotating the axes Ox, Oy through an angle α and translating the origin O to some point O' (Fig. 16) cannot influence physical laws. Such laws must, therefore, take the same form in either of two coordinate systems $\{x,y\}$ and $\{x',y'\}$ connected by the relations

$$
\begin{aligned}
x' &= x\cos\alpha + y\sin\alpha + a, \\
y' &= -x\sin\alpha + y\cos\alpha + b,
\end{aligned}
\tag{6}
$$

with $\alpha = \angle xCx'$, the angle between the axes Ox and $O'x'$, and (a,b) the coordinates of O in the new coordinate system. This means that for a proposition to have mechanical significance it must retain its form under the transformations (6). In addition, Galileo's principle of relativity asserts that the description of all mechanical processes relative to coordinates (x,y) or (x',y') is unaffected if the origin and axes of the coordinate system

---

[18]We are dealing, essentially, with plane-parallel motions of objects (cf. p. 16 and Fig. 15a). Since plane-parallel motion of a rigid body Φ is completely determined by that of its plane section F, it suffices to consider the points of Φ in the plane figure F.

$\{x',y'\}$ move uniformly with respect to the coordinate system $\{x,y\}$ or, equivalently, the origin and axes of the old coordinate system $\{x,y\}$ move uniformly with respect to the new coordinate system $\{x',y'\}$. Now if the origin O of $\{x,y\}$ moves with velocity v along the line l which forms an angle β with the axis $O'x'$ (cf. Fig. 16), then the coordinates $a(t)$ and $b(t)$ of O relative to $\{x',y'\}$ at time t are

$$a(t) = a + v\cos\beta \cdot t,$$
$$b(t) = b + v\sin\beta \cdot t$$

(here a and b are the coordinates of O relative to $\{x',y'\}$ at $t=0$). Hence the relation between the coordinates (x',y') and (x,y) of a point A relative to $\{x',y'\}$ and $\{x,y\}$ is

$$\begin{aligned} x' &= x\cos\alpha + y\sin\alpha + (v\cos\beta)t + a, \\ y' &= -x\sin\alpha + y\cos\alpha + (v\sin\beta)t + b. \end{aligned} \tag{11}$$

It follows that all phenomena which have mechanical significance must be expressible by means of formulas whose form is unaffected by the transformations (11) [or, as mathematicians put it, are *invariant under the transformations* (11)].

We can supplement formulas (11) to some extent. We observe that formulas (10), which describe the motion of a point $A = A(x,y)$, and formulas (11) involve the time t in addition to the coordinates (x,y). This is as it should be, for the basic difference between mechanical and geometric motions is that the time factor is considered in the former and ignored in the latter. Obviously, the choice of a "time origin" must not affect the form of physical laws. Thus, for example, most people today count time from the birth of Christ. On the other hand, the ancient Greeks counted time from the first Olympiad (July 1, 776 B.C.), the ancient Romans counted time from the founding of Rome (April 21, 753 B.C.), Moslems usually count time from the date of Mohammed's flight from Mecca to Medina ("hegira"; July 16, 622 A.D.), and so on. Nevertheless, in all of these time systems (whose origin is often linked to some mythical event) physical laws take the same form. To put it in simple terms, the choice of one calendar or another must have no effect on the content of physics. It is therefore appropriate to add to (11) the transformation

$$t' = t + d,$$

which describes a shift in the time origin; here d is the time of the old time origin (i.e., the moment $t=0$) in the new time system. We thus have

$$\boxed{\begin{aligned} x' &= (\cos\alpha)x + (\sin\alpha)y + (v\cos\beta)t + a, \\ y' &= -(\sin\alpha)x + (\cos\alpha)y + (v\sin\beta)t + b, \\ t' &= t + d. \end{aligned}} \tag{12}$$

We box formulas (12) to emphasize their fundamental importance. The mathematical meaning of Galileo's principle of relativity is that *all properties of* (plane-parallel) *motions which have mechanical significance are ex-*

2. What is mechanics?

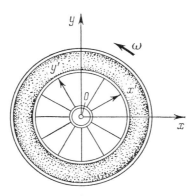

Figure 17

pressible in terms of formulas which are invariant under the transformations (12).

Formulas (12), just like formulas (6) in the previous section, can be looked at in two ways. One way (the alibi interpretation) is to look at (12) as a transition from event $A(x,y,t)$ [characterized by giving the position $A(x,y)$ and the time t] to event $A'(x',y',t')$, where the point $A'(x',y')$ moves uniformly with velocity v with respect to the point $A(x,y)$; here α is the angle at $t=0$ between $A'O'$ [where $O'=O'(a,b)$] and AO, β is the angle between the path of A' and Ox, and d is the value of t' at $t=0$. From this point of view, formulas (12) define a transformation in physical space–time. Such transformations are called *Galilean transformations*. Another viewpoint (the alias interpretation) is to look at formulas (12) as the connection between two reference frames $\{x,y,t\}$ and $\{x',y',t'\}$. Then all reference frames linked by relations of the form (12) (with different parameters α,β,v,a,b,d) are equivalent. Also, relative to such reference frames, the laws of mechanics take a particularly simple, "natural" form. The latter statement requires clarification. Let $\{x,y\}$ be the initial coordinate system, and let $\{x',y'\}$ be a new coordinate system moving with respect to $\{x,y\}$ not uniformly but in some other, more complicated manner; for example, let $\{x',y'\}$ rotate about the origin O of $\{x,y\}$ with angular velocity ω (Fig. 17). Then objects at rest relative to the moving coordinate system $\{x',y'\}$ are acted upon by external forces due to the motion of $\{x',y'\}$. Such forces are called *inertial forces*. In our example, such a force is the "centrifugal force" directed outward from the center of rotation. We experience such a force whenever the car we ride in negotiates a turn (with the car there is associated a coordinate system relative to which we are at rest). The reference frames in which there are no inertial forces, and in which physical laws are therefore not complicated by the need to consider the "extraneous" inertial forces, are called **inertial reference frames**; there are infinitely many such reference frames, and transition from one to another is effected by a transformation (12).[19]

---

[19] In connection with these matters, we suggest that the reader consult [17].

It is now easy to explain why such seemingly simple and basic notions as the path of a moving point or its velocity have no physical meaning and cannot, therefore, appear in a statement of a physical law. In fact, the path of a moving point depends on the choice of an inertial reference frame (we recall that inertial reference frames move uniformly with respect to each other); for example, the tip of a pen wielded by an artist on a fast-moving boat traces an extremely involved path when viewed by an observer on the boat, but a nearly linear path when viewed by a stationary observer on shore.[20] Similarly, the velocity of a point depends on the choice of an inertial reference frame; an object at rest relative to a boat moving uniformly may be moving fast relative to the land. Things are different when we consider the *relative velocity* of one object with respect to another. For example, the velocity with which a runner approaches the ribbon at the finish line is a mechanical quantity, since it does not depend on the inertial reference frame in which we consider runner and ribbon. Whether our coordinate system is linked to the earth or to the fixed stars (in which case both runner and ribbon move with tremendous, "cosmic," velocities), the relative velocity of the runner, which depends on how fast the distance between runner and ribbon is decreasing, is the same. Of even greater importance is the fact that the *acceleration* of a moving point has "absolute" significance, i.e., its value is independent of the choice of an inertial coordinate system. This is not so surprising if we bear in mind the fact that acceleration involves the difference of velocities (cf. p. 17 above). Since transition from one inertial reference frame to another has the effect of changing all velocities by the same amount, the *difference* between two velocities remains unchanged.

Now the velocity of a point is represented by a vector, i.e., velocity is a vector quantity. Moreover, the velocity $\mathbf{v}$ of a point A relative to the old coordinate system $\{x,y\}$ (its so-called *absolute velocity*) differs from the velocity $\mathbf{v}'$ of A relative to the new coordinate system $\{x',y'\}$ (its so-called *relative velocity*) by a vector $\mathbf{a}$ which equals the velocity of the origin O' of $\{x',y'\}$ relative to $\{x,y\}$ (the so-called *transport velocity* of $\{x',y'\}$ relative to $\{x,y\}$). Thus

$$\mathbf{v} = \mathbf{v}' + \mathbf{a}$$

(cf. pp. 48–49). It follows that if $\mathbf{v}_1$ and $\mathbf{v}_2$ are the velocities of two objects (for example runner and ribbon) in one reference frame, and $\mathbf{v}'_1, \mathbf{v}'_2$ are their velocities in another reference frame, then

$$\mathbf{v}_1 = \mathbf{v}'_1 + \mathbf{a} \quad \text{and} \quad \mathbf{v}_2 = \mathbf{v}'_2 + \mathbf{a},$$

so that

$$\mathbf{v}_1 - \mathbf{v}_2 = \mathbf{v}'_1 - \mathbf{v}'_2.$$

Things are much the same if $\mathbf{v}_1$ and $\mathbf{v}_2$ are the velocities of an object at two instants t_1 and t_2. Since $\{x',y'\}$ is supposed to move uniformly relative to $\{x,y\}$, i.e., the vector $\mathbf{a}$ is assumed constant, it follows, just as before, that

$$\mathbf{v}_1 = \mathbf{v}'_1 + \mathbf{a}, \qquad \mathbf{v}_2 = \mathbf{v}'_2 + \mathbf{a}, \quad \text{and} \quad \mathbf{v}_1 - \mathbf{v}_2 = \mathbf{v}'_1 - \mathbf{v}'_2.$$

---

[20]Cf. Galileo's "Dialogues" [15], pp. 171–172.

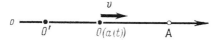

Figure 18

But then the average acceleration $\mathbf{w}_{av} = (\mathbf{v}_1 - \mathbf{v}_2)/(t_1 - t_2)$, and the instantaneous acceleration $\mathbf{w}_{inst} = \lim_{t_1 \to t_2} \mathbf{w}_{av}$ do not depend on the choice of an inertial coordinate system.[21]

So far we have concerned ourselves with two-dimensional (more accurately, plane-parallel) motions affecting points $A(x,y)$ of some plane xOy. However, nothing prevents us from restricting ourselves to even simpler, *rectilinear motions*, where we need only consider motions of points $A = A(x)$ of some fixed line o. Then, if $\{x\}$ and $\{x'\}$ are two inertial reference frames, the origin O of the coordinate system $\{x\}$ moves relative to the coordinate system $\{x'\}$ with constant velocity v (Fig. 18), i.e., at time t, the coordinate $a(t)$ of the (moving) point O relative to the coordinate system $\{x'\}$ is given by

$$a(t) = a + vt,$$

where t is time and a is the $\{x'\}$-coordinate of O at time $t = 0$. Since the relation between the coordinates x and x' of a point A in the coordinate system $\{x\}$ and the coordinate system $\{x'\}$ with origin O' is

$$x' = x + a(t)$$

(cf. Fig. 18),[22] it follows that

$$x' = x + vt + a.$$

By adding the relation

$$t' = t + b,$$

which expresses the possibility of shifting the time origin, we arrive at the formulas

$$\boxed{\begin{aligned} x' &= x + vt + a, \\ t' &= t + b, \end{aligned}} \qquad (13)$$

which give *the relation between two inertial coordinate systems in the case of rectilinear motions* (Galilean transformations for rectilinear motions).

True rectilinear motions do not occur in physics very often. Consequently the mechanics of such motions is of less interest than that of motions in the plane or in three-space. However, it is natural to begin the

---

[21] We suggest that the reader familiar with elements of the calculus check that if $\mathbf{r} = \mathbf{r}(x,y)$ is the position vector of a (moving) point $A(x,y)$, then its acceleration $\ddot{\mathbf{r}} = d^2\mathbf{r}/dt^2$ is not affected by a transition from $\{x,y,t\}$ to $\{x',y',t'\}$, provided that the two reference frames are related by the formulas (12).

[22] Here we are considering only right-handed coordinate systems, in which the positive direction on the x-axis is fixed beforehand (for example, by selecting the positive direction on o, supposed horizontal, to be the direction to the right of O; cf. footnote 6 of Sec. 1).

study of mechanics with the simplest case, that of rectilinear motion, where the difference between mechanical and purely geometric considerations is particularly clear and not obscured by technical difficulties which arise when it is necessary to take into consideration the vector character of the basic mechanical magnitudes (velocities, accelerations, forces). Also, the study of rectilinear motions, which includes in particular the study of (small) displacements of material points under gravity (vertical displacements), has definite "applied" interest. Finally, comparison of formulas (13) and (12) [and even more so, of (13) and (16)] amply illustrates the convenience and simplicity of presentation resulting from the restriction to rectilinear motions.

One more argument in favor of the restriction to rectilinear motions is that the resulting geometry is only two-dimensional, and hence can be more easily visualized. Indeed, in studying rectilinear motions, what interests us is those facts which are described relative to a coordinate system $\{x,y\}$ by formulas invariant under the Galilean transformations

$$\begin{aligned} x' &= x + vt + a, \\ t' &= t + b. \end{aligned} \tag{13}$$

If we represent the position of a point $A(x)$ on a line o at time t by means of an auxiliary point $A(x,t)$ of a (two-dimensional) plane xOt with coordinates x and t (Fig. 19a,b), then we obtain a kind of "geometry" in which the only facts of geometric (or rather mechanical) significance are those facts which can be expressed by means of formulas invariant under the transformations (13). Hence these transformations play the role of "motions" of our geometry, which do not change the properties of figures of interest to us. It is this geometry which we will study in Chapters I and II. While it would be proper to refer to it as "the geometry of Galileo's principle of relativity" we shall use the simpler name "**Galilean geometry.**"

Our study of Galilean geometry will compare it with **Euclidean geometry**, named for the great Greek mathematician EUCLID, who lived and taught in Alexandria in the third century B.C., and to whom we owe the first (excellent) textbook on geometry, from which many generations

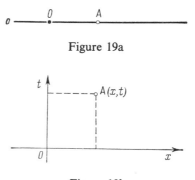

Figure 19a

Figure 19b

learned this subject.[23] The difference between Euclidean geometry and Galilean geometry is that the motions of Euclidean geometry are given by formulas (6) and those of Galilean geometry by formulas (13). To simplify the comparative study of these two remarkable geometries, we shall use the usual Euclidean coordinate symbols x and y in the Galilean plane; specifically, we shall use the letter y (not x) to denote the coordinate of a moving point A on a line o, and x (not t) to denote time. Then we say that Euclidean geometry is the study of properties of figures in the coordinate plane $\{x,y\}$ that are invariant under the transformations (6), and Galilean geometry is the study of properties of figures in the coordinate plane $\{x,y\}$ that are invariant under the transformations

$$\boxed{\begin{aligned} x' &= x \quad\;\;\, + a, \\ y' &= vx + y + b \end{aligned}} \tag{13a}$$

[cf. (13a) and (13), and observe that the letters a and b are reversed in passing from (13) to (13a)]. While we propose to carry out a comparative study of the two geometries, and thus to emphasize the geometric approach to Galilean relativity, we urge the reader not to overlook the mechanical interpretation of the subject.

Finally, we observe that every Galilean transformation (13a) can be composed of the transformation

$$\left. \begin{aligned} x_1 &= x, \\ y_1 &= vx + y, \end{aligned} \right\} \text{ or } \left. \begin{aligned} x_1 &= x + vt, \\ t_1 &= \quad\;\; t, \end{aligned} \right\} \tag{14a}$$

which describes the uniform motion with velocity v of the origin O_1 of the moving coordinate system $\{y_1\}$ on o,[24] and the transformation

$$\left. \begin{aligned} x' &= x_1 + a, \\ y' &= y_1 + b, \end{aligned} \right\} \text{ or } \left. \begin{aligned} x' &= x_1 + b, \\ t' &= t_1 + a, \end{aligned} \right\} \tag{14b}$$

which describes a shift of O_1 to some point O', together with a shift of the time origin. The geometric sense of (14b) [or of (8); see p. 12 above] is that of a *translation* (see Fig. 12b). A transformation (14a) represents a *shear* with coefficient v in the direction of the y-axis. This shear leaves the points of the y-axis fixed, and translates all other points in a direction parallel to it. The magnitude of the translation is proportional (with proportionality constant v) to the distance from the point in question to the y-axis. The direction of the shift is reversed when we cross the y-axis (see Fig. 20). Thus, to check whether a certain quantity is invariant under the transformations (13a), it suffices to see whether it is *invariant under the shears* (14a) *and the translations* (14b). This fact will frequently be of help to us in what follows.

---

[23]For an English translation, see [1]–[3].
[24]We assume that at time $t=0$, O_1 coincides with O.

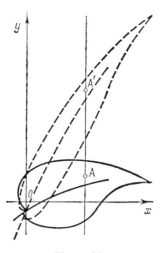

Figure 20

Similarly, the study of mechanics of plane-parallel motions reduces to the study of a geometry of three-space with coordinates $\{x,y,t\}$ whose motions are given by formulas (12). This geometry can be called *three-dimensional Galilean geometry*. It is of considerable interest but it is decidedly more complicated than two-dimensional Galilean geometry. Each motion (12) can be split into three motions: a *rotation*

$$\begin{aligned} x_1 &= x\cos\alpha + y\sin\alpha, \\ y_1 &= -x\sin\alpha + y\cos\alpha, \\ t_1 &= t \end{aligned} \quad (15a)$$

about the t-axis (Fig. 21); a *shear*

$$\begin{aligned} x_2 &= x_1 + (v\cos\beta)t_1, \\ y_2 &= y_1 + (v\sin\beta)t_1, \\ t_2 &= t_1 \end{aligned} \quad (15b)$$

in the direction of the vector $\mathbf{v}=(v\cos\beta, v\sin\beta, 0)$, under which the plane xOy remains pointwise fixed while each plane π parallel to it is translated a distance vt in the direction $\mathbf{v}$;[25] and a *translation*

$$\begin{aligned} x' &= x_2 + a, \\ y' &= y_2 + b, \\ t' &= t_2 + c \end{aligned} \quad (15c)$$

determined by the vector (a,b,c) (Fig. 21c). Hence the propositions of three-dimensional Galilean geometry must be invariant under the rotations (15a), the shears (15b), and the translations (15c).

Four-dimensional Galilean geometry, which studies all properties invariant under motions of objects in space, is even more complex. It can be described more

---

[25]If $vt<0$ (i.e., if the coordinate t which characterizes the plane π is negative), then the direction of the translation is opposite to that of $\mathbf{v}$.

2. What is mechanics?

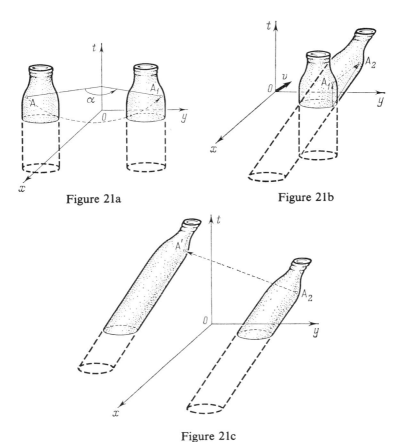

Figure 21a

Figure 21b

Figure 21c

precisely as the study of those properties of four-dimensional space[26] with coordinates $\{x,y,z,t\}$ that are invariant under the general Galilean transformations

$$\begin{aligned}
x' =\ & (\cos\beta\cos\alpha - \cos\gamma\sin\beta\sin\alpha)x + (\sin\beta\cos\alpha + \cos\gamma\cos\beta\sin\alpha)y \\
& + (\sin\gamma\sin\alpha)z + (v\cos\delta_1)t + a, \\
y' =\ & -(\cos\beta\sin\alpha + \cos\gamma\sin\beta\cos\alpha)x + (-\sin\beta\sin\alpha + \cos\gamma\cos\beta\cos\alpha)y \\
& + (\sin\gamma\cos\alpha)z + (v\cos\delta_2)t + b, \quad\quad\quad (16)\\
z' =\ & (\sin\gamma\sin\beta)x - (\sin\gamma\cos\beta)y + (\cos\gamma)z + (v\cos\delta_3)t + c, \\
t' =\ & t + d,
\end{aligned}$$

with $\cos^2\delta_1 + \cos^2\delta_2 + \cos^2\delta_3 = 1$ [cf. formulas (9)]. These transformations are the motions of the geometry. We are not in a position to study this remarkable geometry in depth.

We close this section with a few remarks on another subject. Three-dimensional Galilean geometry, whose motions are given by formulas (12), results from an

---

[26]See, for example, [19]–[21].

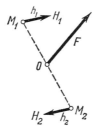

Figure 22

attempt to express the basic facts of plane *kinematics* in geometric language. The "geometrization" of the facts of plane *statics* leads to a different but no less interesting geometry which we propose to describe briefly.

In statics, given a system of forces, we may move the vector of each force along its line of action,[27] add a number of forces applied at the same point using the parallelogram law, or, conversely, decompose a force into the vector sum of several forces applied at the same point. By means of such transformations (which do not change the static properties of the system), any system of forces can be reduced to a *single force* (i.e., a single sliding vector; see footnote) or to a so-called *couple*, i.e., a pair of noncollinear, parallel, oppositely directed forces of equal magnitude. Of even greater interest to us is the fact that *any system of forces can be reduced to a single vector* **F** *applied at a predetermined origin O* (principal vector of the system) *and a couple* $\mathbf{h}_1 = \overline{M_1 H_1}$ *and* $\mathbf{h}_2 = -\mathbf{h}_1 = \overline{M_2 H_2}$, where O is the midpoint of the segment $M_1 M_2$ (Fig. 22).[28] Such a couple is completely determined by its *moment*

$$u = h \cdot \overline{OM_1} \cdot \sin \angle (\overline{OM_1}, \mathbf{h}_1) + h \cdot OM_2 \cdot \sin \angle (\overline{OM_2}, \mathbf{h}_2)$$
$$= (\overline{OM_1} \times \mathbf{h}_1) + (\overline{OM_2} \times \mathbf{h}_2)$$

(principal moment of the system).[29] The symbol "×" in the last formula denotes the *cross product* of two vectors: if the vectors $\mathbf{a} = \overline{QA} = (x_1, y_1)$ and $\mathbf{b} = \overline{QB} = (x_2, y_2)$ are applied at the same point Q, then $\mathbf{a} \times \mathbf{b} = QA \cdot QB \sin \angle AQB = x_1 y_2 - x_2 y_1$, where, as always, the angle AQB is measured counterclockwise from QA to QB (Fig. 23).[30] Moreover, the principal vector of a system of forces $\mathbf{f}_1, \mathbf{f}_2, \ldots, \mathbf{f}_n$ applied at the points $A_1, A_2, \ldots, A_n$ is *the sum of these forces*,

$$\mathbf{F} = \mathbf{f}_1 + \mathbf{f}_2 + \cdots + \mathbf{f}_n,$$

applied at O, and the total moment of the system is *the sum of the moments of the forces* $\mathbf{f}_i$ *about* O:

$$u = (\overline{OA_1} \times \mathbf{f}_1) + (\overline{OA_2} \times \mathbf{f}_2) + \cdots + (\overline{OA_n} \times \mathbf{f}_n).$$

Since the vector **F** is determined by its two coordinates x and y, the set of all possible plane systems of forces is three-dimensional in the sense that each such

---

[27]A vector which is restricted to move along a line is called a *sliding vector*. Thus a system of forces in statics is equivalent to a system of sliding vectors.

[28]See, for example, [23], pp. 241–242.

[29]See, for example, [23], p. 240.

[30]See, for example, [23], pp. 223–224; or [24], pp. 345–358.

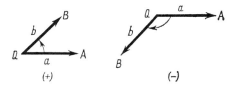

Figure 23

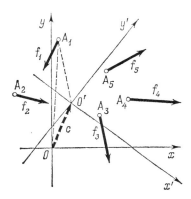

Figure 24

system is determined by three coordinates x, y, and u.[31] If we move the origin of our coordinate system from O to $O'(a,b)$ (where a and b are the components of the vector $\overline{OO'} = \mathbf{c}$; see Fig. 24), then clearly the principal vector $\mathbf{F}$ of the system of forces $\mathbf{f}_1, \mathbf{f}_2, \ldots, \mathbf{f}_n$ remains unchanged, and the new value of the total moment is

$$u' = (\overline{O'A_1} \times \mathbf{f}_1) + (\overline{O'A_2} \times \mathbf{f}_2) + \cdots + (\overline{O'A_n} \times \mathbf{f}_n)$$
$$= \{(\overline{OA_1} - \overline{OO'}) \times \mathbf{f}_1\} + \{(\overline{OA_2} - \overline{OO'}) \times \mathbf{f}_2\} + \cdots + \{(\overline{OA_n} - \overline{OO'}) \times \mathbf{f}_n\}$$
$$= (\overline{OA_1} \times \mathbf{f}_1) + (\overline{OA_2} \times \mathbf{f}_2) + \cdots + (\overline{OA_n} \times \mathbf{f}_n) - \mathbf{c} \times (\mathbf{f}_1 + \mathbf{f}_2 + \cdots + \mathbf{f}_n)$$
$$= u - \mathbf{c} \times \mathbf{F} = u - (ay - bx) = bx - ay + u.$$

(Here we have used the distributive property of the cross product, $(\mathbf{m} + \mathbf{n}) \times \mathbf{p} = (\mathbf{m} \times \mathbf{p}) + (\mathbf{n} \times \mathbf{p})$, and the fact that $\mathbf{c} \times \mathbf{F} = (a,b) \times (x,y) = ay - bx$.) Finally, rotation of the coordinate system through an angle α does not affect the total moment u (since u depends only on the origin O of the coordinate system, not on the orientation of its axes), but changes the coordinates x and y of $\mathbf{F}$ to

$$\left. \begin{array}{l} x' = x\cos\alpha + y\sin\alpha, \\ y' = -x\sin\alpha + y\cos\alpha \end{array} \right\}$$

[cf. formulas (7)].

---

[31] It is not difficult to see that *if the coordinates of a system of forces $\mathbf{f}_1, \mathbf{f}_2, \ldots, \mathbf{f}_n$ are x_1, y_1, u_1, and the coordinates of a system $\mathbf{g}_1, \mathbf{g}_2, \ldots, \mathbf{g}_m$ are x_2, y_2, u_2, then the coordinates of the combined system $\mathbf{f}_1, \mathbf{f}_2, \ldots, \mathbf{f}_n, \mathbf{g}_1, \mathbf{g}_2, \ldots, \mathbf{g}_m$ are $x_1 + x_2$, $y_1 + y_2$, and $u_1 + u_2$.*

In summary, the effect of a change of the coordinate system on the coordinates x, y, u of a (plane) system of forces is given by the formulas

$$\begin{array}{rl} x' = & (\cos\alpha)x + (\sin\alpha)y, \\ y' = & -(\sin\alpha)x + (\cos\alpha)y, \\ u' = & bx \quad - ay + u. \end{array} \tag{17}$$

In addition to this alias interpretation, our formulas also admit an alibi interpretation (cf. footnote 11). We may, therefore, assert that *the study of plane statics reduces to the study of three-space* $\{x,y,u\}$ *under the motions* (17), i.e., to the study of the properties of three-dimensional space that are not affected by the transformations (17).

Since the general theory of systems of forces (reduction to principal vector and total moment) was first developed by the French physicist L. POINCEAU (1777–1859) in his book "Elements of Statics" (1804), the geometry of three-space with the motions (17) should be called the **Poinceau geometry**. In what follows we shall not discuss this geometry, but rather concern ourselves exclusively with Galilean geometry.

PROBLEMS AND EXERCISES

4   The role played by the transformations (13a) in Galilean geometry is similar to that of the direct motions (6) in Euclidean geometry. The totality of (direct and opposite) Galilean motions is given by the formulas

$$\begin{array}{l} x' = \pm x \quad + a, \\ y' = vx \pm y + b, \end{array} \tag{13'a}$$

and the Galilean similitudes are given by the formulas

$$\begin{array}{l} x' = \alpha x \quad + a, \\ y' = vx + \beta y + b, \end{array} \tag{13''a}$$

where $\alpha\beta \neq 0$, and where it is natural to distinguish between direct similitudes (α and β positive) and opposite similitudes (α and/or β negative). What is the mechanical interpretation of the transformations (13'a) and (13''a)? Decompose the transformations (13'a) and (13''a) into simpler ones (cf. formulas (15a–c). Give a classification of the transformations (13'a) and (13''a) (cf. Exercise **3** and Problem **I**).

5   Consider three-dimensional Galilean geometry with the motions (12). We shall find it convenient to write these motions in the form

$$\begin{array}{rl} x' = & (\cos\alpha)x + (\sin\alpha)y + \mu z + a, \\ y' = & -(\sin\alpha)x + (\cos\alpha)y + vz + b, \\ z' = & z + c. \end{array} \tag{12'}$$

In a certain intuitive sense this geometry is intermediate between Euclidean solid geometry and *three-dimensional semi-Galilean geometry*, whose motions are given by the formulas

$$\begin{array}{rl} x' = & x + \lambda y + \mu z + a, \\ y' = & y + vz + b, \\ z' = & z + c. \end{array} \tag{12''}$$

Thus, for example, in Euclidean geometry there are no special directions, in three-dimensional Galilean geometry with the motions (12′) the z-axis is special in the sense that its direction is unchanged by the motions (12′), and in three-dimensional semi-Galilean geometry with the motions (12″) the z-axis and the Oyz plane are special. The geometry with the motions (12″) is the closest three-dimensional analogue of Galilean geometry.

Decompose the motions (12″) into "elementary" motions [in a manner suggested by the decomposition of the motions (12) into the elementary motions (15a–c)].

6   (a) It is natural to think of the motions (12′) as the direct motions of three-dimensional Galilean geometry. How would you define the corresponding opposite motions as well as the corresponding (direct and opposite) similitudes (cf. Exercise 4)? Decompose the motions (12′) into simpler ones [in a manner suggested by decomposition of the motions (12) into the elementary motions (15a–c)]. Classify the motions (12′) (cf. Exercise 3 and Problem I). What is the physical interpretation of these motions? (b) Solve the corresponding problem for the semi-Galilean geometry with the motions (12″).

7   (a) Show that if we start with affine plane geometry whose motions are given by
$$x' = ax + by + e,$$
$$y' = cx + dy + f, \qquad (*)$$
with $ad - bc \neq 0$ (see Chap. 13 of [19]), and single out the motions which preserve a certain direction (say the direction of the y-axis), then we obtain the geometry whose motions are given by (13″a). (b) Describe the geometry whose motions are the motions (*) which preserve *two* directions (say the directions of the x- and y-axes).

VI   Give an analytic description of the Poinceau geometry of three-dimensional statics whose points are systems of forces in space and whose motions are generated by coordinate transformations in the space in which these forces act.

VII   Consider *three-dimensional central Galilean geometry* whose motions are given by
$$x' = (\cos\alpha)x + (\sin\alpha)y + \mu z,$$
$$y' = -(\sin\alpha)x + (\cos\alpha)y + \nu z, \qquad (12'\text{a})$$
$$z' = z,$$
and show that if we take as basic elements *planes* not passing through the origin, then we obtain the Poinceau geometry.

VIII   (a) Classify all (direct and opposite) similitudes of the Euclidean plane (cf. Exercise 3 and Problem I). (b) Classify all (direct and opposite) similitudes of three-dimensional Euclidean space.

IX   (a) Classify and describe all (direct) motions (13a) in the Galilean plane (you will find it helpful to consult subsequent chapters of this book). (b) Classify and describe all (direct and opposite) motions (13′a) in the Galilean plane (cf. Exercise 4 above). (c) Classify and describe all (direct and opposite) similitudes (13″a) in the Galilean plane.

X (a) Characterize the geometric structure of the direct motions (12′) of three-dimensional Galilean geometry. Do the same for the direct and opposite motions of this geometry as well as for its similitudes (cf. Exercises 5–6 above). (b) Solve the corresponding problem for three-dimensional semi-Galilean geometry (cf. Exercises 5–6). (c) Solve the corresponding problem for central three-dimensional Galilean geometry (cf. Problem VII).

XI (a) Solve the analogue of Problem X for the Poinceau geometry with direct motions (17). (b) Solve the analogue of Problem X for the Poinceau geometry of three-dimensional statics (cf. Problem VI).

I. Distance and Angle; Triangles and Quadrilaterals

3. Distance between points and angle between lines

We shall now systematically study Galilean geometry, i.e., the geometry of the plane xOy whose motions are given by the equations

$$\boxed{\begin{aligned} x' &= x + a, \\ y' &= vx + y + b. \end{aligned}} \qquad (1)$$

This means that we shall be interested solely in those properties of figures in the plane xOy that are invariant under the transformations (1) (or, equivalently, under the *shears*

$$\begin{aligned} x' &= x, \\ y' &= vx + y \end{aligned} \qquad (1a)$$

and the *translations*

$$\begin{aligned} x &= x + a, \\ y' &= y + b; \end{aligned} \qquad (1b)$$

cf. p. 25 above); it is only these properties of figures that have geometric significance in this unusual geometry. Also, we shall bear in mind that our geometry arose naturally out of mechanical considerations connected with Galileo's principle of relativity. This implies that, in our case, properties of geometric significance are really properties of mechanical significance; more specifically, facts of one-dimensional kinematics.

Before considering the basic concepts of Galilean geometry we shall find it useful to list certain fundamental properties of the motions (1). We remark that *the transformations* (1) *map*

(a) *lines onto lines*;
(b) *parallel lines onto parallel lines*;
(c) *collinear segments AB, CD onto collinear segments $A'B'$, $C'D'$ with $C'D'/A'B' = CD/AB$*;
(d) *a figure F onto a figure F' of the same area.*

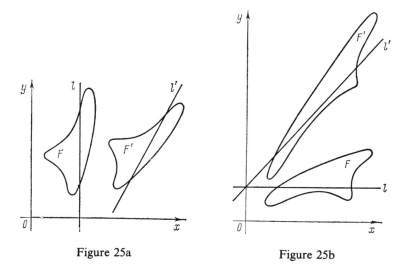

Figure 25a   Figure 25b

(The proofs of these facts are given below.) This means that the concepts of lines, parallel lines, ratios of collinear segments, and areas of figures are significant not only in Euclidean geometry but also in Galilean geometry. Also, it is very important to note that any transformation (1) *takes every line parallel to the y-axis into another line parallel to the y-axis*. Thus, while in Euclidean geometry the term "line parallel to the y-axis" has no geometric significance (since such a line can be carried by a Euclidean motion into an arbitrary line; cf. Fig. 25a), in Galilean geometry lines parallel to the y-axis play a special role different from that of all other lines. On the other hand, lines parallel to the x-axis are not distinguished in Galilean geometry from other "ordinary" lines, i.e., lines not parallel to Oy. This is illustrated in Figure 25b, where a shear (1a) takes a line l parallel to Ox into a line l' not parallel to Ox. In what follows, the term "line" will mean a line not parallel to Oy, whereas parallels to Oy will be referred to as *special lines*.

We now prove that the motions (1) have the properties (a), (b), (c), and (d). Note that since the translations (1b) obviously have these properties (see Fig. 12b), we can restrict the proof to the shears (1a).

(a) A shear takes every parallel to Oy into itself; this follows directly from the definition (1a) of a shear (see p. 25; in particular, Fig. 20). Further, suppose (1a) takes a point A on a line l through the origin O into a point A' (Fig. 26). We denote the line OA' by l', the points where an arbitrary special line m intersects l and l' by M and M', and the points where m and AA' meet the x-axis (which we shall always represent as perpendicular to the y-axis[1]) by Q and P. Since A' is the image of A under

---

[1]This normalization is convenient simply because we are used to rectangular coordinate systems. However, it is important to remember that perpendicularity of lines is significant in Euclidean geometry but not in Galilean geometry.

3. Distance between points and angle between lines   35

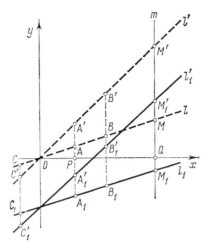

Figure 26

the shear (1a), it follows that
$$\frac{A'P}{AP} = \frac{OP \cdot v + AP}{AP} = \frac{OP}{AP} \cdot v + 1.$$
Looking once more at Figure 26 (where $M'Q/MQ = A'P/AP$ and $OP/AP = OQ/MQ$), we see that
$$\frac{M'Q}{MQ} = \frac{A'P}{AP} = \frac{OP}{AP} \cdot v + 1 = \frac{OQ}{MQ} \cdot v + 1 = \frac{OQ \cdot v + MQ}{MQ},$$
i.e.,
$$M'Q = OQ \cdot v + MQ.$$
Thus the shear takes every point M on l to a point M' on l'. Briefly, *the shear takes the line l onto the line l'*.

Let l_1 be a line parallel to l and not passing through O (Fig. 26). Let d be the common (Euclidean) length of the vertical segments AA_1, BB_1, $CC_1,\ldots$ between l and l_1. The restriction of the shear (1a) to a special line m is a translation along m (by an amount depending on m.). Hence the segments $AA_1 = BB_1 = \cdots = d$ are mapped onto the segments $A'A_1', B'B_1',\ldots$ of the same length,
$$A'A_1' = B'B_1' = C'C_1' = \cdots = d.$$
Since the endpoints $A', B', C',\ldots$ belong to the image l' of l, the endpoints $A_1', B_1', C_1',\ldots$ belong to the image l_1' of l_1. Hence l_1' is a *line parallel to l'*. This proves that the shear (1a) maps *the line l_1 onto the line l_1'*.

(b) The proof of (b) is implicit in the proof of (a). We can also say that a shear maps lines parallel to a given line l onto lines parallel to its image l'.[2]

---

[2] It is easy to show that if l and l' form (Euclidean) angles α and α', respectively, with the x-axis, then $\tan\alpha' = \tan\alpha + v$.

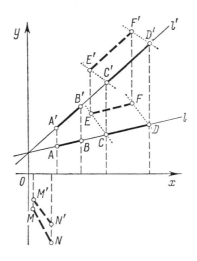

Figure 27

(c) The equality $C'D'/A'B' = CD/AB$ follows directly from Figure 27; its proof is left to the reader. We observe that if AB and EF are parallel segments and the (parallel) segments $A'B'$ and $E'F'$ are their images under a shear, then

$$\frac{E'F'}{A'B'} = \frac{EF}{AB}$$

(cf. Fig. 27, where $EF = CD$, and $E'F' = C'D'$, since the parallelogram $CDFE$ is mapped onto the parallelogram $C'D'F'E'$). Thus the concept of the *ratio of the lengths of parallel segments* is meaningful in Galilean geometry. However, if $AB \not\parallel MN$, then a shear may map these segments onto segments $A'B'$ and $M'N'$ such that $M'N'/A'B' \neq MN/AB$ (cf. Fig. 27, where $AB < A'B'$ but $MN > M'N'$).

(d) The area of a figure F is approximated by the sum of the areas of the squares in the interior of F formed by two families of lines parallel to the coordinate axes with neighboring lines in each family separated by a small distance e (i.e., the area is approximately equal to the number of such squares multiplied by the area e^2 of one square; see Fig. 28). Actually, the area of F is defined as the limit (assuming it exists) of a sequence of these approximations as e decreases to zero. A shear (1a) takes the figure F to a figure F', and every square S of the net to a parallelogram S' (see Fig. 28). Since a shear induces a translation on every parallel to Oy, the sides of S' parallel to Oy have length e, and the perpendicular distance between them is e. Hence area $(S') = e^2 =$ area (S). It is plausible (and can easily be proved) that the area of F' is the limit as e approaches zero of the number of parallelograms in its interior multiplied by the area e^2 of each of them. It follows readily from this that

$$\text{area } F' = \text{area } F.$$

3. Distance between points and angle between lines

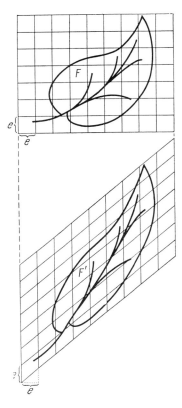

Figure 28

After these preliminaries, we review the concepts of "distance between points" and "angle between lines" in Euclidean geometry and then discuss their analogues in Galilean geometry.

In Euclidean geometry, the *distance* d_{AA_1} between two points $A(x,y)$ and $A_1(x_1,y_1)$ is defined by the formula

$$d_{AA_1} = \sqrt{(x_1-x)^2 + (y_1-y)^2} \; ; \qquad (2)$$

if the distance between the points vanishes, the points coincide. The *angle* δ_{ll_1} between two lines l and l_1 with equations

$$y = kx + s \quad \text{and} \quad y = k_1 x + s_1$$

is defined by the formula

$$\tan \delta_{ll_1} = \frac{k_1 - k}{kk_1 + 1} \; ; \qquad (3)$$

this is suggested by the fact that if φ and φ_1 are the angles formed by l and

l_1 and the x-axis, then $\tan\varphi = k$, $\tan\varphi_1 = k_1$, and

$$\tan\delta_{ll_1} = \tan(\varphi_1 - \varphi) = \frac{\tan\varphi_1 - \tan\varphi}{1 + \tan\varphi_1 \tan\varphi} = \frac{k_1 - k}{1 + k_1 k}$$

(see p. 11 and Fig. 11b above).

If $k = k_1$, then l and l_1 are parallel, and formula (3) assigns the value zero to the angle δ_{ll_1} between the lines. We define the *distance* d_{ll_1} *between parallel lines* by the formula

$$d_{ll_1} = \frac{|s_1 - s|}{\sqrt{k^2 + 1}} ; \qquad (4)$$

this is suggested by the fact that $|s_1 - s| = SS_1$ is the length of the segment on the y-axis between l and l_1, and $\sqrt{k^2 + 1} = \sqrt{\tan^2\varphi + 1} = 1/\cos\varphi$, where φ is the angle formed by each of the lines l and l_1 and the x-axis (cf. p. 10 above; in particular, Fig. 11a). We call the reader's attention to the fact that although δ_{ll_1} and d_{ll_1} both measure the "deflection" of l and l_1, these quantities are radically different; angles are measured in angular units (degrees or radians) and distances in units of length (inches or centimeters). Consequently *these quantities are not comparable*; knowing that two intersecting lines form an angle of 30°, and two parallel lines are 15 cm apart (cf. Fig. 29), we cannot say that one of these two deflections is larger than the other. We also note that the *distance* (4) *between lines is defined only if the angle* (3) *between them is zero*, and that two lines coincide if and only if they form an angle δ equal to zero and the distance d between them is zero.

In Galilean geometry, the *distance* d_{AA_1} between two points $A(x,y)$ and $A_1(x_1,y_1)$ is defined by the formula

$$\boxed{d_{AA_1} = x_1 - x;} \qquad (5)$$

it equals the signed length of the projection PP_1 of the segment AA_1 on the x-axis (Fig. 30a). Note that distances can be negative; in fact, $d_{A_1 A} = -d_{AA_1}$. Since the x coordinate of a point A transforms under a motion (1) in accordance with the formula

$$x' = x + a,$$

it is clear that the difference $x_1 - x$ of the abscissas of two points A_1 and A

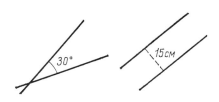

Figure 29a   Figure 29b

3. Distance between points and angle between lines

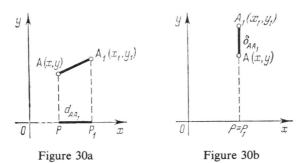

Figure 30a   Figure 30b

is invariant under this motion. [This also follows readily from the geometric properties of a shear (1a) and a translation (1b).]

If the distance d_{AA_1} between A and A_1 is zero, i.e., $x_1 = x$, then A and A_1 belong to the same special line (parallel to the y-axis; cf. Fig. 30b). For such points it makes sense to define the *special distance*

$$\delta_{AA_1} = y_1 - y. \tag{6}$$

In fact, if the abscissas of $A(x,y)$ and $A_1(x_1,y_1)$ coincide ($x_1 = x$), then a motion (1) takes these points to points $A'(x',y')$ and $A'_1(x'_1,y'_1)$, with $x' = x'_1 = x + a$ and

$$y' = vx + y + b,$$
$$y'_1 = vx + y_1 + b.$$

Hence

$$y'_1 - y' = (vx + y_1 + b) - (vx + y + b) = y_1 - y.$$

Thus the difference $y_1 - y$ is unchanged by a motion, and so has geometric significance in the Galilean plane. On the other hand, if the distance $d_{AA_1} = x_1 - x$ between A and A_1 is not zero, then the difference $y_1 - y$ of their ordinates is not preserved by a motion, for, in that case,

$$y'_1 - y' = (vx_1 + y_1 + b) - (vx + y + b) = y_1 - y + v(x_1 - x) \neq y_1 - y.$$

This is not surprising. After all, a shear (1a) keeps the origin O fixed and changes the ordinate of any point M not on the y-axis. But then it cannot preserve the difference of the ordinates of M and O.

It is clear that two points A and B of the Galilean plane coincide if and only if their distance d_{AB} and their special distance δ_{AB} both vanish.

By a **circle** S in the Galilean plane we mean the *set of points $M(x,y)$ whose distances from a fixed point Q have constant absolute value r*; the point $Q(a,b)$ is called the center of S and the (nonnegative) number r its radius. Since

$$d_{QM} = x - a$$

[cf. formula (5)], the equation

$$d_{QM}^2 = r^2$$

which defines S can be written as
$$(x-a)^2 = r^2,$$
or
$$x^2 + 2px + q = 0, \tag{7}$$
where
$$p = -a, \quad q = a^2 - r^2. \tag{7a}$$

It is clear that the circle S with center Q and radius r consists of the points on two special lines whose Euclidean distance from Q is r (Fig. 31a); if r is zero the two special lines coincide (Fig. 31b). We note that while a Galilean circle S has a definite radius (equal to half the Euclidean distance between its two component special lines), it has infinitely many centers, namely the points of the special line through Q (see Fig. 31).

It is natural to define the *angle* δ_{ll_1} between lines l and l_1 intersecting at a point Q as the length of the circular arc NN_1 cut off by l and l_1 from the unit circle S centered at Q (cf. the Euclidean configuration 32a and the Galilean configuration 32b; naturally, by the "length of the arc" NN_1 of the Galilean circle S we mean the special distance δ_{NN_1} between N and N_1).

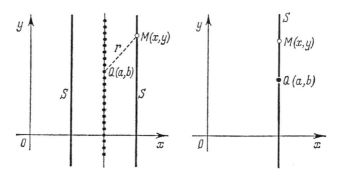

Figure 31a   Figure 31b

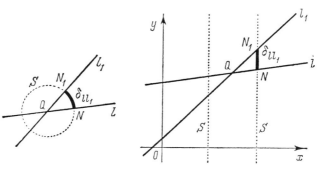

Figure 32a   Figure 32b

3. Distance between points and angle between lines

The angular measure δ_{ll_1} is meaningful in Galilean geometry. In fact, a motion (1) takes the intersection point Q of l and l_1 onto the intersection point Q' of their images l' and l'_1, and the unit circle S and its arc NN_1 onto the unit circle S' (centered at Q') and its arc $N'N'_1$ (Fig. 33). This definition of angle in Galilean geometry can also be phrased as follows. *To determine the angle δ_{ll_1} between the lines l and l_1, draw the special line m one unit to the right of their intersection Q. If N and N_1 are the points where m meets l and l_1, respectively, then*

$$\delta_{ll_1} = \overline{NN_1} = \delta_{NN_1},$$

the signed length of NN_1. If the equations of l and l_1 are $y = kx + s$ and $y = k_1 x + s_1$, and Q has coordinates (x_0, y_0), then the equation of m is

$$x = x_0 + 1.$$

Hence N and N_1 have coordinates

$$(x_0 + 1, k(x_0 + 1) + s) \quad \text{and} \quad (x_0 + 1, k_1(x_0 + 1) + s_1).$$

Consequently

$$\delta_{ll_1} = \delta_{NN_1} = [k_1(x_0 + 1) + s_1] - [k(x_0 + 1) + s]$$
$$= [(k_1 x_0 + s_1) - (k x_0 + s)] + k_1 - k = k_1 - k$$

[for $Q(x_0, y_0)$ lies on both l and l_1, and this implies that $k x_0 + s = k_1 x_0 + s_1$]. In summary,

$$\boxed{\delta_{ll_1} = k_1 - k.} \tag{8}$$

This formula is definitely simpler than formula (3), which defines the (directed) angle between lines in Euclidean geometry.

If we rotate l_1 counterclockwise about Q so that it tends to the special line m through Q, then the angle δ_{ll_1} increases beyond all bounds (Fig. 34). This goes very much against our Euclidean intuition.

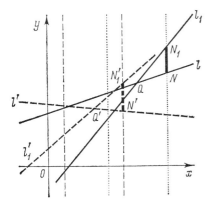

Figure 33

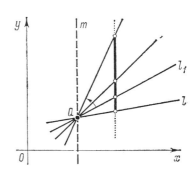

Figure 34

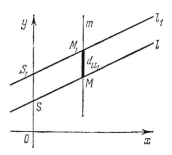

Figure 35

If the lines l and l_1 are parallel, then by formula (8) the angle δ_{ll_1} between them is zero. In that case, we can define the *distance* d_{ll_1} *between the* (parallel) *lines* l *and* l_1 as the (special) length of the directed segment MM_1 between l and l_1 belonging to a special line (the special line is arbitrary; cf. Fig. 35). This definition makes sense because the motions (1) map special lines onto special lines. If the equations of the lines l and l_1 are $y = kx + s$ and $y = kx + s_1$, then clearly

$$d_{ll_1} = s_1 - s. \qquad (9)$$

This formula is also far simpler than the corresponding formula (4) in Euclidean geometry.[3]

Finally we define the *distance* d_{Ml} *from a point M to a line l* as the (special) distance from M to the point of intersection of l with the special line through M (Fig. 36b). This definition is suggested by the following considerations. In Euclidean geometry, the distance from a point M to a line l is defined as the distance from M to the point P on l nearest to M (Fig. 36a). In Galilean geometry, the point P on l nearest to M is at distance zero from M, that is, $d_{MP} = 0$. This prompts us to measure the distance from M to l by the special distance from M to P: $d_{Ml} = \delta_{MP}$. If the equation of l is $y = kx + s$, then the coordinates of P [which is on l and on the special line through $M(x_0, y_0)$] are $(x_0, kx_0 + s)$. Hence

$$-\overline{MP} = -\delta_{MP} = y_0 - (kx_0 + s),$$

and therefore

$$-d_{Ml} = y_0 - kx_0 - s. \qquad (10)$$

Thus, apart from sign, the distance from M to l is the result of substituting the coordinates of M in the left side of the equation $y - kx - s = 0$ of l.[4] It

---

[3] We wish to stress that in defining d_{AA_1}, δ_{AA_1}, δ_{ll_1}, d_{ll_1} we ordered the points A, A_1 and the lines l, l_1. This is reflected in formulas (5), (6), (8), and (9), from which it is apparent that the quantities in question can be positive, negative, or zero.

[4] Cf. p. 234.

3. Distance between points and angle between lines   43

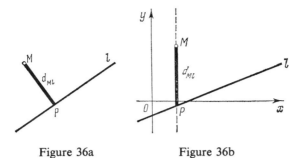

Figure 36a   Figure 36b

is clear that the distance d_{ll_1} between two parallels l and l_1 can be thought of as the distance d_{Ml_1}, where M is any point on l.

The definitions of distance from a point M to a line l and of distance between parallel lines l and l_1 indicate that *in Galilean geometry the special lines play the roles of perpendiculars to a line* (cf. Figs. 37a and 37b, which refer to Euclidean and Galilean geometry, respectively). In what follows we shall frequently return to the analogy between perpendiculars to a line and the special lines.

We shall now give the mechanical interpretation of the concepts introduced in this section, i.e., we shall interpret concepts of Galilean geometry in terms of concepts of kinematics on a line o. Clearly, the kinematic counterpart of a point $M(x,y)$ of the Galilean plane is an *event* (t,x) determined by the position x on o and the time t (Fig. 38); here the x coordinate of $M(x,y)$ corresponds to the time coordinate t of (t,x), and the y coordinate of $M(x,y)$ corresponds to the space coordinate x of (t,x). The counterpart of the distance d between points A and A_1 in the Galilean

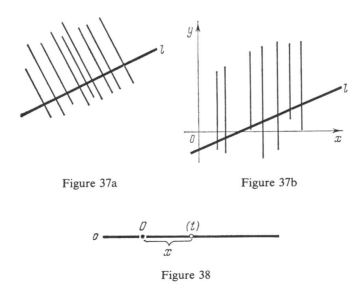

Figure 37a   Figure 37b

Figure 38

plane is the *time interval* $t_1 - t$ between the events corresponding to A and A_1. If this interval is zero, i.e., if the two events are simultaneous, then they are separated by a *space interval* equal to the distance between the points on the line o which correspond to the events A and A_1. This space interval is the special distance δ between the two points in the Galilean plane corresponding to the two events. We note that the space interval δ can be defined only for simultaneous events, for in the case of nonsimultaneous events, the distance between the corresponding points depends on the choice of inertial reference frame, and therefore has no meaning in Galilean geometry. Thus, in a reference frame associated with the earth, the distance between the school to which you go in the morning and the home where you have supper in the evening is small. On the other hand, in a reference frame associated with the fixed stars, the school to which you go in the morning is at an enormous distance from the home where you have supper in the evening, for you and the earth have traveled very far in the time which elapsed between the two events. Also, it should be pointed out that the ordinary distance d between two events and the special distance δ between (simultaneous!) events are not comparable, since they are measured in different units; the first in units of time (seconds, years, centuries) and the second in units of length (inches, kilometers, etc.).

A line l in the Galilean plane with equation $y = kx + s$ corresponds to the *uniform motion*

$$x = kt + s$$

along a line o with velocity k, where, as we know, the velocity depends on the choice of inertial reference frame, and therefore has no physical significance. On the other hand, in the case of two lines with equations $y = kx + s$ and $y = k_1 x + s_1$, that is, in the case of points L and L_1 on the line o which move uniformly in accordance with the equations

$$x = kt + s \quad \text{and} \quad x = k_1 t + s_1,$$

the difference

$$\delta_{ll_1} = k_1 - k \qquad (8)$$

of the velocities of the two points does have physical significance; it is the *relative velocity* of one moving point with respect to the other. The quantity δ_{ll_1}—the angle between the lines l and l_1 in the Galilean plane—can also be described as the velocity of the second point relative to a reference frame in which the first point is at rest. In fact, if by means of a motion (or coordinate transformation) (1) we bring the line l into coincidence with the x-axis, i.e., if we make a fixed point L_0 of o correspond to l, then the difference $k_1 - k$ takes the form

$$\delta_{ll_1} = k_1 - k = k_1' - k' = k_1' - 0 = k_1'$$

(since the velocity k' of the first point in the new frame of reference is zero). But then δ_{ll_1} is indeed equal to the velocity k_1' of the second point relative to a reference frame in which the first point is at rest. The

3. Distance between points and angle between lines 45

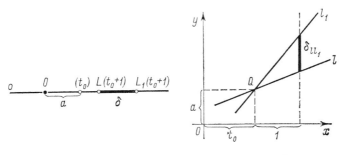

Figure 39a   Figure 39b

following is yet another definition of the angle δ_{ll_1}, regarded as the relative velocity of two uniformly moving points L and L_1 on the line o: If at time t_0 the two points occupy the same position (a) on o [so that the point $Q(t_0, a)$ of the Galilean plane is on both lines l and l_1], then δ_{ll_1} is the distance δ between L and L_1 at time $t_0 + 1$ (cf. Figs. 39a and 39b).

Two parallel lines in the Galilean plane correspond to two uniformly moving points L and L_1 whose velocities (relative to any inertial reference frame) are the same. These points move in accordance with equations

$$x = kt + s \quad \text{and} \quad x = kt + s_1,$$

and their relative velocity δ_{ll_1} is zero. For two such moving points (say two points in Moscow moving in space together with the earth) it is possible to find a reference frame in which both are at rest. It then makes sense to speak of the *distance d_{ll_1} between such moving points*, or the distance between parallel lines. This distance is equal to the difference

$$s_1 - s$$

of the coordinates of the points on the line o at time $t = 0$ (or any other fixed moment). Finally, it is natural to define the distance from a point (event) M to a line l (that is, a point L moving uniformly on o) as the *distance from M to L measured at the time of the event M*.

A special line m in the Galilean plane whose equation is $x = a$ corresponds to a definite *moment in time*

$$t = a.$$

Since it consists of *all* points of the line o at this time, it can also be thought of as a "motion with infinite velocity." This accords with the fact that the slope of a special line is the tangent of the angle between m and the x-axis (i.e., $\tan 90°$) and must be supposed infinite. Finally, a circle in the Galilean plane with center Q and radius r corresponds to the *set of events whose distance from a fixed event $Q(t, a)$ is, in absolute value, equal to the* (time) *interval r*, i.e., the set of events $(t - r, x)$ and $(t + r, x)$ with arbitrary r.

We can summarize our discussion in the form of a dictionary in which the left column contains various concepts of Galilean geometry and the right column contains their counterparts in kinematics on a line o.

Geometric concepts	Mechanical concepts
Point	Event
Line (ordinary)	Uniform motion
Special line	Moment in time
Distance d between points	Time interval between events
Special distance δ between points	Space interval between simultaneous events
Angle δ between lines	Relative velocity of points in uniform motion
Distance d between parallel lines	Distance between points of o at rest with respect to each other
Distance from point to line	Distance (at a given time) from a fixed point of o to a point in uniform motion
Circle of radius r	The set of events which occur r time units before or after a fixed event

We now give a kinematic interpretation of the concept of area of a figure which, as we saw earlier, has significance in Galilean geometry. Let F be a figure, i.e., a set of points in the Galilean plane. To this set of points there corresponds a *set of events*, each of which is determined by prescribing its position (on the line o) and time. We shall now define a quantitative measure of this set of evens. Such a measure must take into consideration the spatial as well as temporal extent of the set F. To define it, we place in F "elementary event sets" of fixed spatial extent e and fixed duration τ. Clearly, the counterpart in the Galilean plane of such an elementary event set E is a rectangle with sides e and τ. We define the *space–time content* of E to be $e\tau$. Now cover F with a net of nonoverlapping rectangles of x-mesh τ and y-mesh e (Fig. 40; cf. Fig. 28). Form the sum of the space–time content of those event sets of the net which lie in F. The limit of these sums as e and τ approach zero (if it exists) is called the space–time content of F. Clearly, it is just the area of the figure F in the Galilean plane.

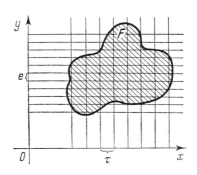

Figure 40

PROBLEMS AND EXERCISES

1. Without relying on drawings, give an analytic proof of the properties (a)–(c) of the Galilean motions (1).

2. Give analytic proofs of the invariance under (1) of the quantity $\delta_{l l_1}$ in (8) associated with a pair of intersecting lines; the quantity $d_{l l_1}$ in (9) associated

4. The triangle

with a pair of parallel lines (for such lines $\delta_{ll_1} = 0$); and the quantity d_{Ml} in (10) associated with a point and a line [cf. the proof, given on pp. 38–39, of the invariance under (1) of the quantities d_{AA_1} and δ_{AA_1} in (5) and (6) associated with pairs of points].

3 Let xOy be the familiar coordinate system in the Galilean plane, where Oy is a special line and $Ox = o$ is an ordinary line. Let S be the (Galilean) unit circle with center O, and M a typical point on S with coordinates x and y. Denote the line OM by l and the angle δ_{ol} by α. Define the *Galilean cosine* $\cos g\, \alpha$, and the *Galilean sine* $\sin g\, \alpha$ by the equations $x = \cos g\, \alpha$, $y = \sin g\, \alpha$. Prove that for all α, $\cos g\, \alpha = 1$ and $\sin g\, \alpha = \alpha$.[5] Prove, for example, that

$$\sin g(\alpha + \beta) = \sin g\, \alpha \cos g\, \beta + \cos g\, \alpha \sin g\, \beta,$$

and that if ABC is a "right triangle" with acute angle $\delta_{AB,AC} = \alpha$ (we recall that this means that $BC \perp AC$ in the sense of Galilean geometry, i.e., BC is a special line; cf. p. 43, in particular, Fig. 37b), then $d_{AC} = d_{AB} \cos g\, \alpha$, $\delta_{BC} = d_{AB} \sin g\, \alpha$. Develop the suggested parallelism between Euclidean and Galilean trigonometry.

I Define the distance between two points, the angle between two lines, and the angle between two planes in three-dimensional Galilean geometry with the motions (12′). Define analogues of the concepts of circle and sphere. Discuss the physical significance (in terms of *plane* kinematics) of the various concepts.

II Define the same concepts in three-dimensional semi-Galilean geometry with the motions (12″).

III (a) Define the same concepts in the Poinceau geometry with the motions (17). Can you give a mechanical interpretation (in terms of plane statics) of these concepts?(b) Define the same concepts in the Poinceau geometry of three-dimensional statics (cf. Problem **VI** in the Introduction). Can you give a mechanical interpretation (in terms of three-dimensional statics) of these concepts?

IV Establish the connections between the quantities in Problems **I** and **III** (a) implied by the results of Problem **VII** in the Introduction.

4. The triangle

The simplest polygon in the Galilean plane is a *triangle* ABC formed by three points A, B, C and three (ordinary) lines $BC \equiv a$, $CA \equiv b$, and $AB \equiv c$ (Fig. 41). Just as in Euclidean geometry, the letters a, b, c are also used to denote the lengths of the sides of the triangle, i.e., the (positive[6]) distances $|d_{BC}| = a$, $|d_{CA}| = b$, $|d_{AB}| = c$, and the letters A, B, C stand not only for the vertices of the triangle but also for the (positive[6]) magnitudes of its angles:

---

[5]Compare these results with the "asymptotic equalities" $\cos \alpha \sim 1$ and $\sin \alpha \sim \alpha$, which hold for small angles α. [The precise meaning of the latter relations is that $\lim_{\alpha \to 0} \cos \alpha = 1$ and $\lim_{\alpha \to 0}(\sin \alpha / \alpha) = 1$; in this connection, see Sec. 13 of the Conclusion.]

[6]Cf. footnote 3.

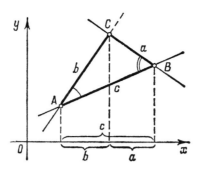

Figure 41

$|\delta_{bc}| = A$, $|\delta_{ca}| = B$, $|\delta_{ab}| = C$. Occasionally, we shall denote the positive length $|d_{AB}|$ of a segment AB by the same letters AB. (Such notation will appear, for example, in some equations involving the lengths of the sides of a triangle.)

The lengths a, b, and c of the sides of a triangle in the Galilean plane and the magnitudes A, B, and C of its angles are connected by very simple relations. If c is the largest side, then

$$\boxed{a + b = c} \qquad (11)$$

(Fig. 41). The mechanical interpretation of this equality is obvious: if A, B, and C are three arbitrary events, then the time interval between the first and last of them is equal to the interval between the first and second plus the interval between the second and last (additivity of time intervals).

From Figure 41, we can easily read off another relation. Suppose C is the largest angle of $\triangle ABC$. If the slopes of BC, CA, and AB are k_1, k_2, and k_3, respectively, then $A = k_2 - k_3$, $B = k_3 - k_1$, and $C = k_2 - k_1$. Hence

$$\boxed{A + B = C.} \qquad (12)$$

To obtain the mechanical interpretation of this relation, we select an (inertial) reference frame in which the side AC of the triangle corresponds to the state of rest.[7] Then $-A$ and $-C$ are the velocities of the uniform motions represented in the Galilean plane by the lines c and a, respectively, while $-B$ is the relative velocity of the motion a with respect to a moving reference frame determined by the motion c (in other words, the velocity of a relative to a frame of reference in which c is the state of rest). Thus, with proper labeling of the uniform motions represented by the lines a, b and c, formula (12) gives the classical law of composition of velocities: *The "absolute velocity" of a motion, i.e., its velocity with respect to a fixed*

---

[7]In other words, we perform a motion (1) to make AC lie on the x-axis.

4. The triangle

reference frame, is the sum of its "relative velocity," i.e., its velocity with respect to a moving reference frame, and its "transport velocity," i.e., the velocity of the moving reference frame. Take, for example, the case of a passenger walking from the rear to the front of a moving train. His velocity with respect to the tracks (his absolute velocity) is the sum of his velocity relative to the train (his relative velocity) and the velocity of the train (the transport velocity which characterizes the motion of the moving reference frame with respect to which we compute the relative velocity).

The mechanical interpretation of Eq. (12) could also be stated: *The largest of the three relative speeds determined by three uniform motions is the sum of the remaining two.*

The "angle formula" (12) can also be deduced from the "side formula" (11) by means of the relation

$$\boxed{\frac{a}{A} = \frac{b}{B} = \frac{c}{C},} \tag{13}$$

which is the analogue, in Galilean geometry, of the law of sines in Euclidean geometry. Relation (13) states that in any triangle of the Galilean plane, the sides are proportional to the opposite angles:

$$a = \lambda A, \qquad b = \lambda B, \qquad c = \lambda C \tag{14}$$

(the significance of the proportionality factor λ will be discussed in Sec. 9 below). To obtain (12) from (11) we need only divide both sides of (11) by λ.

It remains to prove (13). To do this, it suffices to draw the "altitudes" AP, BQ, and CR of $\triangle ABC$, i.e., the segments of the special lines through the vertices of the triangle cut off by the opposite sides (cf. Fig. 42a). [The discussion on p. 42 and Fig. 37 justify calling AP, BQ, and CR the altitudes, in Galilean geometry, of $\triangle ABC$.] We denote the lengths of the

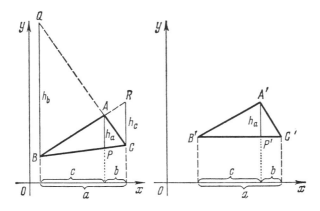

Figure 42a   Figure 42b

altitudes by h_a, h_b, and h_c:
$$h_a = AP = |\delta_{AP}|,$$
$$h_b = BQ = |\delta_{BQ}|,$$
$$h_c = CR = |\delta_{CR}|.$$

The definition of angle in Galilean geometry implies that
$$h_c = B \cdot a = A \cdot b \qquad (15)$$
(Fig. 42a; we recall that $BR = BC = a$ and $AR = AC = b$). Hence
$$\frac{a}{A} = \frac{b}{B}.$$
Similarly, using the altitudes h_b and h_a we see that
$$\frac{a}{A} = \frac{c}{C}$$
and
$$\frac{b}{B} = \frac{c}{C}.$$

We shall now derive another consequence of relations (15). We first show that if S is the area of $\triangle ABC$ (recall that area is a meaningful concept in Galilean geometry; cf. p. 36) then
$$S = \tfrac{1}{2} a h_a = \tfrac{1}{2} b h_b = \tfrac{1}{2} c h_c. \qquad (16)$$
For proof, apply a Galilean motion (1) to $\triangle ABC$ so that its image is a triangle $A'B'C'$ with $B'C'$ parallel to the x-axis (Fig. 42b). Then the Galilean lengths of $B'C'$ and of the altitude h'_a coincide with their Euclidean lengths. Hence
$$S' = \tfrac{1}{2} a' h'_a,$$
where S' is the area of $\triangle A'B'C'$ (note that S' is the Euclidean as well as the Galilean area of $\triangle A'B'C'$; the area of a figure is the same in both geometries). Since the quantities a, h_a and S are invariant under Galilean motions, we have $a' = a$, $h'_a = h_a$, and $S' = S$ (cf. Figs. 42a and 42b). But then
$$S = \tfrac{1}{2} a h_a.$$
A similar argument is used to prove the two remaining relations in (16).

In the third formula of (16), we substitute $h_c = Ab$ from (15). This gives $S = \tfrac{1}{2} bcA$. By symmetry, we obtain
$$S = \tfrac{1}{2} abC = \tfrac{1}{2} acB = \tfrac{1}{2} bcA. \qquad (17)$$
Formulas (17) are the Galilean analogues of the Euclidean relations
$$S = \tfrac{1}{2} ab \sin C = \tfrac{1}{2} ac \sin B = \tfrac{1}{2} bc \sin A. \qquad (17')$$

Consider the Euclidean proposition: *If the sides a and b of $\triangle ABC$ are congruent, then so are the angles A and B opposite them, and conversely.* Formulas (13) imply that this result also holds in Galilean geometry (for

4. The triangle

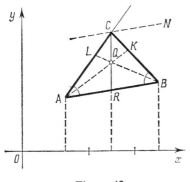 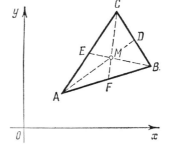

Figure 43   Figure 44

proof, note that $a/A = b/B$; cf. Fig. 43.) Thus a triangle is *isosceles* (i.e., has two equal sides) if and only if it is *isogonal* (i.e., has two equal angles). It is also clear that *the altitude CR of an isosceles triangle ABC bisects the side AB opposite to C* (cf. Fig. 43, where $AR = AC = b$, $BR = BC = a$). On the other hand, CR does not bisect the angle C, since a special line cannot possibly bisect an angle.[8] This implies that in Galilean geometry the angle bisectors need not intersect in a point[8a]; in Figure 43, the bisectors AK and BL of A and B intersect in the midpoint Q of the altitude CR, but, as already pointed out, CR is not the angle bisector CN of C. A Euclidean proposition which does remain valid in Galilean geometry is that *the medians AD, BE, and CF of a triangle ABC intersect in a point M which divides each median, beginning with the vertex, in the ratio* 2:1; $AM:MD = BM:ME = CM:MF = 2:1$. This follows from the fact that the medians of a triangle in Galilean geometry coincide with its Euclidean medians, and the ratio of the segments into which their intersection point divides them is the same in both geometries (Fig. 44; cf. p. 34 above).

We note that in Galilean geometry there are no equilateral triangles; if two sides of a triangle are congruent, then the third side is twice as long [by Eq. (12)]. A similar result holds for angles.

Let ABC be a triangle in the Galilean plane. If, instead of the positive quantities a,b,c,A,B,C (cf. pp. 47–48) we introduce the signed quantities $\bar{a} = d_{BC}$, $\bar{b} = d_{ca}$, $\bar{c} = d_{AB}$, $\bar{A} = \delta_{AB,AC}$, $\bar{B} = \delta_{BC,BA}$, and $\bar{C} = \delta_{CA,CB}$, then formulas (11)–(13) of this section take the following simple form:

$$\bar{a} + \bar{b} + \bar{c} = 0, \quad (11')$$

$$\bar{A} + \bar{B} + \bar{C} = 0, \quad (12')$$

$$\frac{\bar{a}}{\bar{A}} = \frac{\bar{b}}{\bar{B}} = \frac{\bar{c}}{\bar{C}}. \quad (13')$$

---

[8] If the sides meeting at C have slopes k_1 and k_2, the angle bisector must have a slope k satisfying $k_1 - k = k - k_2$, so that $k = (k_1 + k_2)/2$. Since k_1 and k_2 are assumed to be finite, so is k. (Translator's note.)

[8a] It can be shown that in Galilean geometry the angle bisectors in a triangle *never* intersect in a point; see Figure 48b.

Thus here, too, we arrive at relations analogous to (14), where now the sign of the proportionality constant λ has definite geometric significance (cf. footnote 22 of Sec. 9).

We now consider *congruence criteria* for triangles in Galilean geometry. First of all, two triangles ABC and $A'B'C'$ with congruent sides or angles need not be congruent. To see this, we recall that in Galilean geometry the two smallest sides of a triangle determine the third side, and (as in Euclidean geometry) the two smallest angles determine the third angle, but two sides or two angles do not determine a triangle (see Figs. 45a and 45b). Moreover, the familiar congruence criteria of Euclidean geometry, briefly referred to as SAS and ASA, do not hold in Galilean geometry. For example, triangles ABC and A_1BC in Figure 46a have a common angle C enclosed by the common side CB of length a, and the congruent sides CA, CA_1 of length b. However, $AB = a + b$, while $A_1B = a - b$, so triangles ABC and A_1BC are not congruent. Similarly, triangles ABC and A_1BC in Figure 46b have a common side BC enclosed by equal angles, yet $AB > A_1B$. On the other hand, the Euclidean congruence criteria SAS and ASA are valid in Galilean geometry if we deal with directed sides and angles. Specifically, $\triangle ABC \cong \triangle A'B'C'$ if $\bar{a} = d_{BC} = d_{B'C'} = \bar{a}'$, $\bar{b} = d_{CA} = d_{C'A'} = \bar{b}'$, and $\bar{C} = \bar{C}'$; also, $\triangle ABC \cong \triangle A'B'C'$ if $\bar{a} = d_{BC} = d_{B'C'} = \bar{a}'$, $\bar{B} = \bar{B}'$, and $\bar{C} = \bar{C}'$. We leave the proof of these assertions to the reader.

We note that relations (13) imply that *two triangles ABC and $A'B'C'$ with*

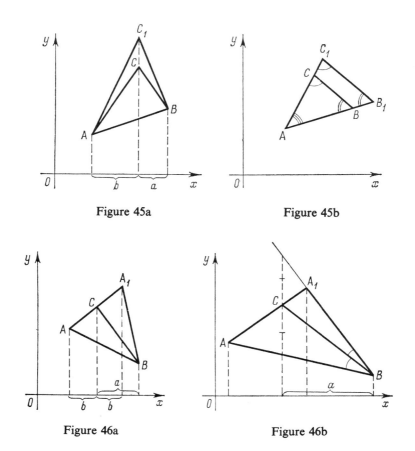

Figure 45a

Figure 45b

Figure 46a

Figure 46b

4. The triangle

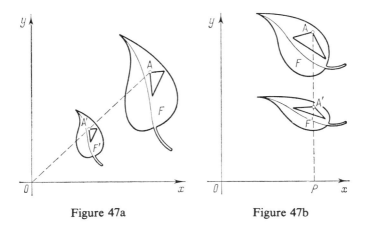

Figure 47a   Figure 47b

congruent angles have proportional sides (cf. Fig. 45b above). Thus

$$\frac{a'}{a} = \frac{b'}{b} = \frac{c'}{c} = k.$$

In this case $\triangle A'B'C'$ can be obtained from $\triangle ABC$ by means of a so-called similitude of the first kind with similitude coefficient k, i.e., by *a mapping of the Galilean plane to itself which preserves* (*the magnitudes of*) *angles, and which multiplies* (*the lengths of*) *segments by a fixed number k*. An example of such a map is a (positive) *central dilatation* with center O, which we define as a transformation of the plane, taking every point A onto the point A' of the ray OA such that

$$\frac{OA'}{OA} = k;$$

cf. Figure 47a. Similarly, *two triangles ABC and $A'B'C'$ with congruent sides have proportional angles* (cf. Fig. 47b). Thus

$$\frac{A'}{A} = \frac{B'}{B} = \frac{C'}{C} = \kappa.$$

In this case, $\triangle A'B'C'$ can be obtained from $\triangle ABC$ by means of a so-called similitude of the second kind with similitude coefficient κ, i.e., by *a mapping of the Galilean plane which preserves segments and multiplies angles by a fixed number κ*. An example of such a transformation is a *compression* with axis Ox and coefficient κ, which takes every point A to the point A' of the ray $PA \| Oy$ such that

$$\frac{PA'}{PA} = \kappa;$$

cf. Figure 47b. We shall not pursue the very interesting study of *similar figures* in the Galilean plane (i.e., figures related by a similitude of the first or second kind, or the product of a similitude of the first kind and one of the second kind).

PROBLEMS AND EXERCISES

4 With the notation of Exercise 3, prove that $a/\sing A = b/\sing B = c/\sing C$, $S_{\triangle ABC} = \frac{1}{2}ab \sing C$, and (after a suitable choice of sides in the triangle) $a^2 = b^2 + c^2 - 2bc \cdot \cosg A$. Develop the analogy between Euclidean and Galilean trigonometry suggested by these relations.

5   Prove the congruence criteria stated on p. 52.

6   (a) Consider a quadrilaterial in the Galilean plane. Find relations involving its sides, angles, the lengths of its diagonals, the angles between the diagonals and between opposite sides, the lengths of its midlines, the length of the segment joining the midpoints of its diagonals, etc. (b) Find relations involving the angles, sides, and diagonals of an arbitrary polygon in the Galilean plane. [*Hint:* You may find it convenient to use directed segments and angles (cf. the remarks in small print on p. 51).]

7   State congruence criteria for quadrilaterals and *n*-gons in the Galilean plane.

8   State criteria for similarity (of the first and second kind; cf. p. 53) of triangles in the Galilean plane.

V   Formulate results involving triangles and tetrahedra in three-dimensional Galilean geometry.

VI Formulate results involving triangles and tetrahedra in three-dimensional semi-Galilean geometry (cf. Problem II).

VII Formulate results involving triangles and tetrahedra in three-dimensional Poinceau geometry (cf. Problem III).

5. The principle of duality; coparallelograms and cotrapezoids

In the previous section we showed that, just as in Euclidean geometry, the medians of a triangle in Galilean geometry intersect in a point (which divides each median in the ratio 2:1 beginning at the vertex). On the other hand, in contrast to Euclidean geometry, the angle bisectors of a triangle in Galilean geometry never intersect in a point (cf. Figs. 48a and 48b[9]). This may disappoint some readers and leave them with the impression that in Euclidean geometry figures are more "regular" and "simple" than in Galilean geometry. However, this is not the case. In fact, Galilean geometry is actually simpler than Euclidean. In particular, the angle bisectors of a triangle in Galilean geometry have a simple and remarkable property which more than makes up for the fact that they are not concurrent. This property will emerge from the principle of duality, an important feature of Galilean geometry which we now proceed to discuss.

It is well known that the properties of lines in Euclidean geometry are often analogous to those of points. Thus, for example, two points determine a unique line (Fig. 49a), while two lines intersect in at most one point (Fig. 49b). The set of points of a line *a* between points *M* and *N* of *a*,

---

[9]Note that it is not quite right to compare Figures 48a and 48b. After all, in Galilean geometry a triangle has precisely three angle bisectors (cf. footnote 8, where it is shown that each angle has only one bisector). On the other hand, in Euclidean geometry a triangle has, strictly speaking, six angle bisectors (three bisectors of the interior angles and three bisectors of the exterior angles). These six angle bisectors of a triangle form a fairly complex configuration: they *intersect three at a time in four points*, the center of the inscribed circle and the three centers of the escribed circles of the triangle. Also, there are *four triples of them whose intersections with the opposite sides of the triangle are collinear.*

5. The principle of duality: coparallelograms and cotrapezoids   55

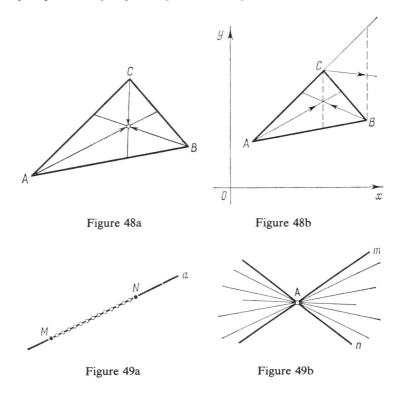

Figure 48a

Figure 48b

Figure 49a

Figure 49b

i.e., the segment MN (Fig. 49a), is analogous to the set of lines passing through a point A and contained between the lines m and n on A, i.e., the angle mAn (Fig. 49b). A triangle can be regarded as a set of three noncollinear points and three segments determined by these points (Fig. 50a), but it can also be thought of as a set of three nonconcurrent lines and the three angles determined by them (Fig. 50b); and so on. Unfortunately, this analogy between points and lines does not go far enough so that one cannot always rely on it in searching for new properties of points or lines. One flaw in the correlation between points and lines is the existence of parallel lines, i.e., lines without a common point. For the analogy between properties of points and properties of lines to be complete, the existence of parallel lines would require the existence of "parallel points," i.e., pairs of points which cannot be joined by lines. But there are no such points in Euclidean geometry. Again, while the distance between points A and B can be arbitrarily large, the size of the angle between two intersecting lines is restricted; if we increase the angle between lines a and b by rotating one of them about their intersection point, then after a rotation ough 180° the rotated line returns to its original position.[10]

---

[10]The angles of arbitrary size employed in trigonometry are not defined merely by a pair of lines, the "sides" of the angle. Consequently, such angles are not strict analogues of segments that are uniquely determined by their endpoints.

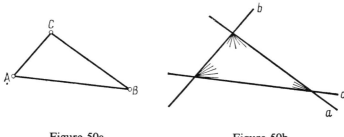

Figure 50a   Figure 50b

In Galilean geometry the situation is different. There we do have "parallel points," i.e., points which cannot be joined by an (ordinary) line. To be sure, such points can be joined by a special line (see, for example, Fig. 51) but in Galilean geometry special lines are radically different from ordinary lines. Again, the angle between two lines in Galilean geometry can be arbitrarily large: starting with a pair of lines a and b intersecting in a point Q, we can pass through Q a line b_1 such that $\angle aQb = \angle bQb_1$, then a line b_2 such that $\angle bQb_1 = \angle b_1Qb_2$, then a line b_3 such that $\angle b_1Qb_2 = \angle b_2Qb_3$, and so on. In this way, we obtain a sequence of lines whose "angular distances" from a increase indefinitely, i.e., a sequence of lines which forms an ever larger angle with a (see Fig. 52b; also, cf. Fig. 52a, which illustrates the corresponding construction for distances between points). We thus have a situation in which the analogy between points and lines is complete. This analogy suggests that *interchanging the words "point" and "line," "distance" and "angle," "lies on" and "passes through" in any theorem of Galilean geometry yields another theorem.*[11]

The italicized assertion (whose proof we discuss in what follows) is known as the **principle of duality**, and theorems related by it are said to be *dual* to each other. This principle sheds new light on some by now familiar facts. Thus in Galilean (and in Euclidean) geometry the angle between two lines is a measure of the "deviation" of one line from the other, and if that angle is zero, i.e., if the lines are parallel, then an additional measure of deviation is the distance between the lines. Similarly, the distance between two points is a measure of the "deviation" of one point from the other, and if that distance is zero, i.e., if the points are on the same special line ("parallel to each other"), then an additional measure of deviation is the special distance, which is measured in units quite different from those of ordinary distance.

The main merit of the principle of duality is that *it enables us to deduce new theorems from known ones.* Thus, for example, it is clear that the formulas

$$a+b=c \quad \text{and} \quad A+B=C,$$

---

[11] Note that in view of the analogy between points and lines, it is obvious but not very important, that our "word transformation" changes a statement of Galilean geometry into another statement. What is important but not obvious is that it changes a *theorem* of Galilean geometry into another *theorem*.

5. The principle of duality: coparallelograms and cotrapezoids   57

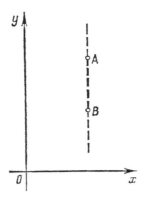

Figure 51

valid for arbitrary triangles, are dual to one another. This means that knowing one of them we could immediately predict the other. In some cases the resulting new theorem coincides with the original one; for example, interchanging sides and angles in the relations $a/A = b/B = c/C$ yields the essentially identical relations $A/a = B/b = C/c$. However, in most cases the dual of a theorem of Galilean geometry is a new theorem which may not be easy to discover without the principle of duality.

It is easy to illustrate the value of the principle of duality by means of examples. We know that in an *isosceles* triangle ABC with $AC = BC$, the altitude (special line) from the vertex C is also the median from C, but it is not the angle bisector of C. Coincidence of the median CF of $\triangle ABC$ and the altitude implies that CF is special, i.e., using the terminology introduced above, *the midpoint F of side AB is parallel to the vertex C* (Fig. 53). The dual of the midpoint F of the side AB is the bisector f of the angle C. Hence the dual of the parallelism of points C and F is the parallelism of lines c and f. We thus arrive at the following assertion: *the bisector f of the angle C at the vertex of an isosceles* (or *isogonal*—cf. p. 51) *triangle ABC is*

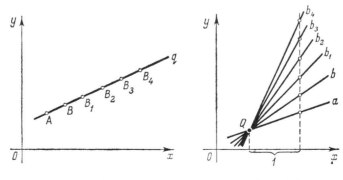

Figure 52a   Figure 52b

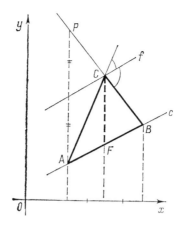

Figure 53

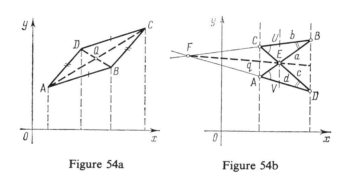

Figure 54a   Figure 54b

parallel to its base $AB = c$. For proof[12] we note that f bisects the vertical segment AP. Moreover, f bisects PB, since $PC = AC = CB$. Thus f is a midline of $\triangle ABP$ and hence $f \| BA$, as asserted.

For other examples, we turn to properties of parallelograms and trapezoids. It is clear that if the opposite sides of a quadrilateral $ABCD$ in the Galilean plane are parallel (in which case we call the quadrilateral a *parallelogram*), then they are congruent (Fig. 54a). Further, the intersection point Q of the diagonals AC and BD of the parallelogram $ABCD$ bisects them (as in Euclidean geometry). The figure in the Galilean plane dual to a parallelogram is a quadrilateral $ABCD$ whose opposite vertices A and C, B and D are parallel, i.e., lie on special lines AC and BD (Fig. 54b). We shall use the term *coparallelogram* for the (necessarily self-intersecting) quadrilateral $ABCD$ with sides $AB = a$, $BC = b$, $CD = c$, $DA = d$. It is clear that *the opposite angles of a coparallelogram are congruent* (and hence $A = C$ and

---

[12] Since we have not yet proved the principle of duality, we can use it only as a guide to the discovery of plausible theorems. If we had a proof of this principle (see Sec. 6), then the duals of known theorems would require no further proof.

5. The principle of duality: coparallelograms and cotrapezoids

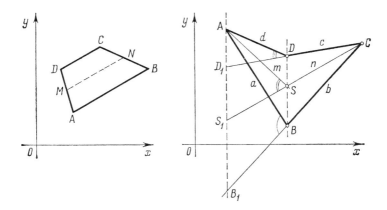

Figure 55a   Figure 55b

$B = D$). This fact, dual to the congruence of the opposite sides of a parallelogram, follows immediately from the definition of angle in Galilean geometry. Further, to the diagonals AC and BD of the parallelogram $ABCD$ there correspond the intersection points E and F of the (pairs of) opposite sides a and c, b and d of the coparallelogram $ABCD$. To the lengths of the diagonals AC and BD of the parallelogram there correspond the magnitudes of the angles aEc and bFd of the coparallelogram. To the intersection point Q of the diagonals AC and BD of the parallelogram there corresponds the line $q = EF$. The dual of the assertion that Q bisects the diagonals AC and BD is the following theorem: *The line EF which joins the points of intersection E and F of the opposite sides of the coparallelogram ABCD bisects the angles at E and F.* For proof, we draw the special line through E, and note that E is the midpoint of its segment UV between b and d (this follows from the congruence of the opposite angles of the coparallelogram). This implies that $\angle bFq = \angle qFd$. But then $\angle aEq = \angle qEc$, for $\angle aEq = \angle bFq + \angle aBb$ and $\angle qEc = \angle qFd + \angle dDc$.

It is natural to define a *trapezoid* as a quadrilateral $ABCD$ with parallel opposite sides AB and DC (Fig. 55a). Since a Galilean trapezoid, thus defined, is also a Euclidean trapezoid, its midline MN (joining the centers M and N of the sides AD and BC) is parallel to the bases AB and CD, and its length is half the sum of their lengths.[13] We call a quadrilateral $ABCD$ in the Galilean plane with sides $\overline{AB} = \bar{a}$, $\overline{BC} = \bar{b}$, $\overline{CD} = \bar{c}$, $\overline{DA} = \bar{d}$ and parallel vertices B and D (this, we recall, means that BD is a special line; see Fig. 55b) a *cotrapezoid*. To the sides BC and DA of the trapezoid $ABCD$ there correspond the vertices C and A of the cotrapezoid $ABCD$, and to the midpoints N and M of the sides of the trapezoid there correspond the bisectors $n = CS$ and $m = AS$ of the angles C and A of the

---

[13] It is clear that the Euclidean relations $MN \| AB \| CD$ and $MN = \frac{1}{2}AB + \frac{1}{2}CD$, i.e., $AB/MN + DC/MN = 2$, remain valid in Galilean geometry.

cotrapezoid with intersection point S. The fact that the midline MN of the trapezoid $ABCD$ is parallel to its bases AB and CD yields the dual assertion that S is parallel to the vertices B and D, i.e., *is on the special line* BD; this follows readily from the definition of angle bisector in Galilean geometry, which implies that the lines AS and CS bisect the segment BD. The assertion that the midline of a trapezoid is half the sum of its bases yields the dual theorem that *the angle mSn is half the sum of the angles aBb and dDc* of the cotrapezoid:

$$\angle mSn = \tfrac{1}{2} \angle aBb + \tfrac{1}{2} \angle dDc \qquad \left(\delta_{mn} = \tfrac{1}{2}\delta_{ab} + \tfrac{1}{2}\delta_{dc}\right).$$

[The latter follows from the fact that (see Fig. 55b)

$$\overline{AS_1} = \overline{AB_1} - \overline{S_1B_1} = \overline{AB_1} - \tfrac{1}{2}\,\overline{D_1B_1}$$
$$= \overline{AB_1} - \tfrac{1}{2}(\overline{AB_1} - \overline{AD_1}) = \tfrac{1}{2}\,\overline{AB_1} + \tfrac{1}{2}\,\overline{AD_1}\,;$$

here $B_1 D_1 \| BD$ is a special line.] A special case of this theorem is the assertion that *in a triangle* ABC (viewed as a "degenerate cotrapezoid" $ABCP$ with BP an altitude) *the bisectors AS and CS of the angles A and C intersect at the midpoint S of the altitude BP, where they form an angle $\tfrac{1}{2}B$* (see Fig. 56a, where $AS_1 = \tfrac{1}{2}AB_1$; compare this result with the theorem about the midline of a triangle, Fig. 56b).

It is now easy to formulate the dual of the theorem on the intersection point of the medians of a triangle. The duals of the midpoints D, E, F of the sides BC, CA, and AB of $\triangle ABC$ are the bisectors $d = AU$, $e = BV$, $f = CW$ of its angles A, B, C (cf. Fig. 57a and 57b). The duals of the lines AD, BE, CF are the intersection points U, V, W of the angle bisectors of $\triangle ABC$ with the opposite sides. Hence the dual of the theorem that the medians AD, BE, CF of a triangle ABC intersect in a point M (Fig. 57a) is

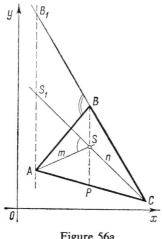

Figure 56a

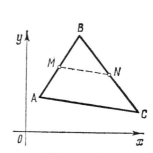

Figure 56b

5. The principle of duality: coparallelograms and cotrapezoids

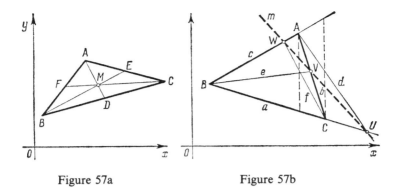

Figure 57a   Figure 57b

the theorem that the points U, V, W lie on a line m (Fig. 57b). Furthermore, the duals of the relations

$$AM:MD = BM:ME = CM:MF = 2:1$$

are the relations

$$\angle aUm : \angle mUd = \angle bVm : \angle mVe = \angle cWm : \angle mWf = 2:1.$$

In summary, *the intersection points of the angle bisectors of a (non-isosceles[14]) triangle ABC with the opposite sides lie on a line m which divides the angles between the bisectors and the opposite sides in the ratio 2:1 beginning with the sides.* This is the property of the angle bisectors of a triangle in Galilean geometry to which we alluded at the beginning of this section.

To prove this theorem, we recall a proof of the theorem about the medians of a triangle.[15] First we show that the intersection point M of the medians AD and BE of $\triangle ABC$ divides the medians in the ratio $AM:MD = BM:ME = 2:1$. To this end, we join the midpoints D and E of BC and CA to the midpoint N of the segment CM (Fig. 58a; so far we cannot claim that CM is part of the median CF of our triangle). Since DN is a midline of $\triangle BMC$, and EN is a midline of $\triangle CMA$, it follows that $DN \| BE$ and $EN \| AD$. But then $MDNE$ is a parallelogram, and so

$$DM = NE = \tfrac{1}{2}MA \quad \text{and} \quad EM = ND = \tfrac{1}{2}MB.$$

This argument holds for any two medians of the triangle. Hence the intersection point of any two medians of the triangle divides each median in the ratio 2:1 beginning with the vertex. This implies that all three medians are concurrent, since both BE and CF intersect AD in the unique point M such that $AM:MD = 2:1$.

We now dualize our argument by interchanging points with lines, and segments with angles. We denote by m the line UV, where U and V are the

---

[14]*If $\triangle ABC$ is isosceles with $AC = BC$, then the line KL passing through the intersection points of the bisectors of the angles A and B with the opposite sides is parallel to the base AB of the triangle and to the bisector of the angle at C (cf. p. 51 above; in particular, Fig. 43).*

[15]This proof works in both Euclidean and Galilean geometry. (Translator's note.)

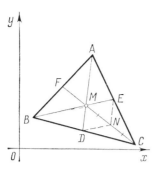
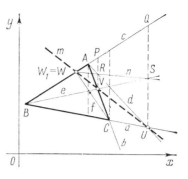

Figure 58a   Figure 58b

intersection points of the angle bisectors $d = AU$ and $e = BV$ with the opposite sides. Let n be the bisector of the angle mW_1c formed by the line m with the side $AB = c$ of $\triangle ABC$ (Fig. 58b; at this point we cannot claim that W_1 coincides with the point W where the bisector f of angle C meets the side c). Let R and S be the intersection points of n with the angle bisectors d and e. Since R is the intersection of the angle bisectors d and n of $\triangle AVW_1$, it lies on the altitude VP (cf. p. 60 and Fig. 56a). Similarly, S, as the intersection of the angle bisectors e and n of $\triangle BUW_1$, lies on the altitude UQ. Hence the (self-intersecting) quadrilateral $RUVS$ is a coparallelogram. But then $\angle mUd = \angle eSn$ and $\angle mVe = \angle dRn$. Now applying to the triangles AVW_1 and BUW_1 the theorem on p. 60, we see that

$$\angle dUm = \angle nSe = \tfrac{1}{2} \angle mUa \quad \text{and} \quad \angle eVm = \angle nRd = \tfrac{1}{2} \angle mVb,$$

i.e., *m divides the angles formed by d and e with the opposite sides of the triangle in the ratio* $2:1$ *beginning with the sides*. Now define W to be the intersection point of the bisector $f = CW$ of the angle C with the side $AB = c$. Then, by symmetry, the line $UW = m_1$ has the property $\angle aUm_1 : \angle m_1Ud = 2 : 1$. Hence m_1 coincides with m, i.e., *the points U, V, W are on m, and*

$$\angle aUm : \angle mUd = \angle bVm : \angle mVe = \angle eWm : \angle mWf = 2 : 1.$$

This proof of the "theorem on angle bisectors" of a triangle is very instructive because it is the exact dual of the proof of the theorem on the medians of a triangle. The latter fact sheds additional light on the principle of duality, for it shows that we can use verbal transformations (interchanging "point" with "line," and so on) not only to formulate the duals of known theorems, but also to prove them. To understand the reason behind this phenomenon, we must examine the process of proving a theorem in geometry. Consideration of this process will not only deepen our understanding of the factors underlying the principle of duality, but will also show what must be done to justify it. Note that, in spite of their suggestiveness, our examples of the use of the principle of duality do not constitute a proof of its validity.

5. The principle of duality: coparallelograms and cotrapezoids

We recall that to prove a theorem in some mathematical system, one proceeds as follows. By means of purely logical steps the theorem in question is reduced to simpler theorems. In turn, these are reduced to still simpler theorems, and so on. This process continues until we arrive at simple propositions called *axioms* which we accept without proof. In the final analysis it is these unproved propositions (axioms) on which we base the proofs of the theorems in the system. In actual proofs of geometric propositions, we carry the reduction process back to previously established theorems rather than to the axioms. This does not affect the general scheme of proof outlined above; it merely means that we leave out established portions of the deductive chain. Now our discussion of Galilean geometry suggests that the basic properties of points and lines, distances and angles are entirely analogous in the sense that *by interchanging the words "point" and "line," and so on, in any axiom of Galilean geometry, we obtain a valid proposition* (see Sec. 6). Thus, for example, the axiom "*two lines have at most one point in common*" (Fig. 59a) yields the assertion "*there is at most one* (ordinary) *line joining two points*" (Fig. 59b). Again, the parallel axiom which asserts that *through a point A not on a line a there passes a unique line not intersecting a* yields the assertion that *on every* (ordinary) *line a not passing through a point A there is a unique point L which cannot be joined to A by an* (ordinary) *line* (Fig. 59d); and so on. Had we drawn up a list of axioms and proved their duals as theorems, then we could claim that the principle of duality (cf. p. 56) holds in our

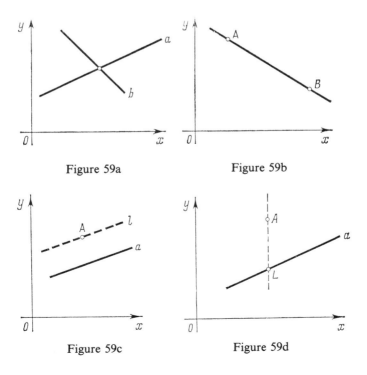

Figure 59a

Figure 59b

Figure 59c

Figure 59d

geometry; in fact, the dual of a theorem could then be proved by interchanging the words "point" and "line," and so on, in the proof of the original theorem. (It should be pointed out that the proof thus obtained would not involve the same axioms as the proof of the original theorem, but rather their duals.) Thus the proofs of duals of theorems of Galilean geometry given above would be superfluous.

The reason that we do not use this approach to establish the principle of duality is that it would require a list of the axioms of Galilean geometry. In the elementary part of our book such a list would be out of place; indeed, many of our readers are probably not even familiar with a complete set of axioms of Euclidean geometry (not usually given in elementary texts).[16] More advanced readers will find it profitable to study the (three) proofs of the principle of duality presented in Section 6.

Here is another proof of the theorem formulated on p. 60 about the intersections of the angle bisectors of a triangle (in the Galilean plane) with the opposite sides. It is well known that the concurrence of the medians of a triangle (valid in Euclidean and in Galilean geometry) follows directly from Ceva's theorem, which asserts that *three lines AU, BV, CW passing through the vertices of a triangle ABC and intersecting the opposite sides in points U, V, W (cf. Fig. 60a) are concurrent or parallel if and only if*

$$\frac{\overline{AW}}{\overline{WB}} \cdot \frac{\overline{BU}}{\overline{UC}} \cdot \frac{\overline{CV}}{\overline{VA}} = 1;$$

cf. p. 25.[17, 18] [Ceva's theorem also implies that the Euclidean angle bisectors of a triangle are concurrent. Indeed, if AU, BV, and CW are the Euclidean angle bisectors of $\triangle ABC$ with sides of lengths $\overline{AB} = \bar{c}$, $\overline{BC} = \bar{a}, \overline{CA} = \bar{b}$, then $\overline{AW} : \overline{WB} = \bar{b} : \bar{a}$, $\overline{BU} : \overline{UC} = \bar{c} : \bar{b}$, $\overline{CV} : \overline{VA} = \bar{a} : \bar{c}$, and

$$\frac{\overline{AW}}{\overline{WB}} \cdot \frac{\overline{BU}}{\overline{UC}} \cdot \frac{\overline{CV}}{\overline{VA}} = \frac{\bar{b}}{\bar{a}} \cdot \frac{\bar{c}}{\bar{b}} \cdot \frac{\bar{a}}{\bar{c}} = 1$$

implies that AU, BV, and CW are concurrent.] Similarly, the collinearity of the points U, V, W in Figure 57b follows from the theorem of Menelaus, which asserts that *points U, V, W on the sides BC, CA, AB of a triangle ABC (Fig. 60b) are collinear if and only if*[19, 20]

$$\frac{\overline{AW}}{\overline{BW}} \cdot \frac{\overline{BU}}{\overline{CU}} \cdot \frac{\overline{CV}}{\overline{AV}} = 1.$$

In fact, if in Figure 57b, $\overline{AB} = d_{AB} = \bar{c}$, $\overline{BC} = d_{BC} = \bar{a}$, $\overline{CA} = d_{CA} = \bar{b}$ (where $\bar{a} + \bar{b} + \bar{c} = 0$), then, in view of relations (13′) on p. 51, consideration of triangles ABU and ACU yields

$$\overline{BU} : \overline{AU} = \angle BAU : \angle ABU = \delta_{cd} : \delta_{ca} \quad \text{and} \quad \overline{CU} : \overline{AU} = \angle UAC : \angle UCA$$
$$= \delta_{db} : \delta_{ab}.$$

---

[16]For complete sets of axioms of plane Euclidean geometry see, for example, [5]–[7] or [4].

[17]For Ceva's theorem see, for example, Problem 27(b) in [11] or Chap. VIII of [33].

[18]We have already noted that the ratio of the Galilean lengths of two collinear segments is the same as the ratio of their Euclidean lengths. Since Ceva's theorem only involves such ratios, its Galilean analogue follows immediately. (Translator's note.)

[19]Cf., for example, Problem 27(a) in [11] or Chap. VIII of the book [33].

[20]The Galilean analogue of Menelaus's theorem follows from the same reasoning as was given in the footnote 18. (Translator's note.)

5. The principle of duality: coparallelograms and cotrapezoids

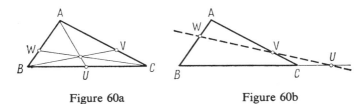

Figure 60a   Figure 60b

Dividing the first of these equations by the second, we obtain

$$\overline{BU} : \overline{CU} = \delta_{cd}\delta_{ab} : \delta_{ca}\delta_{db}.$$

But $\delta_{cd} = \angle BAU = \angle UAC = \delta_{db}$ (since AU bisects angle A). Hence

$$\overline{BU} : \overline{CU} = \angle UCA : \angle ABU = \delta_{ab} : \delta_{ca} = \angle C : \angle B = \bar{c} : \bar{b}$$

(cf. (13′) of Section 4). A similar argument shows that

$$\overline{CV} : \overline{AV} = \bar{a} : \bar{c} \quad \text{and} \quad \overline{AW} : \overline{BW} = \bar{b} : \bar{a}.$$

Now the equality

$$\frac{\overline{AW}}{\overline{BW}} \cdot \frac{\overline{BU}}{\overline{CU}} \cdot \frac{\overline{CV}}{\overline{AV}} = \frac{\bar{b}}{\bar{a}} \cdot \frac{\bar{c}}{\bar{b}} \cdot \frac{\bar{a}}{\bar{c}} = 1$$

implies that *the points U, V, W are collinear.*

We conclude this section with a few remarks about the mechanical meaning of the principle of duality. We arrived at Galilean geometry by taking as points events of one-dimensional kinematics characterized by a location on the line o and a moment in time. The motions of our geometry are Galilean transformations. The "distance" between two events is the time interval separating them. For simultaneous events, for which this distance is zero, we define a new "distance," namely, the space interval between the points on o corresponding to our events. Further, two (nonsimultaneous) events determine a uniform motion along o which "joins" them and this motion plays the role of a line in our geometry. This, then, is the foundation of a theory which, as we saw, resembles Euclidean geometry. However, we are free to try to construct other geometric systems by taking as our points not events but other entities of mechanics. In particular, we could take as the "points" of a new geometric system all uniform motions on the line o, and as its motions, as before, the Galilean transformations which determine the transitions from one inertial reference frame to another. There is a natural metric which can be imposed on our set of uniform motions; specifically, we can take as the distance between two motions their relative velocity—which, as we know, has "absolute" significance (i.e., significance independent of the frame of reference). If both motions have the same velocity (i.e., if their relative velocity is zero), then we can speak of the usual distance between moving points, which in this case does not change with time. Further, two uniform motions (with different velocities) determine a unique event which belongs to both of them and this event plays the role of a line in the "geometry of uniform motions." The principle of duality asserts the essential equivalence of the "geometry of uniform motions" just described and the "geometry of events" treated in the main part of this book.

PROBLEMS AND EXERCISES

9   The midlines of $\triangle ABC$ form $\triangle A'B'C'$ whose angles are congruent to those of $\triangle ABC$ and whose sides are half the size of those of $\triangle ABC$. $\triangle A'B'C'$ is obtained from $\triangle ABC$ by means of a dilatation with center M, which

coincides with the centroid of △*ABC*, and coefficient −1/2. State and prove a theorem of Galilean geometry dual to that just stated.

10 The midpoints A', B', C', D' of the sides of any quadrilateral *ABCD* form the vertices of a parallelogram. Its sides are parallel to the diagonals of *ABCD* and are half their size. The angles of $A'B'C'D'$ are congruent to the angles between the diagonals of *ABCD*, and the area of $A'B'C'D'$ is half the area of *ABCD*. State and prove a theorem of Galilean geometry dual to that just stated. [*Hint*: What concept of Galilean geometry is dual to the concept of area of a figure?]

11 (a) What theorem of Galilean geometry is the dual of Ceva's theorem? (Cf. p. 64.) (b) What theorem of Galilean geometry is the dual of Menelaus' theorem? (Cf. p. 64.)

12 In Euclidean as well as in Galilean geometry, the line joining the intersection point *P* of the diagonals of a trapezoid[21] to the intersection point *Q* of its nonparallel sides bisects its bases. State and prove the dual of this theorem in Galilean geometry. [*Hint*: To prove the original theorem, note that the dilatation γ_1 with center *Q* and coefficient $k_1 = QD/QA$ takes *AB* to *DC*. This implies that γ_1 carries the midpoint *M* of *AB* to the midpoint *N* of *DC*. But then *M*, *N*, *Q* are collinear. To prove the collinearity of *M*, *N*, *P* we consider the dilatation γ_2 with center *P* and coefficient $k_2 = -PC/PA$.]

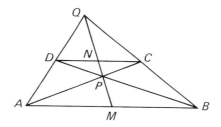

13 Give other examples of pairs of dual theorems of Galilean geometry.

VIII Discuss the principle of duality in three-dimensional semi-Galilean geometry (see Exercise 5). Give examples of pairs of dual theorems of this geometry. [*Hint*: In this geometry the dictionary which enables us to go from a theorem to its dual includes the following pairs of concepts.

Point	Plane
Line	Line
Plane	Point
The point *A* is in the plane π	The plane α contains the point *P*
The point *A* is on the line *c*	The plane α contains the line *c*
The line *a* is in the plane π	The line *a* contains the point *P*
Distance between points	Angle between planes
Angle between lines	Angle between lines
Angle between planes	Distance between points

and so on.]

---

[21]Assumed to be not a parallelogram. (Translator's note.)

6. Proofs of the principle of duality

Our discussion of the principle of duality in Galilean geometry on pp. 62–64 made the principle seem plausible but did not rigorously prove it. In this section, we present a number of proofs of the principle.

We pointed out earlier that the principle of duality does not hold in Euclidean geometry (however, see Problem IX below). It was first discovered in *projective geometry*.[22] The founder of projective geometry, the French engineer and geometer J. V. PONCELET (1787–1867), deduced the (projective) principle of duality from the fact that polarities of the projective plane (cf. Chap. 1, Sec. 4 of [12]) interchange points and lines and thus transform the diagram associated with a given projective theorem into the diagram of its dual (whose validity follows from properties of polarities). At about the same time, another French geometer, J. D. GERGONNE (1771–1859), also considered the principle of duality. Gergonne wrote theorems and their duals in two columns. Transition from one column to the other required interchanging the terms "line" and "point," "passes through the point" and "lies on the line," and so on. Gergonne was familiar with the concept of a polarity but did not use it to establish the principle of duality. He relied, instead, on an argument analogous to that used in Section 5 of this book to make the Galilean principle of duality plausible. In a sense, Gergonne may be said to have anticipated the derivation of the principle of duality from a list of axioms of projective geometry (of course, at that time mathematicians had no such list and were not aware of the need for one). A third proof of the (projective) principle of duality is due to a contemporary of Poncelet and Gergonne, the German geometer A. F. MÖBIUS (1790–1868) and is based on the use of "point coordinates" and "line coordinates." These could be used to translate an analytic proof of a projective theorem into an analytic proof of its dual (with "line coordinates" replacing the ordinary coordinates of a point).

In this section we give three proofs of the principle of duality in Galilean geometry. The first proof uses an approach analogous to that of Möbius, the second echoes Poncelet, and the third is in the spirit of Gergonne.

I. Analytic proof of the principle of duality (the approach of Möbius)

Since every (ordinary) line l in the Galilean plane is uniquely determined by the coefficients k and s of its equation $y = kx + s$, we refer to the pair of numbers (k,s) as *the coordinates of the line l* (the **tangential coordinates** of l). The mapping $(x,y) \to (x',y')$ defined by Eq. (1) of Section 3 sends each point (x,y) of the Galilean plane to another point (x',y'). It induces a mapping $(k,s) \to (k',s')$ which sends each line (k,s) of the Galilean plane to

---

[22]See, for example, Chap. 14 in [19], Chap. 7 in [31], Chap. 5 in [32], and [14].

another line (k',s'). To describe the latter mapping, we invert

$$x' = x \quad + a,$$
$$y' = vx + y + b, \qquad (1)$$

and obtain

$$x = x' - a,$$
$$y = -v(x'-a) + y' - b,$$

or

$$x = \quad x' \quad - a,$$
$$y = -vx' + y' + av - b. \qquad (1')$$

This means that the mapping (1) carries the line l, i.e., the set of points (x,y) satisfying the equation $y = kx + s$, to the set of points (x',y') satisfying the equation

$$-vx' + y' + av - b = k(x' - a) + s,$$

or

$$y' = (k+v)x' + (-ak + s + b - av). \qquad (18)$$

Putting

$$k' = \quad k \quad + v,$$
$$s' = -ak + s + (b - av), \qquad (19)$$

we see that (18) can be written as

$$y' = k'x' + s' \qquad (18')$$

and thus represents a line l', the image of l under (1).

We have defined Galilean geometry to be the study of properties of figures [sets of points (x,y)] invariant under all the transformations (1). If a figure is defined to be a collection of lines rather than of points, then we must describe Galilean geometry as the study of figures [sets of lines (k,s)] invariant under all the transformations (19). If we rewrite (19) in the form

$$k' = \quad k + A,$$
$$s' = Vk + s + B, \qquad (20)$$

where $V = -a$, $A = v$, and $B = b - av$, then it becomes clear that the difference between Galilean geometry of points and Galilean geometry of lines is just a matter of terminology.

For example, the fundamental invariant of a pair of points $A(x,y)$ and $A_1(x_1,y_1)$ under all the transformations (1) is their distance

$$d_{AA_1} = x_1 - x, \qquad (5)$$

or if $d_{AA_1} = 0$, their special distance

$$\delta_{AA_1} = y_1 - y \qquad (6)$$

(cf. pp. 38–39). Similarly, the fundamental invariant of a pair of lines (k,s)

and (k_1, s_1) under all the transformations (20) is the angle

$$\delta_{ll_1} = k_1 - k \tag{8}$$

between l and l_1, or if $\delta_{ll_1} = 0$, the "special angle"

$$d_{ll_1} = s_1 - s \tag{9}$$

(or distance) between l and l_1. Again, from the point of view of the geometry of points, a line q with tangential coordinates (k,s) is the set of points (x,y) satisfying the equation

$$y = kx + s \tag{21}$$

(cf. Fig. 52a). On the other hand, from the point of view of the geometry of lines, a point $Q(x,y)$ can be characterized by the pencil of lines (21), where x and y are fixed and k and s are variable (cf. Fig. 52b). Thus in the geometry of lines the (ordinary) coordinates (x,y) of a point arise as the coefficients in the equation (21), which we rewrite for the sake of symmetry as

$$s = (-x)k + y. \tag{21'}$$

Starting with this equation we can again establish the invariance of the distance between points; the distance in question is given by (5) if the value of (5) is different from zero, and by (6) otherwise.

The complete symmetry between the study of points (x,y), lines (21) and invariants of points and lines under the mappings (1), and the study of lines (k,s), points (21) [or (21')] and invariants of lines and points under the mappings (20) establishes the principle of duality.

II. The "transformation" proof of the principle of duality (the approach of Poncelet)

We denote by $l = l(k, t)$ a line in the Galilean plane whose equation in "symmetric" form is

$$y = kx - t \quad \text{or} \quad t + y = kx, \tag{22}$$

where $-t$ is the y-intercept s (cf. the first proof of the principle of duality). We define the distance between parallel lines l and l_1 whose equations are $y = kx - t$ and $y = kx - t_1$ to be

$$d_{ll_1} = -d_{l_1 l} = t_1 - t. \tag{23}$$

Then we consider the **polarity** $\pi : M \leftrightarrow m$ which interchanges the point $M(X, Y)$ and the line $m(X, Y)$ (whose equation is $y = Xx - Y$):

$$M(X, Y) \leftrightarrow m(X, Y). \tag{24}$$

The geometric interpretation of the map π is apparent from Fig. 61: π interchanges the point $M(X, Y)$ and the line $m(X, Y)$ with y-intercept $-Y$ which makes a Galilean angle X with the axis of abscissas. (This implies in particular that in Fig. 61 we have $OP = Y/X$ and $NQ = X^2$.)

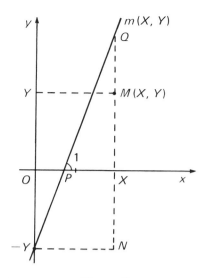

Figure 61

It is easy to see that the polarity π has the following properties:

(i) If $\pi(M) = m$ [so that $\pi(m) = M$] and $\pi(N) = n$, then $d_{MN} = \delta_{mn}$.
(ii) If M and N are parallel (i.e., lie on a special line), *then* $m \| n$ *and* $\delta_{MN} = d_{mn}$ [where $\pi(M) = m$ and $\pi(N) = n$].
(iii) If $m = \pi(M)$, $N = \pi(n)$, then $d_{Mn} = d_{Nm}$.
Property (iii) implies property
(iv) If M is on n, then $\pi(M) = m$ passes through $N = \pi(n)$ (i.e., if $d_{Mn} = 0$, then $d_{Nm} = 0$).

[In fact, if $M(x,y) \in n(k,t)$, then the coordinates (x,y) of M and the coordinates (k,t) of n are connected by the symmetric relation (22). But then the coordinates (x,y) of $m = \pi(M)$ and the coordinates (k,t) of $N = \pi(n)$ are connected by the same relation. Hence $N(k,t) \in m(x,y)$.]

Property (iv) implies that π interchanges the pencil of (ordinary) lines through M and the points of the line $m = \pi(M)$.

The existence of the mapping π with properties (i)–(iv) implies the principle of duality. This follows from the fact that π interchanges the diagram of a Galilean proposition and the diagram of its dual.

III. Axiomatic proof of the principle of duality (the approach in the spirit of Gergonne)

In Supplement B of the present book, Galilean geometry is described in terms of points and vectors. Specifically, to each pair of points A and B there is associated a vector $\overline{AB} = \mathbf{a}$; the vectors satisfy the groups of axioms I, II, III[(2)] and IV[(G)], and the correspondence $\overline{AB} = \mathbf{a}$ between points and vectors satisfies the axioms in group V. Also, Supplement B contains a sketch of the development of plane Galilean geometry based on these five

6. Proofs of the principle of duality

groups of axioms. (While the extensive development of Galilean geometry presented in Sections 3–5 is ultimately based on these axioms, it seemed pointless to dwell on this issue in those sections.) Here we assume familiarity with Galilean geometry and prove that it satisfies not only the axioms listed in Supplement B but also their duals, which are the result of interchanging in the axioms the terms "point" and "line" and of linguistic adjustments necessitated by this interchange. In this way the principle of duality is established, for as soon as we can deduce a theorem of Galilean geometry from the axioms, we can immediately deduce its dual from the duals of those axioms.

We define a *vector* in the Galilean plane as an ordered pair of points (A, B) and call the line AB its *carrier*. We say that the vectors (A, B) and (C, D) (or, using earlier notation, $\overline{AB}$ and $\overline{CD}$) are *equal* if their carriers are parallel (or coincident) and their lengths are the same: $d_{AB} = d_{CD}$ or, in the case of "special vectors" whose carriers are special lines, $\delta_{AB} = \delta_{CD}$ (Fig. 62a). [This rule is equivalent to the *parallelogram law*: If $\overline{AB} = \overline{CD}$ and their carriers are distinct, then the points A, B, C, D are vertices of a parallelogram.] Similarly, we define a **doublet** in the Galilean plane as an ordered pair (a, b) or $\overline{ab}$ of ordinary lines, and a *special doublet* as an ordered pair of parallel lines. We call the lines a and b the *beginning* and *end* of the doublet, respectively. We call the intersection point of the lines

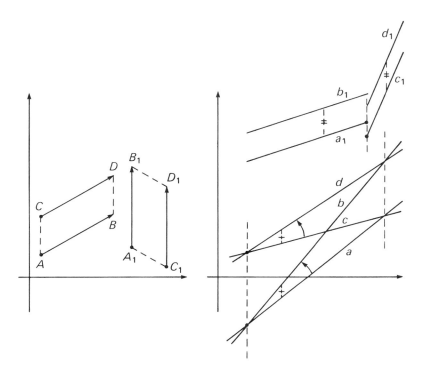

Figure 62a   Figure 62b

of an ordinary doublet, and any point on the beginning of a special doublet, its *vertex*. We call the angle δ_{ab} of a doublet $\overline{ab}$ its *angle*, and the distance d_{ab} associated with a special doublet $\overline{ab}$ its *special angle*. We say that the doublets $\overline{ab}$ and $\overline{cd}$ are *equal* if their vertices are parallel (i.e., lie on the same special line) or coincident and their angles, or special angles, are equal (Fig. 62b). We shall use boldface capital letters to denote doublets. [Our definition of equality of doublets is equivalent to the following assertion: If $\overline{ab} = \overline{cd}$ and the vertices of the doublets are distinct, then the lines a, b, c, d are the sides of a coparallelogram. (Can you prove this?)] We define all *zero* doublets, i.e., doublets with the same beginning and end, to be equal. We denote a zero doublet $\overline{aa}$ (and a zero vector $\overline{AA}$) by **0**. Our definition of equality of doublets implies that any doublet is equal to some doublet with a preassigned beginning a.

Addition of vectors is defined by the "triangle law": $\overline{AB} + \overline{BC} = \overline{AC}$ (Fig. 63a). This definition implies the commutativity of vector addition (**a** + **b** = **b** + **a** follows from the parallelogram law: $\overline{AB} + \overline{AD} = \overline{AC}$ if $AB \| DC$ and $AD \| BC$; see Fig. 63a) and its associativity [i.e., the fact that (**a** + **b**) + **c** = **a** + (**b** + **c**)] follows readily from the equality $(\overline{AB} + \overline{BC}) + \overline{CD} = \overline{AB} + (\overline{BC} + \overline{CD}) = \overline{AD}$. Similarly, the definition $\overline{ab} + \overline{bc} = \overline{ac}$ of doublet addition (Fig. 63b) implies that this operation is commutative (the equality **A** + **B** = **B** + **A** can be deduced from the "coparallelogram law" for addition of doublets: $\overline{ab} + \overline{ad} = \overline{ac}$ if the vertices of the doublets $\overline{ab}$ and $\overline{dc}$, $\overline{ad}$ and $\overline{bc}$ are parallel; cf. Fig. 63b) and associative [since clearly $(\overline{ab} + \overline{bc}) + \overline{cd} = \overline{ab} + (\overline{bc} + \overline{cd}) = \overline{ad}$]. Further, it is obvious that for each doublet **A** we have **A** + **0** = **A**, and that the doublet $\overline{ba}$ is the *additive inverse* of $\overline{ab}$ in the sense that $\overline{ab} + \overline{ba} = \mathbf{0}$. Thus addition of doublets satisfies all the axioms of group I. Addition of doublets also satisfies the axioms of group V with lines playing the role of points: the definition of addition of doublets states that $\overline{ab} + \overline{bc} = \overline{ac}$, and the definition of equality of doublets

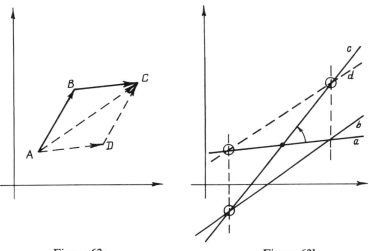

Figure 63a   Figure 63b

6. Proofs of the principle of duality

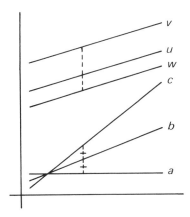

Figure 64

implies that given a line a there is just one doublet with beginning a equal to a given doublet **A**.

We now define multiplication of a doublet by a number: $\overline{ac} = \alpha \overline{ab}$ if the doublets $\overline{ac}$ and $\overline{ab}$ have a common vertex and $\delta_{ac} = \alpha \delta_{ab}$ or $d_{ac} = \alpha d_{ab}$ (see Fig. 64, where $\overline{ac} = 2\overline{ab}$ and $\overline{uw} = -\frac{1}{2}\overline{uv}$). From this definition it follows readily that $1\mathbf{A} = \mathbf{A}$ for any doublet **A**, that $\alpha(\beta \mathbf{A}) = (\alpha\beta)\mathbf{A}$, and that $(\alpha + \beta)\mathbf{A} = \alpha \mathbf{A} + \beta \mathbf{A}$ for all numbers α and β and all doublets **A**. With a little effort we can prove that $\alpha(\mathbf{A} + \mathbf{B}) = \alpha \mathbf{A} + \alpha \mathbf{B}$ for all numbers α and all doublets **A, B** (cf. Exercise **19** below).

Addition of doublets and multiplication of doublets by numbers can be used to assign *coordinates* to doublets. Let **I** denote any doublet $\overline{oe}$ where, for the sake of simplicity, we take o to be a horizontal line and $\delta_{oe} = 1$. Let **O** be the special doublet $\overline{of}$ with $d_{of} = 1$ (Fig. 65). If $\mathbf{A} = \overline{oa}$ is a doublet with beginning o, end a, and vertex M, then by drawing the line $u \| a$ through the point O we obtain the decomposition $\overline{oa} = \overline{ou} + \overline{ua} = \overline{ou} + \overline{ov}$, where the line $v \| o$ is determined by the condition $d_{ov} = d_{ua}$. It follows that $\overline{ou} = \xi \overline{oe}$ and $\overline{ov} = \eta \overline{of}$, where the numbers $\xi = \delta_{oa}$ and $\eta = d_{OM} \cdot \delta_{oa}$ are the coefficients in the decomposition

$$\mathbf{A} = \xi \mathbf{I} + \eta \mathbf{O} \qquad (25)$$

and are called the *coordinates* of the doublet **A** (or of the line a[23]). The decomposition (25) implied that doublets satisfy axiom $\text{III}_1^{(2)}$ [i.e., if **A, B, C** are any three doublets, then there exist numbers α, β, γ not all zero, such that $\alpha \mathbf{A} + \beta \mathbf{B} + \gamma \mathbf{C} = \mathbf{0}$ (proof?)] as well as axiom $\text{III}_2^{(2)}$. The latter assertion follows from the fact that a special doublet is never a multiple of an ordinary doublet, so that an ordinary doublet and a special doublet (say, the pair **I, O**) are always linearly independent.

---

[23] More appropriately, the *rectangular coordinates* of a; it is natural to call the numbers $\rho = \delta_{oa}$ and $\varphi = d_{OM}$ the *polar coordinates* of a. The connection between the rectangular and polar coordinates of a line is given by the relations $\xi = \rho = \rho \cos \varphi$, $\eta = \rho \varphi = \rho \sin \varphi$ (cf. Exercise 3 of Sec. 3).

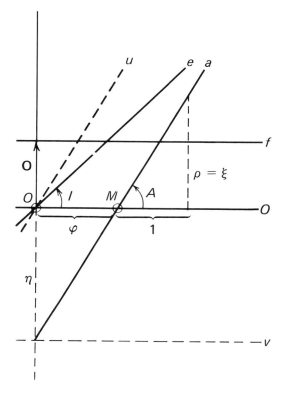

Figure 65

The formulas $\xi = \delta_{oa}$ and $\eta = d_{OM} \cdot \delta_{oa}$ and the definition of multiplication of a doublet by a number imply readily that if **A** has coordinates (ξ, η) [for which we shall write $\mathbf{A} = (\xi, \eta)$ or $\mathbf{A}(\xi, \eta)$], then $\alpha \mathbf{A}$ has coordinates $(\alpha\xi, \alpha\eta)$. It is somewhat more difficult to prove that $\mathbf{A}(\xi, \eta) + \mathbf{B}(\xi_1, \eta_1) = \mathbf{C}(\xi + \xi_1, \eta + \eta_1)$; an assertion which follows, for example, from the coparallelogram law (see Fig. 66). It is clear that the equalities $\alpha(\xi, \eta) = (\alpha\xi, \alpha\eta)$ and $(\xi, \eta) + (\xi_1, \eta_1) = (\xi + \xi_1, \eta + \eta_1)$ imply all the rules for operating with doublets.

The *scalar product* of doublets is defined by the formula

$$\overline{ab} \cdot \overline{cd} = \delta_{ab} \cdot \delta_{cd} \tag{26}$$

or

$$\mathbf{AB} = |\mathbf{A}||\mathbf{B}|, \tag{26a}$$

where the *absolute value* or *modulus* $|\mathbf{A}|$ of an ordinary doublet **A** is its angle, and the modulus of a special doublet is defined to be zero. It is clear that the scalar product of the doublets $\mathbf{A}(\xi, \eta)$ and $\mathbf{B}(\xi_1, \eta_1)$ is equal to

$$\mathbf{AB} = \xi\xi_1. \tag{26b}$$

Equation (26b) shows that scalar multiplication of doublets satisfies axioms $\text{IV}_1 - \text{IV}_5^{(G)}$: $\mathbf{AB} = \mathbf{BA}$, $(\alpha\mathbf{A})\mathbf{B} = \alpha(\mathbf{AB})$, $(\mathbf{A} + \mathbf{B})\mathbf{C} = \mathbf{AC} + \mathbf{BC}$, $\mathbf{AA} = \mathbf{A}^2 > 0$ for a nonspecial doublet **A**, and $\mathbf{O}^2 = 0$ for the special doublet **O**. Finally, the "supplementary" axiom $\text{IV}_6^{(G)}$ (see p. 252) holds because of the

6. Proofs of the principle of duality

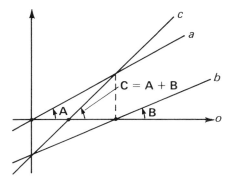

Figure 66

special angle d_{ab} associated with each special doublet $\overline{ab}$ (i.e., each doublet $\overline{ab}$ with $a \| b$; such doublets **O** satisfy **AO**=0 for all **A**). The "supplementary" scalar product $(\overline{ab}\cdot\overline{mn})_1$, where $a\|b$ and $m\|n$, is defined by the relation

$$(\overline{ab}\cdot\overline{mn})_1 = d_{ab}\cdot d_{mn}. \tag{26'b}$$

We have shown that our system of lines and doublets satisfies the axioms in groups I, II, III$^{(2)}$, IV$^{(G)}$, and V and have thus established the principle of duality.

Let $\overline{ab}$ and $\overline{cd}$ be doublets with vertices P and Q. In addition to the scalar product $\overline{ab}\cdot\overline{cd} = \delta_{ab}\cdot\delta_{cd}$, we can define the *cross product*

$$\overline{ab}\times\overline{cd} = \delta_{ab}\cdot\delta_{cd}\cdot d_{PQ}. \tag{27}$$

If $\mathbf{A}=\mathbf{A}(\xi,\eta)$ and $\mathbf{B}=\mathbf{B}(\xi_1,\eta_1)$, then it is easy to see that

$$\mathbf{A}\times\mathbf{B} = \xi\eta_1 - \eta\xi_1. \tag{27a}$$

Formula (27a) implies the following relations:

$$\mathbf{A}\times\mathbf{B} = -(\mathbf{B}\times\mathbf{A}),$$
$$(\alpha\mathbf{A})\times\mathbf{B} = \alpha(\mathbf{A}\times\mathbf{B}),$$
$$(\mathbf{A}+\mathbf{B})\times\mathbf{C} = (\mathbf{A}\times\mathbf{C}) + (\mathbf{B}\times\mathbf{C})$$

(cf. p. 252, Supplement B). The cross product of doublets yields a formula for the *distance between the intersection points* P *and* Q *of the lines* a,b *and* c,d, *respectively*:

$$d_{PQ} = \frac{\overline{ab}\times\overline{cd}}{|\overline{ab}|\cdot|\overline{cd}|}. \tag{28}$$

If the doublets $\overline{ab}$ and $\overline{cd}$ have coordinates (ξ,η) and (ξ_1,η_1), then

$$d_{PQ} = \frac{\xi\eta_1 - \eta\xi_1}{\xi\xi_1}. \tag{28a}$$

PROBLEMS AND EXERCISES

14 Describe in geometric terms: (a) A line translation
$$k' = k + A,$$
$$s' = s + B,$$

where k and s are tangential coordinates and A and B are arbitrary (fixed) numbers. **(b)** A line shear

$$k' = k,$$
$$s' = Vk + s.$$

(c) A general transformation (20).

15 Give specific examples of analytic proofs of theorems of Galilean geometry and of "dual" proofs of their duals (the dual proofs involve tangential coordinates).

16 Prove properties (i)–(iii) of polarities (p. 70).

17 Consider specific examples of theorems of Galilean geometry and their transforms under a polarity π.

18 **(a)** Describe the set of points M in the Galilean plane for which $\pi(M) \ni M$ (where π is a polarity), and the set of lines m for which $\pi(m) \in m$. **(b)** Use Exercise **18(a)** and property (iv) (p. 70) of a polarity to give a new geometric description of a polarity.

19 Show that $\alpha(\mathbf{A} + \mathbf{B}) = \alpha \mathbf{A} + \alpha \mathbf{B}$ for all numbers α and all doublets $\mathbf{A}, \mathbf{B}$.

20 Show that $\overline{oa} \times \overline{ob} = 2S_{oab}$, where S_{oab} is the oriented area of the triangle with sides o, a, b (just as $\overline{OA} \times \overline{OB} = 2S_{OAB}$, where S_{OAB} is the oriented area of $\triangle OAB$). What is the meaning of the term "oriented" used above? [$\overline{OA} \times \overline{OB}$ is the cross product of two vectors (cf. p. 252 of Supplement B), and $\overline{oa} \times \overline{ob}$ is the cross product of two doublets (cf. p. 75).]

21 Give an example of a proof of a theorem of Galilean geometry using the axioms listed in Supplement B and a "dual" proof of its dual using the properties of lines and doublets given above.

IX Consider the polarity $\pi: M \leftrightarrow m$ of the Euclidean plane which maps the point $M(r, \varphi)$ with polar coordinates $r (\neq 0)$ and φ to the line $m(1/r, \varphi)$ with "normal coordinates" $p = 1/r$ and $\alpha = \varphi$; here the term "normal coordinates" refers to the numbers α and p in the "normal equation" $x \cos\alpha + y \sin\alpha - p = 0$ of the line cf. Eq. (16) of Supplement A]. Make up a dictionary associated with π starting with

point M	line m
line m	point M
point M on line n	line m passing through point N
parallel lines m, n	points M, N on a line through the origin O
angle between lines m and n	angle MON
the distance MN	the quantity $M'N'/(OM' \cdot ON')$, where M', N' are the projections of O onto m, n

and give examples of dual (i.e., related by π) Euclidean theorems.

X Justify the principle of duality in three-dimensional semi-Galilean geometry (cf. Problem **VIII**, Sec. 5) using **(a)** the analytic; **(b)** transformational; **(c)** axiomatic approach.

II. Circles and Cycles

7. Definition of a cycle; radius and curvature

In elementary geometry we study properties of figures in the Euclidean plane bounded by line segments and circular arcs. Examples of such figures are triangles, circles, and circular segments.

In the Euclidean plane, a circle is usually defined as *the set* (locus) *of points which are at a fixed distance r from a given point Q* (Fig. 67a). An equivalent but less common definition of a circle is *the set of points from which a given segment AB is seen at a constant directed*[1] *angle α* (Fig. 67b). In Galilean geometry these two definitions yield different sets. We saw that a *circle s* in the Galilean plane defined as *the set of points which are at a fixed distance r from a given point Q* was a figure of little interest, namely, a pair of special lines (Fig. 68a). Of greater interest is *the set Z of points M from which a given ordinary segment AB* (i.e., a segment on an ordinary line) *is seen at a constant* (directed) *angle α*. The latter set is called a **cycle**.

The common school practice is to speak of (positive) angles between lines rather than of directed angles between lines. If we follow this practice, then (in Euclidean geometry) the locus of points M from which a given segment AB can be seen at a constant angle α is a pair of circular arcs (Fig. 67c) rather than a circle. (Note that the directed angle between the lines M_1A and M_1B in Figure 67c is the negative of the directed angle between the lines MA and MB in that figure.) This is a convincing demonstration of the appropriateness of the notion of directed angle between lines.

It is easy to deduce the equation of a cycle. Let $A(a_1, a_2)$ and $B(b_1, b_2)$ be two points in the Galilean plane, and let $M(x,y)$ be a point such that $\angle AMB = \alpha$. The slopes k and k_1 of the lines MA and MB (see Fig. 68b)

---

[1] By the directed angle between an ordered pair of intersecting lines we mean the smallest positive angle through which we must rotate the first line in order to bring it into coincidence with the second.

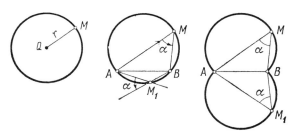

Figure 67a   Figure 67b   Figure 67c

are given by

$$k = \frac{y - a_2}{x - a_1} \quad \text{and} \quad k_1 = \frac{y - b_2}{x - b_1}.$$

In view of Eq. (8) in Section 3, we have

$$\alpha = k_1 - k = \frac{y - b_2}{x - b_1} - \frac{y - a_2}{x - a_1}.$$

Consequently, the coordinates of all points $M(x,y)$ with $\angle AMB = \alpha$ satisfy the equation

$$\frac{y - b_2}{x - b_1} - \frac{y - a_2}{x - a_1} = \alpha$$

or[2]

$$\alpha(x - b_1)(x - a_1) - (x - a_1)(y - b_2) + (x - b_1)(y - a_2) = 0.$$

The latter equation can be written as

$$(b_1 - a_1)y = \alpha x^2 + [(b_2 - a_2) - \alpha(b_1 + a_1)]x + \alpha a_1 b_1 - a_1 b_2 + a_2 b_1$$

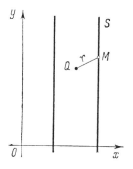
Figure 68a

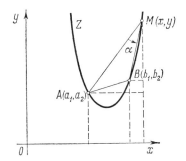
Figure 68b

---

[2]Note that MA and MB are ordinary lines, so that $x - b_1 \neq 0$ and $x - a_1 \neq 0$; there is no useful notion of Galilean angle between two lines, one of which is special.

7. Definition of a cycle; radius and curvature

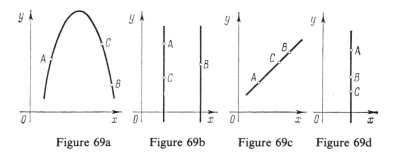

Figure 69a   Figure 69b   Figure 69c   Figure 69d

or as

$$y = ax^2 + 2bx + c,\qquad(1)$$

with

$$a = \frac{\alpha}{b_1 - a_1} \neq 0, \quad b = \frac{b_2 - a_2 - \alpha(b_1 + a_1)}{2(b_1 - a_1)}, \quad c = \frac{\alpha a_1 b_1 - a_1 b_2 + a_2 b_1}{b_1 - a_1}\qquad(1a)$$

(AB is an ordinary line, so that $b_1 - a_1 \neq 0$). Eq. (1) is the familiar equation of a (Euclidean) *parabola*.

Thus *the cycles of the Galilean plane are the parabolas* (1). Sometimes the term *cycle* is applied to the set of points given by the equation

$$ax^2 + 2b_1 x + 2b_2 y + c = 0.\qquad(2)$$

Equation (2) is more general than Eq. (1) and reduces to Eq. (1) when $b_2 \neq 0$, $a \neq 0$. The "special" cycles included in Eq. (2) are circles (pair of special lines), special lines, and ordinary lines. To obtain a circle we put $b_2 = 0$, $a \neq 0$, $b_1^2 - ac > 0$. To obtain a special line we put $b_2 = 0$, $a \neq 0$, $b_1^2 - ac = 0$[3] or $a = b_2 = 0$, $b_1 \neq 0$. To obtain an ordinary line we put $a = 0$, $b_2 \neq 0$. [Similarly, the equation of a Euclidean circle (Eq. (2') of Section 1 of the Introduction) can be written in the form

$$a(x^2 + y^2) + 2b_1 x + 2b_2 y + c = 0,\qquad(2')$$

which includes circles as well as lines (Eq. (2') represents a line if $a = 0$).] If "cycle" is taken to mean "a curve described by an equation of the form (2)," then we can assert that *three points of the Galilean plane determine a unique cycle* [i.e., an ordinary cycle (1), a circle, a special line, or an ordinary line; see Figs. 69a–d]. Nevertheless, in what follows we shall mean by "cycle" only a curve (1) (a Euclidean parabola).

A cycle Z is characterized by the property that *its inscribed angles subtended by a fixed segment AB have a constant value α* (cf. Figs. 70a and

---

[3] If $b_2 = 0$, $a \neq 0$, then Eq. (2) reduces to the quadratic equation $ax^2 + 2b_1 x + c = 0$. If $b_1^2 - ac > 0$ (i.e., if the roots x_1 and x_2 of this equation are real and distinct), then this equation is equivalent to the pair of equations $x = x_1$ and $x = x_2$, and so represents a circle. If $b_1^2 - ac = 0$ (i.e., if the roots of the equation coincide), then this equation represents a special line (a circle of radius zero). If $b_1^2 - ac < 0$ (i.e., if the equation has imaginary roots), then the equation represents the empty set.

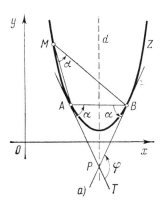 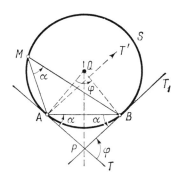

Figure 70a   Figure 70b

70b, which refer to a Galilean cycle Z and a Euclidean circle S). This implies that *the angle between the chord AB of the cycle Z and a tangent to Z at an endpoint of AB is* also *equal to* α. For proof, note that if M tends to A in Figure 70, then the chord AM tends to the tangent AT at A and the chord MB tends to the chord AB. Since all pairs of lines AM and MB form the same angle α, the tangent AT and the chord AB also form angle α. This result implies that *the tangents PA, PB from a point P to a cycle are equal*; in fact, the triangle PAB has equal base angles (at the base AB) and so is isosceles (see p. 51).

We shall now try to define the "radius" of a cycle Z. The usual definition of the radius of a Euclidean circle as the distance from its center to one of its points cannot be carried over to our centerless cycle in the Galilean plane. The same is true of the definition of the radius r of a circle S as the ratio

$$r = \frac{s}{\varphi} \quad (3)$$

of the length s of an arc AB of S to the corresponding central angle $\varphi = \angle AQB$ measured in radians[4] (Fig. 70b). However, we can rephrase the latter definition by replacing the central angle φ by the inscribed angle $\alpha = \varphi/2$ and say that *the radius r of a circle S is the ratio of the length of an arbitrary arc $s = $ arc AB of S to twice the inscribed angle α subtended by s*:

$$r = \frac{s}{2\alpha}. \quad (3a)$$

If we wish to adapt the above definition of radius to a cycle Z, then we must show that *for a given cycle Z and any arc AB of Z, the ratio s/α of the length s of arc AB to the inscribed angle α subtended by arc AB is constant*.

---

[4]To be sure, the usual high school definition of radian measure relies on the equation $\varphi = s/r$, which is equivalent to (3). However, there are other definitions of radian measure. Thus, for example, the radian measure of an angle can be defined by the condition $\lim_{\varphi \to 0}(\sin\varphi/\varphi) = 1$.

7. Definition of a cycle; radius and curvature

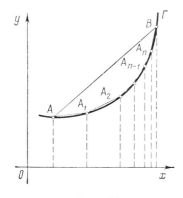
Figure 71a

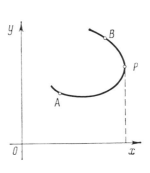
Figure 71b

But first we must define the length of an arc AB of a curve Γ in the Galilean plane. It is natural to rely on the corresponding Euclidean definition, and to say that *the length of an arc AB of a curve Γ is the limit of the length $AA_1 + A_1A_2 + \cdots + A_{n-1}A_n + A_nB$ of a polygonal line $AA_1A_2\ldots A_nB$* (Fig. 71a) *inscribed in Γ as the longest link in the polygonal line tends to zero*. Since the Galilean length of any polygonal line $AA_1A_2\ldots A_nB$ inscribed in the arc AB is equal to the length of the chord AB,[5] it is natural to say that *the length of an arc AB of a curve is equal to the length $s = d_{AB}$ of the chord AB*.

We can now evaluate s/α. Let AB be a chord of a cycle Z with endpoints $A(a_1, a_2)$, $B(b_1, b_2)$, and let α be the inscribed angle subtended by AB. In view of Eq. (1a), we have

$$\frac{s}{\alpha} = \frac{d_{AB}}{\alpha} = \frac{b_1 - a_1}{\alpha} = \frac{1}{a}.$$

Thus s/α is indeed constant and, as in the case of a Euclidean circle, we can define the "radius" r of the cycle Z as *half the ratio of the length s of the arc AB* (or, equivalently, *the chord AB*) *to the inscribed angle α subtended by AB*. In algebraic terms,

$$r = \frac{1}{2}\frac{s}{\alpha} = \frac{1}{2}\frac{b_1 - a_1}{\alpha} = \frac{1}{2a}, \qquad (3b)$$

where a is the coefficient of x^2 in the equation (1) of the cycle. Unlike the radius of a circle S, the radius of a cycle Z can be positive or negative according as the (Euclidean) parabola representing the cycle opens upwards (Fig. 72a) or downwards (Fig. 72b).

---

[5]We assume that the arc AB of the curve Γ does not "go in reverse" (relative to the positive direction of the x-axis which, as we know, is the direction of increasing time), i.e., does not look like the arc in Figure 71b. [If a curve in the Galilean plane is smooth in the sense of having an ordinary tangent line at each point, then it cannot look like the curve in Figure 71b (since the tangent at P is special and special lines are excluded from the class of lines).]

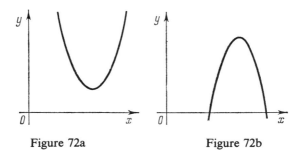

Figure 72a   Figure 72b

In trigonometry, we speak of signed angles. If this usage is introduced into Euclidean geometry, and the radius r of a circle S is defined by formula (3) [or (3a)], then r is positive or negative according as S (and thus an arc of S) is traversed counterclockwise or clockwise (see Figs. 73a and 73b); this convention pertaining to the sign of the radius of a directed circle is often useful (see, for example, [37], pp. 489–493; or [13], p. 260). In the case of a cycle Z in the Galilean plane, matters are different in the sense that there is a "preferred" way of traversing the cycle determined by the positive direction of the x-axis (the direction of increasing time values). This preferred way of traversing the cycle determines the sign of its radius

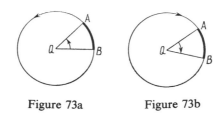

Figure 73a   Figure 73b

It is clear that if the radius r of a Euclidean circle S increases beyond all bounds, but S remains tangent to a fixed line l, then S tends to l (Fig. 74a). (That, of course, is why we regard straight roads on the surface of the earth as rectilinear in spite of the fact that they belong to circles of large radius on the spherical surface of the earth.) On the other hand, if the radius of the circle decreases, then the circle tends to be less linelike and more "curved" (Fig. 74b). That is why we call the reciprocal ρ of the radius r of

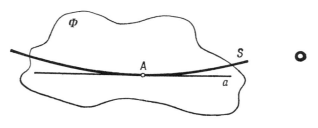

Figure 74a   Figure 74b

7. Definition of a cycle; radius and curvature

the circle its *curvature*; in view of (3a) we have

$$\rho = \frac{1}{r} = \frac{2\alpha}{s}. \tag{4}$$

Similarly, in the case of a cycle Z we call $\rho = 1/r$ the *curvature* of Z; in view of (3b) we have

$$\rho = \frac{1}{r} = 2\frac{\alpha}{s} = 2a. \tag{4'}$$

Here, too, we can justify calling ρ the curvature of the cycle by noting that as $r \to \infty$ (that is, $\rho \to 0$) the cycle tends to a line (cf. Figs. 75a and 75b, which represent cycles of large and small radius, respectively).[6] We note that just as in the case of two circles of equal radius, *it is always possible to map a cycle of radius $r = 1/(2a)$ onto any other cycle of the same radius by means of a suitable translation* (since all parabolas $y = ax^2 + bx + c$, a fixed, can be obtained from the parabola $y = ax^2$ by means of translations).

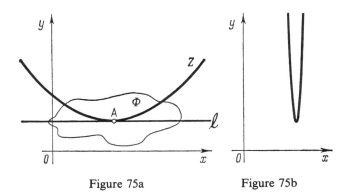

Figure 75a   Figure 75b

We shall now give the mechanical interpretation of the concept of a cycle. A curve Γ in the Galilean plane given by the equation

$$y = f(x) \tag{5}$$

corresponds to the motion of a (material) point P along the line o described by the equation

$$x = f(t), \tag{5'}$$

where x denotes the position of P on o at time t.[7]

---

[6]This terminology has the following precise meaning: Let S be a circle (or Z a cycle) passing through a point A, and let l be its tangent at A. Let Φ be any bounded region containing A. If the radius $r \to \infty$, then $S \cap \Phi$ ($Z \cap \Phi$) tends to $l \cap \Phi$ in the sense that the largest distance from a point in $S \cap \Phi$ to l tends to zero (see Figs. 74a and 75a).

We note that if $r \to 0$ (i.e., $\rho \to \infty$), then the circle S tends to a point (Fig. 74b) and the cycle Z tends to a ray of a special line (Fig. 75b).

[7]Here we again stipulate that the curve Γ never "goes in reverse"; see footnote 5.

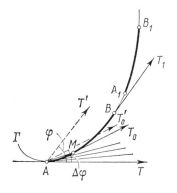

Figure 76

The form of (5′) corresponding to a cycle Z is

$$x = at^2 + 2bt + c. \qquad (1')$$

Such equations characterize motions with constant acceleration (or deceleration). In fact, the velocity v of the motion (1′) at time t is given by

$$v = \lim_{\Delta t \to 0} \frac{\Delta x}{\Delta t} = \lim_{\Delta t \to 0} \frac{\left[a(t+\Delta t)^2 + 2b(t+\Delta t) + c\right] - (at^2 + 2bt + c)}{\Delta t}$$

$$= \lim_{\Delta t \to 0} \frac{2at\Delta t + a(\Delta t)^2 + 2b\Delta t}{\Delta t} = \lim_{\Delta t \to 0} (2at + a \cdot \Delta t + 2b) = 2at + 2b,$$

and its acceleration w is given by

$$w = \lim_{\Delta t \to 0} \frac{\Delta v}{\Delta t} = \lim_{\Delta t \to 0} \frac{\left[2a(t+\Delta t) + 2b\right] - (2at + 2b)}{\Delta t} = \lim_{\Delta t \to 0} \frac{2a \cdot \Delta t}{\Delta t} = 2a.$$

Thus *Eq. (1′) describes the motion of a point on a line o with constant acceleration*

$$w = 2a$$

equal to the curvature of the corresponding cycle Z in the Galilean plane. Specifically, cycles with positive curvature (Fig. 72a) correspond to motions with constant positive acceleration, and cycles with negative curvature (Fig. 72b) correspond to motions with constant negative acceleration (i.e., constant deceleration).

It is possible to define the curvature ρ (and the radius of curvature r) for any curve Γ in the Euclidean or Galilean plane. To do this, we first define the tangent to a curve Γ at a point A of Γ as the limiting position of the secant AM as M on Γ tends to A (Fig. 76; the assertion that the line AM tends to the line AT means that $\angle AMT \to 0$). (This definition of a tangent line to a curve was used above in the proof of the theorem on p. 80 about the angle between a chord of a cycle and the tangent to the cycle at an endpoint of the chord.) It is natural to think of the tangent as giving the "direction" of Γ at A. Support for this idea comes from mechanics (see p. 89 below): If a (material) point P is induced by certain forces to

7. Definition of a cycle; radius and curvature

move along a (Euclidean) curve Γ, and if when P reaches A these forces cease to act, then P moves "by inertia" along the line AT.[8]

We base the concept of curvature on that of the tangent. The intuitive notion of "straightness" of a line l is that *its direction is the same at all points*, i.e., that all of its tangents are the same (and coincide with l itself). Other curves have nonzero "curvature," i.e., their direction is not the same at all points. A numerical measure of curvature is the angular rate of change of the tangent. More precisely, if A and B are two points on Γ such that the arc length of arc AB is s and the angle between the tangents AT and BT_1 at A and B is φ, then we define the *average curvature* ρ_{av} of Γ on the arc AB as

$$\rho_{av} = \frac{\varphi}{s}, \quad \text{where } s = \text{arc } AB \text{ and } \varphi = \angle TAT' \quad (AT' \| BT_1) \quad (6)$$

(Fig. 76).[9] Intuitively, portions of Γ with large average curvature are more curved than those with small average curvature. (The curve Γ in Fig. 76 has larger average curvature between A and B than between A_1 and B_1.) *The curvature ρ of Γ at A is defined as the rate of change of the tangent at A*, i.e., the curvature ρ of Γ at A is the limit of the average curvature ρ_{av} on the arc AM as M tends to A along Γ:

$$\rho = \lim_{\Delta s \to 0} \frac{\Delta \varphi}{\Delta s}, \quad \text{where } \Delta S = \text{arc } AM \text{ and } \Delta \varphi = \angle TAT'_0 \quad (AT'_0 \| MT_0) \quad (6a)$$

(Fig. 76; see also p. 17 above).[10] The *radius of curvature r of Γ at A is defined as the reciprocal of its curvature at A*:

$$r = \frac{1}{\rho}. \quad (7)$$

Now we consider a (Euclidean) circle S (Fig. 77a). The circle S "has the same structure at all of its points" (the precise meaning of this intuitive statement will be considered in the next section) and it is therefore natural to expect that *the curvature ρ has the same value at all points of a circle S* (and is inversely proportional to its radius; see p. 83 above and Figs. 74a and 74b). In more rigorous terms, Eq. (6) implies that the average curvature ρ_{av} of a circle S of radius r on an arc AB is φ/s, where (see Fig. 70b above) $\varphi = \angle TAT' = \angle TPT_1$ (here $AT' \| BT_1$). Since φ is an exterior angle of the isosceles triangle APB, we have $\varphi = 2 \angle PAB = 2\alpha$; using this, we obtain $s = \text{arc } AB = r(\angle AQB) = r(2 \angle AMB) = r \cdot 2\alpha = r\varphi$. Thus the average curvature

$$\rho_{av} = \frac{\varphi}{s} = \frac{\varphi}{r\varphi} = \frac{1}{r}$$

of a circle S of radius r is the same for each arc AB. It follows that at each point A of the circle we have

$$\rho = \frac{1}{r}$$

[see Eq. (4) above, which was used to define the curvature of a circle].

---

[8] In other words, if P moves along a curve Γ in accordance with the law $\mathbf{r} = \mathbf{r}(t)$, where $\mathbf{r} = \overline{OM}$ is the position vector of P and t is time, then the instantaneous velocity $\mathbf{v} = d\mathbf{r}/dt$ of P at time t is directed along the tangent to Γ.

[9] We assume that as we move from A to B along the arc AB of Γ, the tangent to Γ turns always in the same direction.

[10] Here and in what follows we restrict ourselves to sufficiently smooth curves Γ. Such a curve is assumed to have a tangent AT (a requirement which rules out corners) and a definite curvature ρ at each of its points.

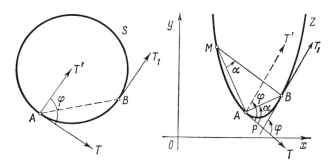

Figure 77a Figure 77b

We now turn to Galilean geometry. We define the average curvature ρ_{av} on an arc AB of a curve Γ and the curvature ρ of Γ at A by Equations (6) and (6a) above, except that we now use Galilean arc length s and angle φ (in particular, the length s of the arc AB is equal to the length d_{AB} of the chord AB[11]). In the special case where Γ is a cycle Z (see Fig. 77b), the average curvature ρ_{av} on the arc AB is φ/s, where $s = \text{arc } AB$ and $\varphi = \angle TAT' = \angle TPT_1$ (here $AT' \| BT_1$). From the "angle formula" (12), Section 4, applied to $\triangle ABP$, it follows readily that

$$\varphi = \angle APB = 2\angle PAB = 2\alpha,$$

where, as we know, $\alpha = \angle PAB = \angle ABP$ is equal to the inscribed angle of the cycle Z subtended by the arc AB. Hence

$$\rho_{av} = \frac{\varphi}{s} = \frac{2\alpha}{s} = 2a$$

[see (1a) in Sec. 7], where a is the coefficient of x^2 in the equation (1) of the cycle Z. It is clear from this that all arcs AB of the cycle Z have the same average curvature $\rho_{av} = 2a$. This implies that the curvature ρ of Z has the same value

$$\rho = 2a \tag{7a}$$

at each point A of Z.

If, as in (7), we define the reciprocal r of the curvature as the radius of curvature of the cycle, then we have

$$r = \frac{1}{\rho} = \frac{1}{2a}. \tag{7b}$$

Our definitions of curvature of a (Euclidean) circle and a (Galilean) cycle give us a new approach to the concepts of curvature ρ and radius of curvature r at any point of a curve Γ (in the Euclidean or Galilean plane). Thus consider all circles (cycles) passing through the point A of Γ and having the same tangent $AT = l$ at A as Γ. From these circles (cycles), we select the circle S_0 (cycle Z_0) which is closest to Γ (Figs. 78a and 78b), in the sense that for any other circle S (cycle Z) the distance MM' between points on Γ and S (or Γ and Z) sufficiently close to A which project to the same point N on a is larger than the distance between the corresponding points M and M_0' of Γ and S_0 (or Γ and Z_0). The circle S_0 (cycle Z_0) is called the *osculating circle* (*osculating cycle*) of Γ at A, its curvature $\rho = 1/r$ (or $\rho = 2a$) is called the *curvature of Γ at A*, and the radius r of the circle S_0 (cycle Z_0) is called the *radius of curvature of Γ at A*.[12]

---

[11] Cf. footnote 5.

[12] If none of the circles S (cycles Z) is closer to Γ than the tangent $AT = a$, then the line l plays the role of the osculating circle S_0 (osculating cycle Z_0) at A_0. In that case, the curvature ρ of Γ at A is zero (and its radius of curvature r is infinite).

7. Definition of a cycle; radius and curvature

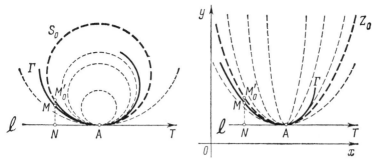

Figure 78a   Figure 78b

It is not difficult to obtain analytical expressions for the curvature ρ and the radius of curvature r of a curve $y = f(x)$ at a point $A(x,y)$. The slope of the tangent to the curve $y = f(x)$ at the point $A(x,y)$ is given by the derivative $f'(x)$ of the function $f(x)$. Let AT and A_1T_1 be the respective tangents to our curve at its neighboring points A and A_1. Let φ and φ_1 be the Euclidean or Galilean angles which AT and A_1T_1 make with the x-axis. Then $\Delta\varphi = \varphi_1 - \varphi$ is the angle $\angle TPT_1$ in Figure 79a. Let Δs be the Euclidean or Galilean length of the arc AA_1; our problem is to compute $\rho = \lim_{A_1 \to A}(\Delta\varphi/\Delta s)$.

In the Galilean case, this is quite easy. Let $A_1 = (x + \Delta x, y + \Delta y)$. Since the Galilean angle between two lines is equal to the difference of the slopes, we have

$$\Delta\varphi = f'(x + \Delta x) - f'(x).$$

And since the Galilean length of an arc is equal to the difference between the abscissas of its endpoints, we have

$$\Delta s = (x + \Delta x) - x = \Delta x.$$

Hence

$$\frac{\Delta\varphi}{\Delta s} = \frac{f'(x + \Delta x) - f'(x)}{\Delta x},$$

from which it follows that

$$\rho = \lim_{\Delta x \to 0} \frac{f'(x + \Delta x) - f'(x)}{\Delta x} = f''(x). \qquad (8_1)$$

For the radius of curvature r we have

$$r = \frac{1}{\rho} = \frac{1}{f''(x)}. \qquad (8_2)$$

The Euclidean case is somewhat more difficult. Here we have $\tan\varphi = f'(x)$ and $\tan\varphi_1 = f'(x + \Delta x)$. Hence

$$\tan\Delta\varphi = \tan(\varphi_1 - \varphi) = \frac{\tan\varphi_1 - \tan\varphi}{1 + \tan\varphi_1 \tan\varphi}$$

$$= \frac{f'(x + \Delta x) - f'(x)}{1 + f'(x + \Delta x)f'(x)}.$$

Therefore,

$$\lim_{\Delta x \to 0} \frac{\tan\Delta\varphi}{\Delta x} = \lim_{\Delta x \to 0} \frac{[f'(x + \Delta x) - f'(x)]/\Delta x}{1 + f'(x + \Delta x)f'(x)} = \frac{f''(x)}{1 + f'(x)^2}.$$

From the definition of arc length, it is easy to see that for small arcs,

$\Delta s \approx \sqrt{(\Delta x)^2 + (\Delta y)^2} \approx \sqrt{1+f'(x)^2}\, \Delta x$ (see Fig. 79b). More precisely, we have

$$\lim_{\Delta x \to 0} \frac{\Delta s}{\Delta x} = \sqrt{1+f'(x)^2}\,.$$

Hence

$$\lim_{\Delta x \to 0} \frac{\Delta \varphi}{\Delta s} = \lim_{\Delta x \to 0} \frac{\Delta \varphi}{\tan \Delta \varphi} \frac{\tan \Delta \varphi}{\Delta x} \frac{\Delta x}{\Delta s} = \lim_{\Delta x \to 0} \frac{\Delta \varphi}{\tan \Delta \varphi} \lim_{\Delta x \to 0} \frac{\tan \Delta \varphi}{\Delta x} \lim_{\Delta x \to 0} \frac{\Delta x}{\Delta s}$$

$$= 1 \cdot \frac{f''(x)}{1+f'(x)^2} \cdot \frac{1}{\sqrt{1+f'(x)^2}} = \frac{f''(x)}{[1+f'(x)^2]^{\frac{3}{2}}}\,.$$

[Here we have used the fact that $\Delta \varphi \to 0$ as $\Delta x \to 0$, and therefore $\lim_{\Delta x \to 0}(\Delta \varphi / \tan \Delta \varphi) = \lim_{\Delta \varphi \to 0}(\Delta \varphi / \tan \Delta \varphi) = 1$.] We have thus obtained the following formulas for the Euclidean curvature ρ and the radius of curvature r:

$$\rho = \frac{f''(x)}{\left[1+f'(x)^2\right]^{\frac{3}{2}}}, \tag{$8'_1$}$$

$$r = \frac{1}{\rho} = \frac{[1+f'(x)]^{\frac{3}{2}}}{f''(x)}. \tag{$8'_2$}$$

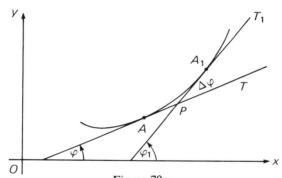

Figure 79a

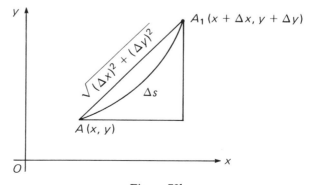

Figure 79b

7. Definition of a cycle; radius and curvature

Equations $(8_1')$ and (8_1) imply that the curvature of a circle $x^2+y^2=r^2$ at each of its points is $1/r$ and the curvature of a cycle $y=ax^2+bx+c$ at each of its points is $2a$. Also, starting with these formulas, we can show that *Euclidean circles and Galilean cycles are the only curves of constant curvature ρ*. For this statement to be accurate we must regard (ordinary) lines as limiting cases of circles (cycles) with constant zero curvature (and "infinite radius of curvature"). Thus, for example, it is clear that if $y=f(x)$ is a curve in the Galilean plane with

$$\rho = f''(x) = \text{const.},$$

then the equation of the curve is

$$y = \tfrac{1}{2}\rho x^2 + c_1 x + c_2$$

with arbitrary constants c_1 and c_2, i.e., the curve is a cycle. Quite generally, *any curve Γ in the Euclidean or Galilean plane is determined* (up to position) *by its curvature*, i.e., by the equation

$$\rho = \rho(s), \qquad (9)$$

where ρ is the curvature of Γ at the (variable) point A and s is the length of the arc QA of Γ measured from some arbitrary but fixed "origin" Q on Γ to A. This statement is the so-called *fundamental theorem on plane curves*.[13] Equation (9) is called the *natural equation* of the curve Γ.

It is clear that in the mechanical interpretation of Galilean geometry, the curvature $\rho = y''(x)$ at the point $A(x,y)$ of the curve Γ defined by the equation $y = f(x)$ corresponds to the acceleration $w = d^2x/dt^2$ (at time t) of a material point moving along a line o in accordance with the equation $x = f(t)$. It is also clear that the osculating cycle Z_0 of Γ at A corresponds to a motion with constant acceleration of a material point whose (constant) acceleration, velocity, and position on o at time t coincide with the acceleration w (given by the curvature of the osculating cycle Z_0, i.e., twice the coefficient of x^2 in the equation of Z_0), velocity v, and position $f(t)$ on o at time t of a point moving in accordance with the equation $x = f(t)$. In this connection, note that the tangent a to the curve Γ at the point A of Γ corresponds to uniform motion with velocity v equal to the velocity at time t. (The value of v equals the slope of the tangent to Γ at A.) It follows that the slope $k = v$ of the tangent to Γ is not invariant under Galilean motions. On the other hand, the curvature $\rho = w$ and the radius of curvature $r = 1/\rho$ of the osculating cycle are Galilean invariants. (This is because the acceleration w is the same in all Galilean reference frames, and therefore has second mechanical significance.)

In view of Newton's second law of motion

$$f = mw,$$

where m is the mass of a material point and f is the force acting on the point, we see that the curvature $\rho = w$ differs from the force only by the constant factor m. Hence the fundamental theorem on curves in Galilean geometry reduces to the assertion that *the motion of a material point on a line o is fully determined by the*

---

[13]See, for example, [36], p. 26. [In Galilean geometry, in view of Eq. (8_1), the curve $y=f(x)$, given by its natural equation $f''(x) = \rho = \rho(s)$ [or $f''(x) = \rho(x)$, since if $Q = (0,b)$, then $s = x$] is determined by the formula

$$y = \int \left[\int \rho(x)\, dx \right] dx = y_0(x) + c_1 x + c_2,$$

where c_1 and c_2 are arbitrary constants. It is clear that all such curves are related by Galilean motions.]

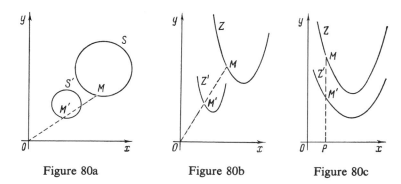

Figure 80a   Figure 80b   Figure 80c

forces f acting on the point (f is proportional to the curvature $\rho = w = f/m$). The significance of this fact in mechanics sheds additional light on the important role played by the fundamental theorem on curves.

We conclude this section by noting that since the curvature $\rho = \lim_{\Delta s \to 0}(\Delta \varphi / \Delta s)$ of a curve Γ involves the ratio of angle to arc length, and the radius of curvature $r = 1/\rho = \lim_{\Delta \varphi \to 0}(\Delta s / \Delta \varphi)$ involves the ratio of arc length angle, it follows that *every transformation of the Euclidean plane which preserves angles and multiplies distances by a factor k, multiplies the curvature of a curve by $1/k$ and its radius of curvature by k.* An analogous statement holds for similitudes of the first kind of the Galilean plane with coefficient k.[14] For example, a dilatation with center O and coefficient k takes a circle S (cycle Z) of radius r to a circle S' (cycle Z') with radius kr (Figs. 80a and 80b). On the other hand, *every similitude of the second kind* [14] *of the Galilean plane with coefficient κ preserves distances and multiples angles by κ; it therefore multiplies the curvature of a curve Γ by κ, and its radius of curvature by $1/\kappa$.* For example, a compression with axis Ox and coefficient κ takes a cycle of radius r into a cycle of radius r/κ (Fig. 80c).

Problems and Exercises

1. In Euclidean geometry, a curve can be described by a vector equation $\mathbf{r} = \mathbf{r}(t)$, where $\mathbf{r} = \overline{OM}$ is the radius vector of a variable point M on the curve, and t is the parameter. The "natural parameter" (arc length) s is defined by the condition $\mathbf{r}' = d\mathbf{r}/ds = \mathbf{t}$, where $\mathbf{t}^2 = 1$. The curvature ρ is defined by the equation $\mathbf{t}' = \rho \mathbf{n}$, where $\mathbf{t}\mathbf{n} = 0$ and $\mathbf{n}^2 = 1$ (this implies that $\mathbf{n}' = -\rho \mathbf{t}$). From these relations we can easily compute the curvature of the curve, find its natural equation and determine all curves of constant curvature (cf. [36]). The development of the elements of plane Galilean differential geometry follows the same pattern. The reader should try to carry out this program.

2. View a curve in the Galilean plane as a "one-parameter family of lines" (tangents to the curve). Let $\mathbf{R} = \mathbf{R}(t)$ be the equation of the curve, where $\mathbf{R} = \overline{om}$ is a variable doublet (cf. Chap. 1, Sec. 6). Develop a theory of curves based on the calculus of doublets and dual to that sketched in Exercise **1**.

3. What can you say about evolutes and involutes of curves (cf. [36] or Chap. 17 of [19]) in Galilean geometry?

---

[14]See p. 53.

8. Cyclic rotation; diameters of a cycle   91

I   In Euclidean space, a curve is characterized by its curvature and torsion (cf. [36] or Chap. 17 of [19]). Develop elements of the differential geometry of curves in three-dimensional Galilean geometry. Investigate curves that are invariant under glide rotations (analogues of helices).

II Solve Problem I in three-dimensional semi-Galilean geometry (cf. Sec. 3, Problem II).

III By a developable surface (in Euclidean as well as in non-Euclidean space) we mean a surface which is generated by lines, and which has a constant tangent plane along each of the generating lines. Such a surface may be thought of as a one-parameter family of planes, the tangent planes to the surface. Develop elements of the differential geometry of developable surfaces in three-dimensional semi-Galilean geometry (cf. Problem II) dual (in the sense of the principle of duality in Problems VIII and X, Chap. I) to the differential geometry of curves in that geometry (cf. Problem II; in this dual theory, the role of the calculus of vectors is played by the calculus of doublets. For some of the issues to be considered, see Exercise 2).

IV Consider the surfaces of three-dimensional Galilean geometry (see Exercise 5 in the Introduction) which are the analogues of Euclidean spheres. Find the equations and form of such surfaces. Sketch various possible approaches to their study (including the approach via the differential geometry of the space under consideration).

V   Consider the same problem for three-dimensional semi-Galilean geometry (cf. Problem II).

VI Which of the problems and exercises in this section are meaningful in the Poinceau geometry (see p. 30)?

8. Cyclic rotation; diameters of a cycle

Euclid's "Elements" (see footnote 23 of the Introduction), whose role in the history of geometry we discussed above, opens with "definitions" of the basic geometric entities such as point, (straight) line, curve, and surface.[15] In particular, Euclid's definition 4 states that

A straight line is a line [i.e., a curve] *which lies evenly with the points on itself.*

As a definition this vague phrase is entirely inadequate, but as a description of the "most important" property of a line it is excellent. After all, a fundamental property of a line is its "homogeneity," i.e., the essential sameness of any two of its segments of equal length. In this sense, no point on a line is different from another. However, this property of a line does

---

[15]Today these definitions are viewed as the least fortunate aspect of the otherwise remarkable work of Euclid. The author of the *Elements* seems not to have been aware of the fact that without any explicit knowledge of geometry it is not possible to give acceptable definitions of anything, and that the initial geometric notions suggested by intuition must remain undefined (just as the initial assertions of geometry, the axioms, must be accepted without proof).

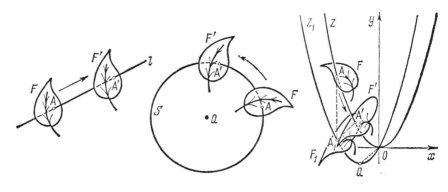

Figure 81a   Figure 81b   Figure 81c

not characterize it, i.e., it does not set lines apart from all other curves in the (Euclidean) plane. Indeed, a circle also "lies evenly with the points on itself," i.e., it is also homogeneous: any two of its arcs of equal length are congruent, so that nothing distinguishes one of its points from another. The homogeneity of a line and of a circle manifests itself in the existence of a "linear glide" and a "circular glide," defined respectively as a motion which takes a point A on a line l to a preassigned point A' on l and maps l to itself, and a motion which takes a point A on a circle S to a preassigned point A' on S and maps S to itself. For a line l, such a motion is a translation along l (Fig. 81a), while for a circle S it is a rotation about the center Q of S (Fig. 81b).

We turn next to Galilean geometry.

Since a translation (Fig. 81a) is a motion of Galilean geometry, it is clear that a line l in the Galilean plane admits a "linear glide." Also, it is not difficult to show that a cycle Z in the Galilean plane admits a "cyclic glide." In fact, the shear

$$x_1 = x,$$
$$y_1 = vx + y \tag{10a}$$

[cf. (14a), Sec. 2], which can be written as

$$x = x_1,$$
$$y = y_1 - vx_1, \tag{10'a}$$

maps the cycle Z given by the equation

$$y = ax^2 \tag{11}$$

(Fig. 81c) to the cycle Z_1 given by the equation

$$y_1 - vx_1 = ax_1^2,$$

which can be written as

$$y_1 = ax_1^2 + vx_1 = a\left(x_1^2 + \frac{vx_1}{a} + \frac{v^2}{4a^2}\right) - \frac{v^2}{4a},$$

8. Cyclic rotation; diameters of a cycle 93

or as

$$y_1 + \frac{v^2}{4a} = a\left(x_1 + \frac{v}{2a}\right)^2. \tag{11a}$$

It is clear that the cycle Z in (11) and the cycle Z_1 in (11a) are congruent. Specifically, Z is the result of the application of the translation

$$\begin{aligned} x' &= x_1 + \frac{v}{2a}, \\ y' &= y_1 + \frac{v^2}{4a}, \end{aligned} \tag{10b}$$

to Z_1. It follows that the transformation

$$\boxed{\begin{aligned} x' &= x + \frac{v}{2a}, \\ y' &= vx + y + \frac{v^2}{4a}, \end{aligned}} \tag{12}$$

which is the composite of the translation (10b) and the shear (10a) maps the cycle Z to itself; (10a) takes a point A of Z to a point A_1 of Z_1 and (10b) takes A_1 to a point A' of Z (Fig. 81c). A motion (12) of the Galilean plane which maps the cycle Z to itself is called a *cyclic rotation* (determined by the cycle Z) with coefficient v (or a "cyclic rotation of the cycle Z"[16]). Since the shear (10a) moves each point A a distance zero, (i.e., it takes A to A_1 such that $d_{AA_1} = 0$) and the translation (10b) moves each point a distance $v/2a$ (i.e., it takes A_1 to A' such that $d_{A_1A'} = v/2a$), it follows that *a cyclic rotation* (12) *moves each point a distance* $v/2a$; it takes a point A to a point A' such that

$$d_{AA'} = \frac{v}{2a}.$$

This implies that by choosing an appropriate value for the coefficient v of a cyclic rotation we can move a point A of a cycle Z an arbitrary (positive or negative) distance d; specifically, we need only put $v/2a = d$ or $v = 2ad$. In other words, *there exists a cyclic rotation* (12) *which takes any point A of a cycle Z to a preassigned point A' of Z.* It can be shown that just as the only curves in the Euclidean plane with this kind of homogeneity are lines and circles, so, too, the only curves in the Galilean plane with such homogeneity are lines and cycles (cf. p. 89).

A rotation of the Euclidean plane which maps a circle S to itself moves each point of the plane along a circle concentric with S (Fig. 82a). Similarly, a cyclic rotation of the Galilean plane determined by the cycle Z in (11) moves each point of that plane along a cycle "parallel" to Z, i.e., a cycle obtained from Z by a "special" translation (meaning a translation in

---

[16]*Every (direct) motion of the Euclidean plane is a rotation or a translation* [see Exercise **3(a)** in the Introduction]. Similarly, it can be shown that *every (direct) motion of the Galilean plane* [see (1), Sec. 3] *is a cyclic rotation or a translation* [see Problem **IX(a)** in the Introduction].

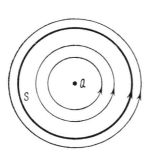

Figure 82a   Figure 82b

the vertical direction). In fact, the shear (10a) [or (10′a)] takes the cycle

$$y = ax^2 + c \tag{11'}$$

parallel to Z to the cycle

$$y_1 - vx_1 = ax_1^2 + c,$$

or, equivalently, the cycle

$$y_1 + \frac{v^2}{4a} = ax_1^2 + vx_1 + \frac{v^2}{4a} + c = a\left(x_1 + \frac{v}{2a}\right)^2 + c$$

[see (11a)]. The latter is taken by the translation (10b) to the cycle

$$y' = ax'^2 + c,$$

i.e., the initial cycle (11′). This shows that *the cyclic rotation* (12) *carries each of the cycles* (11′) (where c is an arbitrary constant), as well as Z, *to itself*. Thus (12) *moves each point of the plane along a suitable cycle* (11′) (Fig. 82b). The analogy between concentric circles and parallel cycles will be carried further in the sequel.

Cyclic rotations are a very convenient tool for proving a variety of properties of cycles. The first theorem which we shall prove using cyclic rotations is the familiar theorem which asserts that *the tangents PA and PB from a point P to a cycle Z are congruent* (see p. 80). We assume that the cycle Z (Fig. 83) is given by Eq. (11)[17]; this is no loss of generality, for whatever the cycle, there always exists a coordinate system in which it is described by Eq. (11). We apply the cyclic rotation (12) which maps Z to itself and takes P to P' on the y-axis. [In the general case, we apply the cyclic rotation which maps the cycle Z to itself and takes P to P' on the axis of symmetry of the (Euclidean) parabola Z.] Our cyclic rotation takes the tangents PA and PB of Z to tangents $P'A'$ and $P'B'$; this follows from the fact that tangency is preserved under Galilean motions. Since the parabola $y = ax^2$ is symmetric with respect to the y-axis, it is obvious that the segments $P'A'$ and $P'B'$ are congruent in the Euclidean as well as in

---

[17] We shall assume that the axes of our coordinate system are Euclidean–perpendicular. This will make our proof "basically non-Galilean" inasmuch as we shall use facts and ideas which have no meaning in Galilean geometry (cf. footnote 1, Chap. I, Sec. 3).

8. Cyclic rotation; diameters of a cycle

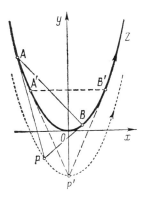

Figure 83

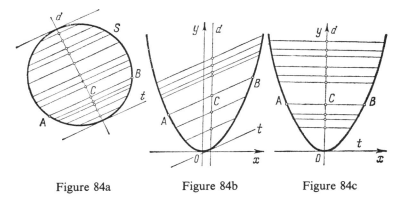

Figure 84a   Figure 84b   Figure 84c

the Galilean sense (the latter means that the projections of $P'A'$ and $P'B'$ on the x-axis are congruent). But the motion (12) leaves the (Galilean) length of segments unchanged. Hence $d_{AP} = d_{A'P} = d_{A'P'} = d_{P'B'} = d_{PB}$.[18]

We give another example of a similar kind. It is well known that *the midpoints of a family of parallel chords of a circle S lie on a line—a diameter d of S* (Fig. 84a). This assertion is just a reformulation of the fact that a circle is symmetric with respect to each of its diameters. Now we consider a family of parallel chords of the cycle Z with equation $y = ax^2$. Such chords belong to lines with a fixed slope k_0 (Fig. 84b). We note that the shear (10a) [or (10'a)] maps the line

$$y = kx + s$$

with slope k to the line $y_1 - vx_1 = kx_1 + s$, or equivalently, the line

$$y_1 = k_1 x_1 + s \qquad (k_1 = k + v).$$

---

[18]Note that $d_{PA} = -d_{PB}$. [We remark that if PA and PB are tangents from a point P to an oriented circle S (i.e., a circle with clockwise or counterclockwise orientation), then it is also natural to put $PA = -PB$.]

with slope $k_1 = k + v$, and the translation (10b) preserves the direction of each line. It follows that the cyclic rotation (12) with coefficient $v = -k_0$ maps the selected chords of Z to chords with slope 0, i.e., parallel to the x-axis (Fig. 84c). Clearly, the midpoints of the latter chords are on the y-axis, which is the axis of symmetry of the parabola $y = ax^2$. It follows that the midpoints of the original chords must lie on a line d. Specifically, d is the preimage of the y-axis under our cyclic rotation. Since the image of d under a Galilean motion is the special line Oy, it follows that d itself must be special. The line d is called a *diameter* of the cycle Z (corresponding to the chords with slope k_0).

Using this terminology, we can say that *the midpoints of a family of parallel chords of a cycle Z lie on a special line d*, a diameter of Z.

It is not difficult to show that every special line is a diameter of Z. In fact, consider the special line l with equation $x = m$. The cyclic rotation (12) with coefficient v satisfying $v/2a = -m$, i.e., with coefficient $v = -2am$, maps l onto the y-axis. This implies that l bisects all chords of the cycle Z which our cyclic rotation maps onto lines parallel to the x-axis. (It is easy to check that the diameter $x = m$ of the cycle $y = ax^2$ bisects all chords of Z that have slope $k = 2am$.)

Now let us move a chord AB of Z which is bisected by a diameter d so that it stays parallel to its original position and its length decreases. Then, in the limit, the points A, B and the midpoint C (on d) of the chord AB coincide. We conclude that *the tangent t to a cycle Z at the end of a diameter d of Z is parallel to the chords bisected by d* (compare Figs. 84b and 84c with Fig. 84a, which refers to a circle).

We can now give a new justification for the use in Galilean geometry of special lines as "perpendiculars" (see p. 43 and Figs. 36 and 37 above). In Euclidean geometry, *a diameter of a circle is perpendicular to the tangents at its endpoints*. It follows that in order to draw a perpendicular to a line a at a point A it suffices to draw a circle S tangent to a at A; the radius QA of S is the required perpendicular (Fig. 85a). By analogy, in Galilean geometry we say that a diameter of a cycle is perpendicular to the tangent at its endpoint, i.e., the role of the perpendicular to a line a at a point A is

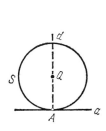
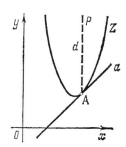

Figure 85a Figure 85b

8. Cyclic rotation; diameters of a cycle   97

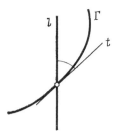

Figure 86

played by the diameter AP of a cycle Z tangent to a at A (i.e., the special line AP through A; Fig. 85b). If we define the angle between a line l and a curve Γ to be the angle between l and the tangent t to Γ at the point of intersection of l and Γ (Fig. 86), then we can say that the diameters of a cycle Z are perpendicular to Z (just as the diameters of a circle S are perpendicular to S; see Figs. 85a and 85b). We note that if PA and PB are the tangents from a point P to a cycle Z, then the diameter d of Z passing through P bisects the chord AB (this follows from the fact that d is an altitude of the isosceles triangle PAB; see Figs. 70a and 70b).

Our discussion sheds new light on the analogy between concentric circles and parallel cycles. Define the *distance from a point A to a curve Γ* to be the distance from A to the point B on Γ nearest to A (in other words, $d_{A\Gamma} = \min d_{AB}$ for all B on Γ). It is possible to prove that in Euclidean geometry this distance (at least for a smooth curve Γ without endpoints) is measured along *a perpendicular AB* to Γ; in other words, the line AB is perpendicular to the tangent to Γ at B (Fig. 87a).[19] In Galilean geometry the distance from A to Γ is generally equal to zero. In such cases, we measure the distance from A to Γ in terms of the special distance δ_{AB} from A to B on Γ (Fig. 87b). We shall say that a curve Γ_1 is *parallel* to a curve Γ (in symbols: $\Gamma_1 \| \Gamma$) if the points of Γ_1 are equidistant from Γ (it can be shown that in that case Γ is parallel to Γ_1, i.e., $\Gamma_1 \| \Gamma \Rightarrow \Gamma \| \Gamma_1$). In Euclidean

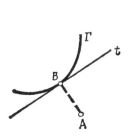

Figure 87a

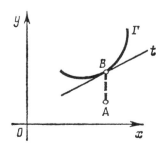

Figure 87b

[19] See, for example, [36], pp. 275–277.

geometry, if Γ and Γ_1 are lines, this is equivalent to the usual definition of parallelism (Fig. 88a). A parallel to a circle S is any circle S_1 concentric with S; in the case of parallel circles S and S_1, the distance from a point A of S_1 to S is measured along the common diameter of S and S_1 through A (Fig. 88b). In Galilean geometry, a parallel to an (ordinary) line l is any line l_1 with $l_1 \| l$ (Fig. 88c), and a parallel to a cycle Z is any cycle Z_1 parallel to Z in our earlier sense of the term (cf. p. 93); in the case of parallel cycles Z and Z_1 the distance from a point A on Z_1 to Z is measured along the common diameter of Z and Z_1 passing through A (Fig. 88d).

The fact that the midpoints of a family of parallel chords of a cycle Z lie on a diameter of Z has a reasonably simple mechanical interpretation.

We select a frame of reference in which the parallel chords are horizontal. Then they correspond to fixed points on the line o (i.e., to states of rest). We assume o to be vertical and Z to open downwards, and consider a part W of Z symmetric with respect to the special line T through the maximum M of Z. Then W may be thought of as a representation (in space–time) of the path on o of an object tossed vertically upward which eventually returns to the launching point. (The acceleration of the moving object has constant absolute value and opposite signs on the two legs of the journey.) The endpoints of each horizontal chord AB are the two events where the moving object is at the same point of o. The two times when it is at this point are symmetric about the moment when it reaches its

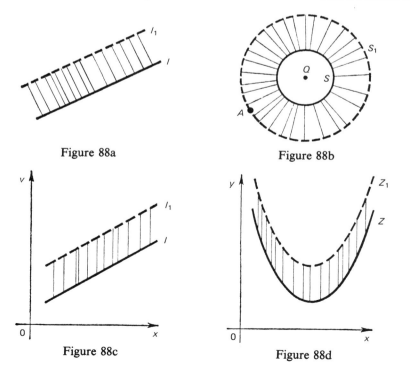

Figure 88a

Figure 88b

Figure 88c

Figure 88d

8. Cyclic rotation; diameters of a cycle

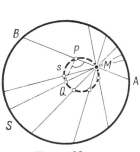
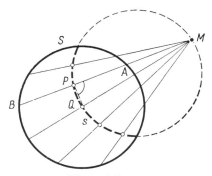

Figure 89a Figure 89b

maximum (represented by M on Z). It follows that this moment is the time coordinate of the points of T. But then T bisects AB.

We shall give another illustration of the use of cyclic rotations to prove a property of cycles.

It is easy to show that *the midpoints of all chords of a Euclidean circle S passing through a fixed point M form a circle (or circular arc) s*. For proof, note that the line QP joining the center Q of S to the midpoint P of a chord AB through M is perpendicular to AB. Therefore, P is a point of the circle s with diameter MQ (Figs. 89a and 89b). This proof relies on the concept of the center of a circle, and therefore cannot be adapted to cycles in the Galilean plane.

However, the above result can be proved differently. On each chord AB of S which passes through M, we select a point N such that $AN = MB$. All such points N lie on the circle S_1 concentric with S and passing through M (since, by symmetry, each chord MN of S_1 intersects S at points A and B such that $AN = MB$; see Figs. 90a and 90b). On the other hand, the midpoint P of the chord AB is also the midpoint of the chord MN. This means that P is the image of N under the central dilatation with center M and coefficient $\frac{1}{2}$. Hence the totality of such points P belongs to the circle s which is the image of the circle S_1 under this dilatation.

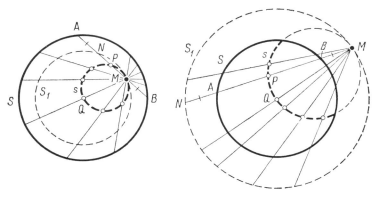

Figure 90a Figure 90b

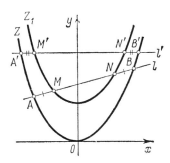

Figure 91

The latter argument can easily be adapted to cycles in Galilean geometry. We note that if a line l intersects the cycle Z with equation $y = ax^2$ at points A and B, and the cycle Z_1 with equation $y = ax^2 + c$ at points M and N (Fig. 91), then $AM = NB$. (To prove this, we perform a cyclic rotation (12) which maps l to a line l' parallel to the x-axis. Then, by symmetry, l' intersects Z and Z_1 at points A', B' and M', N' such that $A'M' = N'B'$.) Now let M be a point and Z a cycle. On each chord AB of Z passing through M we choose a point N such that $AN = MB$ (Figs. 92a and 92b). In view of the result just proved, all such points N belong to a cycle Z_1 through M parallel to Z. Now the center P of AB is also the center of the segment MN. Hence P is the image of N under the central dilatation with center M and coefficient $\frac{1}{2}$. But then *the midpoints of the chords of Z which pass through the point M form a cycle* (*or an arc of a cycle*) z which is the image of the cycle Z_1 under the central dilatation with center M and coefficient $\frac{1}{2}$.

We now return to the principle of duality discussed in Section 5, Chapter I. This principle tells us that to each theorem of Galilean geometry there corresponds another theorem, its dual, which is obtained by replacing in the original theorem the terms "point," "distance," and "parallel lines," respectively, by "line," "angle," and "parallel points" (i.e., points on the same special line), and conversely. In particular, this procedure replaces concepts of Galilean geometry appearing in the original theorem by their duals (for example, parallelograms are replaced by coparallelograms). We shall now explain the concept of Galilean geometry which is dual to the concept of a cycle.

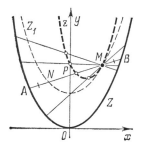
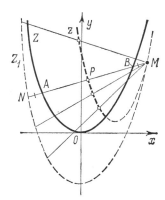

Figure 92a Figure 92b

8. Cyclic rotation; diameters of a cycle

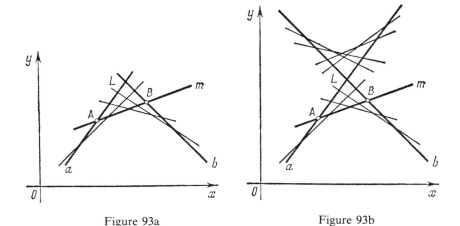

Figure 93a   Figure 93b

We defined a cycle Z as the set of points M from which a given segment AB is seen at a constant angle:

$$\angle AMB = \delta_{MA,MB} = \alpha.$$

The dual of the segment AB is an angle aLb determined by lines a and b. The dual of a point M is a line m. The duals of the lines MA and MB are the points A and B in which m intersects a and b, and the dual of the angle $\delta_{MA,MB}$ (i.e., $\angle AMB$) is the distance $\overline{AB}$ ($=d_{AB}$). We are thus led to consider *the set Σ of lines m on which a given angle aLb cuts off a segment AB of fixed length*

$$\overline{AB} = d_{AB} = d$$

(Fig. 93a; note that if we take the length AB to be always positive, i.e., we work with nondirected distances, then the condition $AB=$ const. defines a pair of cycles —cf. Fig. 93b and 67c).

We know that there exist Galilean motions (cyclic rotations) which map the set of points M (i.e., the cycle Z) to itself. By the principle of duality, there exist Galilean motions which map the set of lines m to itself. (We recall that a motion of the Galilean plane was defined as a mapping of the points and lines of the plane preserving distances between points and angles between lines; therefore, the dual of a motion is a motion.) From this we conclude that *the lines m are the tangents of a cycle*. This result can also be proved without using the principle of duality. To see this, let m be a line which intersects the sides of the angle aLb in points A and B, and let Z be the cycle tangent to a,b,m at the points D,E,F (Fig. 94).[20] Since the tangents from a point to a cycle are congruent, it follows that $DA=AF$ and $FB=BE$. This, and the fact that $DE=DA+AB+BE$ [see (11), Sec. 4] imply[21] that

$$DE = AF + AB + FB = 2AB.$$

Thus *the length AB ($=d$) of the segment of the tangent m to the cycle Z contained between the sides of the angle aLb is equal to half the length of the chord DE of Z.* Hence the sides of the angle aLb determine a segment of constant length d ($=\frac{1}{2}DE$) on every tangent m of Z. It follows that *the set of lines m from which lines a and b cut off a segment of constant length d is the set of tangents to a cycle Z.*

---

[20]Concerning the existence of such a cycle, see pp. 104–105.

[21]Note that in this argument we are using directed (i.e., positive or negative) lengths of segments (see the lines AB and A_1B_1 in Fig. 94).

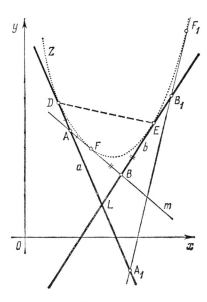

Figure 94

It is now easy to formulate the duals of the properties of cycles established thus far. For example, consider the result that *the tangents AP and PB from a point P to a cycle Z are congruent*. The dual of a cycle Z regarded as a set of points is a cycle Z viewed as the set of its tangents.[22] The dual of the point P is a line p. The duals of the tangents $PA \equiv a$ and $PB \equiv b$ through P are the points A, B of Z on p. Finally, the duals of the distances AP and PB are the angles $\angle aAp$ and $\angle pBb$ (Fig. 95). Hence the dual of our result is the result that *every chord of a cycle Z forms congruent angles with the tangents to Z at its endpoints*. This theorem differs only in form from the theorem stated on p. 80.

Next consider the result that *the midpoints of parallel chords of a cycle lie on a special line*, (i.e., are parallel points). The duals of parallel lines l and l_1 are parallel points L and L_1 (cf. Fig. 96). The duals of the points A, B and A_1, B_1 in which l and

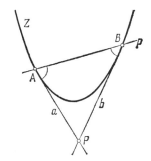

Figure 95

[22]Strictly speaking, the dual of a cycle of radius c is a cycle with curvature c (i.e., with radius of curvature $1/c$). This is so because the concept of radius of a cycle is dual to the concept of its curvature.

8. Cyclic rotation; diameters of a cycle 103

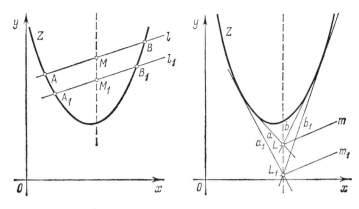

Figure 96

l_1 intersect the cycle Z are the tangents a,b and a_1,b_1 from L and L_1 to Z. Finally, the duals of the midpoints M and M_1 of the segments AB and A_1B_1 are the bisectors m and m_1 of the angles aLb and $a_1L_1b_1$. Hence the dual of our result is that *the bisectors m and m_1 of the angles aLb and $a_1L_1b_1$ are parallel* (cf. Exercise 7 below).

PROBLEMS AND EXERCISES

4 Give a direct proof (without using properties of cyclic rotations) of the theorem on diameters of a cycle, i.e., the theorem that the midpoints of parallel chords of a cycle lie on a special line.

5 Give other examples of the use of cyclic rotations to prove geometric properties of cycles in the Galilean plane.

6 Describe the mapping on the set of ordinary lines in the Galilean plane dual to a cyclic rotation.

7 Give a direct proof (without using the principle of duality) of the dual of the theorem on diameters of a cycle (see pp. 102–103 and Fig. 96b).

8 Formulate the dual of the theorem on the midpoints of concurrent chords of a cycle. Prove this theorem without using the principle of duality (cf. pp. 99–100. In particular, note Figs. 92a and 92b).

9 Give other examples of pairs of dual theorems involving cycles in the Galilean plane.

VII Investigate glides along themselves of surfaces in three-dimensional Galilean geometry which are analogous to Euclidean spheres (in connection with such surfaces, see Sec. 7, Problem IV). Use such glides to obtain geometric properties of these analogues of spheres.

VIII Investigate the question discussed in Problem VII in three-dimensional semi-Galilean geometry (cf. Problem V, Sec. 7).

IX Discuss the application of the principle of duality of three-dimensional semi-Galilean geometry (see Problems VIII and X in Chap. 1) to Problems V and VIII of the present chapter.

9. The circumcycle and incycle of a triangle

Let ABC be a triangle in the Galilean plane with sides $AB=c$, $BC=a$, $CA=b$; as usual, the letters a,b,c denote the (ordinary) lines AB, BC, CA and also the Galilean lengths of the corresponding segments. It is clear that there is a unique cycle through the points A, B, C; it can be defined as, say, the set of points from each of which the segment AB is seen at a Galilean angle equal to C. This cycle Z is called the **circumcycle** of the triangle ABC. A kind of analogue of the circumcycle is the **incycle** z of the triangle ABC, defined as the cycle tangent to the lines a,b,c. We shall now prove the existence and uniqueness of the incycle of a triangle.

It is easy to see that if the incycle z of a triangle ABC exists, then it must touch one side of the triangle (i.e., one of the segments AB, BC, CA) and the extensions of the other sides. Specifically, suppose our cycle is tangent to the side AB at F, and the extensions of the sides CA and CB at E and D (Fig. 97). The triangle DCE is isosceles (since the tangents CD and CE from C to the cycle z in Fig. 97 are congruent). Hence its altitude CP is also a median, so the special line CP through C intersects the segments ED and AB. For this to happen, it is necessary that $AB=c$ be the largest side of the triangle ABC, and therefore $c=a+b$. Since

$$CD+CE=(CB+BD)+(CA+AE)$$
$$=CB+BF+CA+AF=CB+CA+(BF+AF)$$
$$=CB+CA+AB,$$

we have

$$CD=CE=\frac{a+b+c}{2}=c=a+b. \qquad (13)$$

Therefore,

$$AE=CE-CA=a,$$
$$BD=CD-CB=b. \qquad (13a)$$

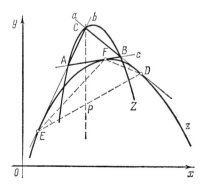

Figure 97

Since $AF=AE$ and $BF=BD$, we have
$$AF=a, \quad BF=b. \tag{13b}$$

Equations (13)–(13b) determine completely the positions of the points D, E, F on the sides of the triangle ABC. We shall now show that *the circumcycle z of the triangle DEF* (which we know exists) *is tangent to the sides AB, BC, CA of the triangle ABC at the points F, D, E*, and is thus the incycle of the triangle ABC. In fact, the inscribed angle $\angle DEF$ of the cycle z is equal to the difference of the angles $\angle DEC$ and $\angle FEA$. We know the angles C and A of the triangles DEC and FEA, and we know that these triangles are isosceles. Using the angle formula (12) of Section 4, we conclude that
$$\angle DEC = \frac{C}{2}, \quad \angle FEA = \frac{A}{2}$$
and, therefore,
$$\angle DEF = \frac{C}{2} - \frac{A}{2} = \frac{C-A}{2} = \frac{B}{2}.$$
On the other hand, an application of the angle formula to the isosceles triangle DFB with known angle B shows that
$$\angle FDB = \frac{B}{2}.$$
It follows that the line $DB \equiv a$, which forms with the chord DF of the cycle z the angle $\angle FDB = B/2$, equal to the inscribed angle subtended by that chord, is tangent to z. Using a similar argument, we can prove that the lines $EA \equiv b$ and $FA \equiv c$ are tangent to the cycle z at the points E and F.

We now consider the radii R and r and the curvatures P and ρ of the circumcycle Z and incycle z of the triangle ABC. We regard all these quantities and the sides and angles of the triangle as positive. In view of the definition of the radius of a cycle, we have
$$\frac{a}{A} = \frac{b}{B} = \frac{c}{C} = 2R; \tag{14}$$
this gives a geometric meaning to the coefficient of proportionality λ in (14), Section 4. Equations (14) enable us to express the radius R of the circumcycle Z of the triangle ABC in terms of its sides and angles. These expressions are simpler than the related expressions known as the "law of sines" of Euclidean geometry:
$$\frac{a}{\sin A} = \frac{b}{\sin B} = \frac{c}{\sin C} = 2R. \tag{14'}$$
(Here R is the radius of the circumcircle of the Euclidean triangle ABC.) Equations (14) and (17) in Section 4 imply that
$$R = \frac{a}{2A} = \frac{abc}{2(Abc)} = \frac{abc}{2 \cdot 2S},$$

where S is the area of $\triangle ABC$. In final form

$$R = \frac{abc}{4S}. \tag{15}$$

Curiously enough, the last equation is identical with the corresponding equation of Euclidean geometry obtained from the relations $R = a/2\sin A$ and $S = \frac{1}{2}bc\sin A$ [cf. (14') and (17) in Sec. 4].

We now turn to the incycle z of the triangle ABC. We saw that this cycle was the circumcycle of the triangle DEF in Figure 97, whose sides $EF = d$, $DF = e$, $DE = f$ and angles $\angle EDF = D$, $\angle DEF = E$, $\angle DFE = F$ can be easily computed. For example, we saw that $E = \angle DEF = B/2$. Similar computations show that

$$D = \frac{A}{2}, \qquad E = \frac{B}{2}, \qquad F = \frac{C}{2}. \tag{16}$$

From the isosceles triangles EAF (where $EA = AF = a$), DBF (where $DB = BF = b$) and DCE (where $DC = CE = c$), we obtain the relations

$$d = 2a, \qquad e = 2b, \qquad f = 2c.$$

Now from Eqs. (14) we have

$$\frac{d}{D} = \frac{e}{E} = \frac{f}{F} = 2r,$$

i.e.,

$$\frac{a}{A} = \frac{b}{B} = \frac{c}{C} = \frac{r}{2}. \tag{14a}$$

We thus arrive at the remarkable relation

$$r = 4R, \tag{17}$$

which implies that

$$\mathrm{P} = 4\rho. \tag{17a}$$

We note that the rather unexpected relations (17) and (17a), without analogues in Euclidean geometry, follow directly from the duality principle. In fact, it is clear that the dual of the circumcycle Z of the triangle ABC, the cycle containing the points A, B, C, is the incycle z of $\triangle ABC$, the cycle with tangents a, b, c. Next, the magnitudes of the angles of a triangle and the lengths of its sides are duals. Finally, the radius $r = \lim_{\Delta\varphi \to 0}(\Delta s/\Delta\varphi)$ of a cycle and its curvature $\rho = \lim_{\Delta s \to 0}(\Delta\varphi/\Delta s)$ are duals. It follows that the dual of the formulas

$$\frac{a}{A} = \frac{b}{B} = \frac{c}{C} = 2R \tag{14}$$

for the radius of the circumcycle of a triangle are the formulas

$$\frac{A}{a} = \frac{B}{b} = \frac{C}{c} = 2\rho$$

for the curvature of the incycle. The latter are equivalent to the relations

$$\frac{a}{A} = \frac{b}{B} = \frac{c}{C} = \frac{r}{2}. \tag{14a}$$

A glance at Eqs. (14) and (14a) yields the relation (17).

9. The circumcycle and incycle of a triangle

In Section 4, Chapter I, we discussed the simple mechanical significance of the side formula (11) and the angle formula (12) connecting the elements of a triangle. We saw that (11) expressed *the law of additivity of time intervals*, and Eq. (12) could be interpreted as *the law of composition of velocities* (cf. pp. 48–49). We are now able to give a mechanical interpretation of Eq. (13), Section 4, which is the last of the three fundamental relations connecting the sides and angles of any triangle. To do this, let $A(x_1,y_1)$, $B(x_2,y_2)$ and $C(x_3,y_3)$ be three points in the Galilean plane which are vertices of a triangle ABC (Fig. 98a). To these points there correspond three events $A(t_1,x_1)$, $B(t_2,x_2)$, and $C(t_3,x_3)$, no two of which are simultaneous. It can be shown (we do this below) that there exists a unique motion with (positive, negative, or zero) constant acceleration w which "joins" these three events. We know that the acceleration w has absolute significance in the sense that it is independent of the choice of an inertial reference frame. We choose an inertial reference frame so that two of our events, say A and B, differ in time but not in their positions on the line o. Also, we take the position of A on o as the origin of the coordinate (x), and the time of A as the origin of the time coordinate. Then the events A, B, C have space–time coordinates $A(0,0)$, $B(t_2,0)$, $C(t_3,x_3)$ (compare Fig. 98b with Fig. 98a, which also shows the axes of the new coordinate system corresponding to the imposed conditions).

We shall now determine the motion

$$x = at^2 + 2bt + c \tag{18}$$

with constant acceleration joining our three events (cf. pp. 83–84). Since the event $x=0$, $t=0$ satisfies (18), we have

$$c = 0. \tag{18a}$$

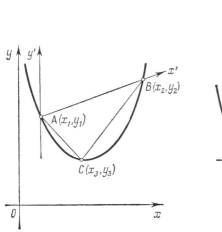
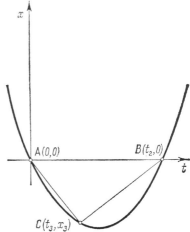

Figure 98a   Figure 98b

Since the event $x=0$, $t=t_2$ also satisfies (18), we have

$$at_2^2 + 2bt_2 = 0, \quad \text{i.e., } 2b = -at_2 \tag{18b}$$

(for $t_2 \neq 0$). To compute the last unknown coefficient a in (18), we substitute the values $x = x_3$ and $t = t_3$. This yields the relation

$$x_3 = at_3^2 - at_2 t_3,$$

which implies that

$$a = \frac{x_3}{t_3^2 - t_2 t_3} = \frac{x_3}{t_3(t_3 - t_2)}.$$

Since

$$\frac{1}{(t_3 - t_2)} - \frac{1}{t_3} = \frac{t_2}{t_3(t_3 - t_2)},$$

we have

$$a = \frac{x_3/(t_3 - t_2) - x_3/t_3}{t_2}. \tag{18c}$$

We have thus determined the required unique motion with constant acceleration. This motion is given by Eq. (18), where the values of the coefficients a, b, c are given by (18a–c). We saw that the acceleration w of the motion (18) is $2a$. Hence

$$\frac{w}{2} = \frac{x_3/(t_3 - t_2) - x_3/t_3}{t_2}.$$

Clearly t_2 is the time interval between the events B and A, i.e., t_2 is equal to the side $AB = c$ of the triangle ABC. Also, the velocity v_2 of the uniform motion of a material point which leaves $x = 0$ at time $t = 0$ and reaches $x = x_3$ at time $t = t_3$ is

$$v_2 = \frac{x_3}{t_3},$$

and the velocity v_1 of the uniform motion of a material point which leaves $x = 0$ at $t = t_2$ and reaches $x = x_3$ at $t = t_3$ is

$$v_1 = \frac{x_3}{(t_3 - t_2)}.$$

It follows that the difference

$$\frac{x_3}{t_3 - t_2} - \frac{x_3}{t_3} = v_1 - v_2$$

is the relative velocity of the second uniform motion relative to an observer moving according to the first uniform motion [see (8), Section 3], i.e., it is equal to the magnitude of the angle C of triangle ABC. We may, therefore, rewrite the formula which expresses the acceleration w in terms of the coordinates t_2, t_3, and x_3 of the events A, B, C as[22a]

$$\frac{|w|}{2} = \frac{C}{c}.$$

By symmetry, we also have

$$\frac{|w|}{2} = \frac{A}{a} \quad \text{and} \quad \frac{|w|}{2} = \frac{B}{b},$$

so that

$$\frac{A}{a} = \frac{B}{b} = \frac{C}{e} \left(= \frac{|w|}{2} \right).$$

Note that (in the cgs system) the time intervals a, b, c are measured in seconds, and the speeds A, B, C are measured in centimeters per second. Hence the ratios $A/a, B/b, C/c$ are measured in cm/sec², in agreement with the fact that these ratios represent (the absolute value of) an acceleration. In other words, since the sides of the triangle ABC have the dimension t of time, and the angles have the dimension l/t of velocity, the ratios of angles to opposite sides have the dimensions l/t^2 of acceleration. Also, in Galilean geometry the area S of a triangle has the dimension lt (cm·sec) (cf. p. 46), so that the ratio $4S/abc$ has the dimension $lt/t^3 = l/t^2$ of acceleration, and the ratio $2ABC/S$ has the dimension $(l/t)^3/lt = l^2/t^4$ of the square of acceleration [cf. Exercise **9(a)** below].

We now consider the circle S_1 passing through the midpoints A_1, B_1, C_1 of the sides BC, CA, AB of a (Euclidean) triangle ABC (Fig. 99a) and a cycle Z_1 passing through the midpoints A_1, B_1, C_1 of the sides of a (Galilean) triangle ABC (Fig. 99b). In Figure 99a, the circle S_1 is circumscribed about the triangle $A_1B_1C_1$ with $B_1A_1 = AB/2$, $C_1B_1 = BC/2$, $A_1C_1 = CA/2$ and angles $\angle B_1A_1C_1 = \angle CAB = A$, $\angle A_1C_1B_1 = C$, $\angle C_1B_1A_1 = B$ (in fact, $\triangle A_1B_1C_1$ is the image of $\triangle ABC$ under the central dilatation with coefficient $-\frac{1}{2}$ and center M at intersection of the medians of $\triangle ABC$). Hence the radius R_1 of S_1 has the value

$$R_1 = \frac{R}{2}, \tag{19}$$

where R is the radius of the circumcircle of triangle ABC. An analogous statement holds for the cycle Z_1. In the case of the cycle Z_1, we have the additional relation $R_1 = r/8$, where r is the radius of the incycle of $\triangle ABC$ [cf. Eq. (17)].

The Euclidean circle S_1 also passes through the feet P, Q, R of the altitudes AP, BQ, CR of the triangle ABC, and might therefore be called *the six-point circle*[23] of $\triangle ABC$. In fact, since the midline A_1B_1 of $\triangle ABC$

[22a]We note that if $\bar{c} = \overline{AB} = d_{AB}$ is the signed length of the side AB and $\bar{C} = \delta_{CA,CB}$ is the signed magnitude of the angle C of the triangle, then $w/2 = \bar{C}/\bar{c}$.

[23]It can be shown that S_1 also passes through the midpoints of the segments joining the intersection point of its altitudes to its vertices. This additional fact justifies calling S_1 *the nine-point circle* of $\triangle ABC$ (cf., for example, [11], Problem **17(a)**, Section 1.7 of [19], or Chap. XI of [33]). [Since the Galilean altitudes of a triangle are not concurrent, this fact is exclusively "Euclidean."]

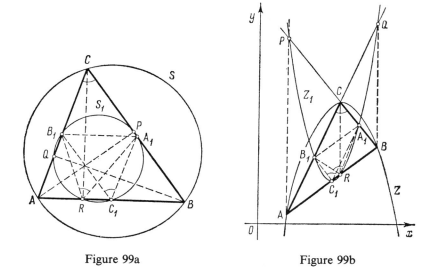

Figure 99a   Figure 99b

bisects its altitude CR, the points C and R are symmetric with respect to the line A_1B_1. But then

$$\angle A_1RB_1 = \angle ACB. \qquad (20)$$

Now $\angle A_1C_1B_1$, which is an inscribed angle of S_1 subtended by the arc A_1B_1, is also congruent to $\angle ACB$.[24] Hence R is on S_1 (Fig. 99a). Similar arguments show that Q and P are on S_1. It is easy to see that our Euclidean argument carries over to Galilean geometry, where the triangles B_1RA_1 and B_1CA_1 are also symmetric with respect to the line A_1B_1 (cf. p. 130) and therefore congruent. Thus here, too, Eq. (20), which implies that the point R is on S_1, remains valid. [In this case we need not rely on symmetry with respect to a line. It is obvious (cf. Fig. 98b) that $B_1C = B_1R$ and $CA_1 = A_1R$, i.e., the triangles B_1CR and A_1CR are isosceles. Hence

$$\angle B_1RA_1 = \angle B_1CA_1.$$

Therefore (just as in the Euclidean case), the point R belongs to the cycle Z_1.] Z_1 is called *the six-point cycle* of triangle ABC.

A remarkable theorem of Euclidean geometry asserts that *the six-point (or nine-point) circle S_1 of a triangle ABC is tangent to its incircle s and its excircles s_a, s_b, s_c* (Fig. 100a). This theorem also carries over to Galilean geometry: *The six-point cycle Z_1 of a triangle ABC is tangent to its incycle z* (Fig. 100b). This fact is an indication of the far-reaching analogy between Euclidean and Galilean geometry.

---

[24] For the arguments to be independent of diagrams we must use directed angles here (and in what follows).

9. The circumcycle and incycle of a triangle 111

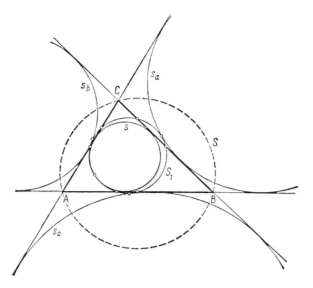

Figure 100a

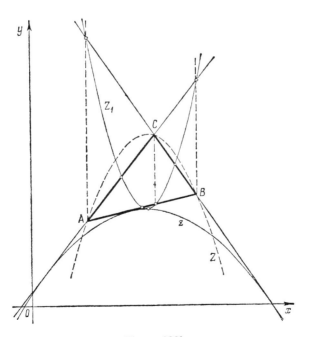

Figure 100b

We now give a proof of the theorem just formulated. Except as noted, the proof holds for Euclidean and Galilean geometry (cf. also Exercise **10** below and pp. 138–141). For definiteness, we use the Euclidean terminology. At one point in the proof we shall find it necessary to use results established in Section 10 below.

Let C_1T be the tangent to the six-point circle S_1 (or the six-point cycle Z_1) at the midpoint C_1 of the side AB (Figs. 101a and 101b). Since

$$\angle AC_1T = \angle AC_1B_1 - \angle TC_1B_1$$

and

$$\angle AC_1B_1 = \angle ABC = B, \quad \angle TC_1B_1 = \angle C_1A_1B_1 = A,$$

we have

$$\angle AC_1T = B - A.$$

We denote the points of contact of the circle s (cycle z) with the sides AB, BC, CA of triangle ABC by F, D, E and the points of contact of the circle s_c with these sides by F_1, D_1, E_1. We draw the diameter CWX of the circles s and s_c through C, (and similarly the diameter CR of the cycle s);[25] here W and R are on AB and X is on the extension of CW beyond W. Through W and R we draw the tangents WK, WK_1 to the circles s, s_c (and similarly the tangent RK to the cycle z). In view of the theorem on tangents from a point to a circle, we have

$$WK = WF \quad \text{and} \quad WK_1 = WF_1$$

(in the Galilean case, $FR = RK$, that is, $d_{FR} = d_{RK}$). In Figure 101a (bearing in mind that WC is a diameter of s and s_c), we have

$$\angle AWK = 180° - 2\angle BWC = 180° - 2\left(180° - B - \frac{C}{2}\right)$$
$$= 2B + C - 180° = 2B + C - (A+B+C) = B - A,$$

and

$$\angle BWK_1 = 180° - \angle AWK_1 = 180° - 2\angle AWX = 180° - 2\angle BWC$$
$$= \angle AWK = B - A,$$

so that KWK_1 is a single line parallel to C_1T.

The proof of the analogous result in the Galilean case is entirely different. Here we can use the fact that $AR = AC = b$ and $AF = CB = a$, so that $FR = RK = b - a$, and thus $FK = FR + RK = 2(b - a)$ (Fig. 101b). Hence the magnitude of the angles inscribed in the cycle z of radius r and subtended by the arc FK is

$$\frac{FK}{2r} = \frac{2(b-a)}{2r} = \frac{b-a}{4R} = \frac{1}{4}\left(\frac{b}{R} - \frac{a}{R}\right) = \frac{1}{4}(2B - 2A) = \frac{B-A}{2}$$

[see definition (3a) of the radius of a cycle and formulas (17) and (14)], so that

$$\angle KFR = \angle RKF = \frac{B-A}{2}.$$

Hence

$$\angle KRB = \angle KFR + \angle RKF = B - A \; (= \angle AC_1T),$$

which implies that $RK \parallel C_1T$.

Finally, we draw the line C_1K, and denote by L its second point of intersection with the incircle s (incycle z). We also draw the line C_1K_1 and denote by L_1 its second point of intersection with the excircle s_c. We claim that *L is on the circle* S_1

---

[25]Thus in the subsequent argument the angle bisector CW of the Euclidean triangle ABC and the altitude CR of the Galilean triangle play, quite unexpectedly, similar roles.

9. The circumcycle and incycle of a triangle

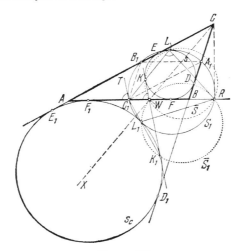

Figure 101a

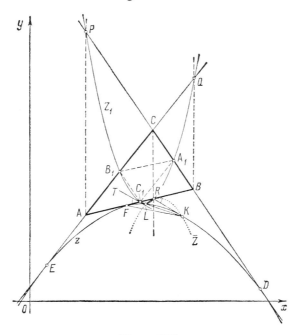

Figure 101b

(*the cycle* Z_1) and that L_1 *is also on* S_1 (we shall soon see that s and S_1, as well as z and Z_1, are tangent at L, and that s_c and S are tangent at L_1).

By the theorem on the secant and tangent from C_1 to the circles s and s_c, we have

$$C_1 K \cdot C_1 L = C_1 F^2 \tag{21a}$$

and

$$C_1 K_1 \cdot C_1 L_1 = C_1 F_1^2. \tag{21b}$$

By the corresponding theorem of Galilean geometry (see pp. 119–120), the equality (21a) also holds in Figure 101b. Now we again have different arguments for the Euclidean and the Galilean cases, and the argument for the Galilean case is far simpler. Since $AF=a$, $AC_1=c/2=(a+b)/2$, and $AR=b$ (Fig. 101b), we have, obviously,

$$FC_1 = \frac{a+b}{2} - a = \frac{b-a}{2} \quad \text{and} \quad C_1R = b - \frac{a+b}{2} = \frac{b-a}{2}.$$

Hence the equality (21a) is equivalent to the relation

$$C_1K \cdot C_1L = C_1R^2, \tag{22}$$

which implies that the circumcycle $\bar{Z}$ of triangle KLR (represented by dots in Fig. 101b) is tangent to AB at R. It follows that

$$\angle KLR = \angle KRB = B - A$$

(an inscribed angle of a cycle is congruent to the angle between the chord and the tangent). Since $\angle AC_1T = B - A$, the latter equality can be rewritten in the form

$$\angle RLC_1 = \angle AC_1T.$$

This last equality implies that L is on the cycle Z_1. (In fact, $\angle RLC_1$ is congruent to $\angle AC_1T$, so that $\angle RLC_1$ is an inscribed angle.)

The Euclidean argument (Fig. 101a) is essentially quite similar to the one just given, but is somewhat more difficult. We denote by R the foot of the perpendicular of triangle ABC passing through the vertex C, and prove that

$$C_1K \cdot C_1L = C_1W \cdot C_1R \tag{22a}$$

and

$$C_1K_1 \cdot C_1L_1 = C_1W \cdot C_1R. \tag{22b}$$

(Note that the role of the two points W and R in Fig. 101a is played in Fig. 101b by the single point R.) As usual, we denote the sides of triangle ABC by a,b,c. Since

$$BF = BD$$

and

$$BD + BF = (a - CD) + (c - AF) = a + c - (CD + AF)$$
$$= a + c - (CE + AE) = a + c - b,$$

we have $BF = (a+c-b)/2$, and therefore

$$C_1F = \frac{c}{2} - \frac{a+c-b}{2} = \frac{b-a}{2}.$$

On the other hand,

$$CE_1 = CD_1,$$

and

$$CE_1 + CD_1 = (b + AE_1) + (a + BD_1)$$
$$= a + b + (AE_1 + BD_1)$$
$$= a + b + (AF_1 + BF_1)$$
$$= a + b + c,$$

so that

$$CE_1 = \frac{a+b+c}{2}, \quad AF_1 = AE_1 = \frac{a+b+c}{2} - b = \frac{a+c-b}{2},$$

9. The circumcycle and incycle of a triangle

and thus
$$C_1F_1 = \frac{c}{2} - \frac{a+c-b}{2} = \frac{b-a}{2}.$$
This means that the products $C_1K \cdot C_1L$ $(= C_1F^2)$ and $C_1K_1 \cdot C_1L_1$ $(= C_1F_1^2)$ are equal, and it therefore suffices to prove one of the equalities (22a) and (22b)—say (22a).

We have
$$BW + AW = c \quad \text{and} \quad \frac{BW}{AW} = \frac{a}{b},$$
since CW is the angle bisector of the angle C of $\triangle ABC$. Therefore,
$$BW = \frac{a}{a+b}c,$$
so that
$$C_1W = \frac{c}{2} - \frac{a}{a+b}c = \frac{(b-a)c}{2(a+b)}.$$
It is somewhat more difficult to compute the length x of the segment C_1R. Since
$$x - \frac{c}{2} = C_1R - C_1B = BR$$
and
$$x + \frac{c}{2} = C_1R + AC_1 = AR,$$
we have
$$a^2 - \left(x - \frac{c}{2}\right)^2 = b^2 - \left(x + \frac{c}{2}\right)^2 \quad (= CR^2),$$
and thus
$$\left(x + \frac{c}{2}\right)^2 - \left(x - \frac{c}{2}\right)^2 = b^2 - a^2,$$
i.e.,
$$2cx = b^2 - a^2.$$
Consequently,
$$C_1R = x = \frac{b^2 - a^2}{2c},$$
and hence
$$C_1W \cdot C_1R = \frac{(b-a)c}{2(a+b)} \cdot \frac{b^2 - a^2}{2c} = \frac{(b-a)^2}{4} = C_1F^2 = C_1K \cdot C_1L, \quad (23)$$
which is what we wished to prove.

We note that Eqs. (22a) and (22b) imply the existence of the circle $\bar{S}$ through K, L, W, R and the circle $\bar{S}_1$ through K_1, L_1, W, R (these circles are represented in Fig. 101a by dots). The existence of the circle $\bar{S}$ implies that
$$\angle RLK = 180° - \angle KWR = 180° - \angle TC_1B = \angle TC_1A,$$
i.e., the angle $\angle RLC_1$ is congruent to the angle between the chord RC_1 of S_1 and its tangent C_1T. This means that $\angle RLC_1$ is an inscribed angle of S_1, i.e., L lies on S_1. From the existence of the circle $\bar{S}_1$, we deduce by a similar argument that L_1 also lies on S_1.

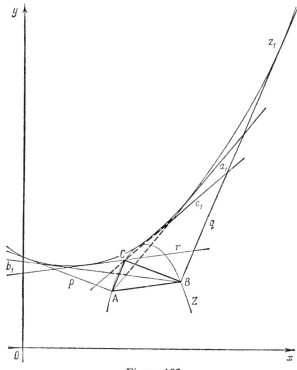

Figure 102

We have now nearly completed the proof. It remains only to show that the circles s and S_1 intersect—in fact, are tangent—at L, and that the circles s_c and S_1 are tangent at L_1. But this follows immediately from the fact that the tangents KW and C_1T to the circles s and S_1 at the endpoints K and C_1 of the chords LK and LC_1 are parallel. Indeed, since the tangents to these circles at the other endpoint L of these chords form the same angles with them as the tangents at K and C_1, it follows that the tangents at L coincide, i.e., S_1 and s are tangent to L. Similarly, the parallelism of the tangents K_1W and C_1T to the circles s_c and S_1 at K_1 and C_1 implies that these circles are tangent at L_1.

It is of interest to formulate the dual of the theorem on the six-point cycle. This theorem asserts that *the incycle of the triangle formed by the angle bisectors a_1, b_1, c_1 of triangle ABC* (which is the dual of the triangle $A_1B_1C_1$ whose vertices are the midpoints of triangle ABC) *is also tangent to the lines p, q, r passing through the vertices A, B, C and parallel to the opposite sides.* This "six-line cycle" z_1 (whose radius is twice the radius of the incycle z of triangle ABC) *is tangent to the circumcycle Z of the triangle* (Fig. 102; cf. Exercise **11** below).

PROBLEMS AND EXERCISES

10 (a) Show that if R and P are the radius and curvature of a circumscribed cycle of a triangle in the Galilean plane with sides a, b, c, angles A, B, C, and area S, then $R^2 = S/2ABC$, $\mathrm{P} = 4S/abc$, and $\mathrm{P}^2 = 2ABC/S$. (b) Without using relations (17) and (17a), derive the connection between the radius (or curvature) of the inscribed cycle of a triangle and its sides, angles, and area.

11 (a) Prove that the inscribed cycle z of $\triangle ABC$ is mapped to its circumscribed cycle Z by the dilatation γ with coefficient $\frac{1}{4}$ and center Q on the special line m through the intersection point M of the medians of triangle ABC with $d_{Qa} = 4d_{Qt}$, where $a = BC$, $t \| a$ is the tangent to Z, and A lies between a and t. (b) Using (a) and the fact that the dilatation γ_1 with center M and coefficient $-\frac{1}{2}$ maps Z to the six-point cycle Z_1, show that z is mapped to Z_1 by the dilatation γ_2 with coefficient $-\frac{1}{8}$ and center at the point L of m with $LM = MQ/3$. (c) Using (b), prove that z and Z_1 are tangent. (*Hint*: Prove that L lies on Z_1.)

12 Without using the principle of duality, prove: (a) the theorem on the six-line cycle of a triangle (cf. p. 116 and Fig. 102); and (b) the theorem that the six-point cycle z_1 is tangent to the circumscribed six-line cycle Z of a triangle.

X Discuss the problem of inscribed and circumscribed "spheres" (surfaces discussed in Sec. 7, Problem **IV**) of a tetrahedron in three-dimensional Galilean geometry. Does an analogue of the theorem on the six-point cycle hold in this geometry?

XI Consider Problem **X** in three-dimensional semi-Galilean geometry (cf. Sec. 7, Problem **V**).

10. Power of a point with respect to a circle or a cycle; inversion

We recall that *if a line l through a point M intersects a* (Euclidean) *circle S in points A and B* (Figs. 103a–c), *then the product $MA \cdot MB$ depends only on M and S, and is independent of l.* The product $MA \cdot MB$ is usually signed. Specifically, $MA \cdot MB$ is taken as positive if the directions of the segments MA and MB coincide, and negative otherwise. The signed product

$$MA \cdot MB \tag{24}$$

is called *the power of the point M with respect to the circle S*. It is clear that the power of M with respect to S is positive when M is outside S (Fig. 103a), zero when M is on S (Fig. 103b), and negative when M is inside S (Fig. 103c). When M is outside S, its power with respect to S is equal to the square of the tangent MT from M to S (for we may take l to be the tangent line MT, in which case the points A and B coincide with the point of tangency T).

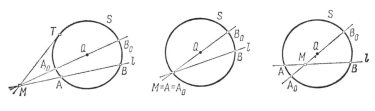

Figure 103a   Figure 103b   Figure 103c

If $MQ = d$, where Q is the center of the circle S with radius r, and the line l passes through Q and intersects S in points A_0 and B_0, then the length of one of the segments MA_0 and MB_0 is $d+r$ and the length of the other segment is $|d-r|$. The sign of the product $MA \cdot MB$ is the same as the sign of the difference $d-r$. Thus in all cases *the power of the point M with respect to the circle S is equal to*

$$d^2 - r^2. \tag{25}$$

In terms of the coordinates (x_0, y_0) of M and (a,b) of Q, the power of M with respect to S is

$$(x_0 - a)^2 + (y_0 - b)^2 - r^2.$$

Since the equation of S is

$$(x-a)^2 + (y-b)^2 - r^2 = 0, \tag{26}$$

or equivalently,

$$x^2 + y^2 + 2px + 2qy + f = 0, \tag{26'}$$

where $p = -a$, $q = -b$, $f = a^2 + b^2 - r^2$ [see Eqs. (2)–(2a), Sec. 1], it follows that *the power of $M(x_0, y_0)$ with respect to S can be obtained by replacing x and y on the left side of the equation of S [(26) or (26')] by the coordinates x_0 and y_0 of M*. This implies that *if S is the circle given by Eq. (26), then the set of points whose power with respect to S is k is given by the equation*

$$(x-a)^2 + (y-b)^2 - r^2 = k,$$

or

$$(x-a)^2 + (y-b)^2 - (r^2 - k) = 0,$$

i.e., *the set in question is a circle concentric with S*. [Incidentally, this follows directly from (25)]. If S is a circle given by Eq. (26'), and S_1 is a circle given by the equation

$$x^2 + y^2 + 2p_1 x + 2q_1 y + f_1 = 0, \tag{26''}$$

then *the set of points with the same power with respect to S and S_1* is given by the equation

$$x^2 + y^2 + 2px + 2qy + f = x^2 + y^2 + 2p_1 x + 2q_1 y + f_1$$

or

$$2(p - p_1)x + 2(q - q_1)y + (f - f_1) = 0.$$

This set *represents a line r, the so-called radical axis* of S and S_1 (Fig. 104).[26] [An exception is the case where $p = p_1$ and $q = q_1$, i.e., when S and S_1 are concentric. In this case, the radical axis is the empty set.] It is not difficult to see that *the radical axes r_1, r_2, r_{12} determined by three circles S, S_1 and S_2 are either parallel or concurrent* (Fig. 105; in fact, if two of the

---

[26]This fact, called the radical axis theorem, is easy to prove without using the equation of a circle (cf. Sec. 1 of [13] or Chap. III of [33]).

10. Power of a point with respect to a circle or a cycle; inversion

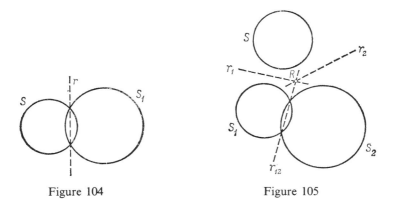

Figure 104   Figure 105

radical axes intersect in a point R, then R has the same power with respect to all three circles, and so belongs to the third radical axis). The intersection point of three radical axes is called the *radical center* of the three circles.

We now turn to Galilean geometry. It is obvious that if l is an (ordinary) line passing through a point M and intersecting a Galilean circle S in points A and B, then *the product $MA \cdot MB$ depends only on M and S, and is independent of l*; indeed, for every line through M the distances MA and MB are the same (Fig. 106). The product

$$MA \cdot MB \qquad (24)$$

(which can be positive, zero, or negative since MA and MB are directed distances) is called *the power of the point M with respect to the circle S*. It is clear that the power of M with respect to S is positive if M is outside S (Fig. 106a), zero if M is on S (Fig. 106b), and negative if M is in the interior of S (Fig. 106c). If $d = MQ$ is the distance from M to a center Q of S (this distance does not depend on the choice of Q) and $r = AQ = QB$ is the radius of S, supposed for convenience to have the same sign as d, then one of the segments MA and MB is $d+r$ and the other is $d-r$. It follows that in Galilean geometry, just as in Euclidean geometry, *the power of a point M with respect to a Galilean circle S is*

$$d^2 - r^2. \qquad (25)$$

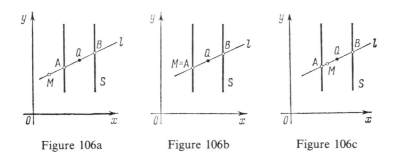

Figure 106a   Figure 106b   Figure 106c

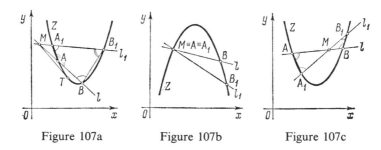

Figure 107a   Figure 107b   Figure 107c

The corresponding argument for cycles is more involved. Let l and l_1 be two lines passing through a point M and intersecting a cycle Z in points A, B, and A_1, B_1 (Fig. 107). It is clear that if M is on Z (Fig. 107b), then $MA \cdot MB = MA_1 \cdot MB_1 = 0$. If M is not on Z, then $\angle ABB_1$ and $\angle AA_1B_1$ are inscribed angles of Z subtended by the same arc and, therefore, congruent. Similarly, $\angle A_1AB = \angle A_1B_1B$. It follows that the angles of the triangles MAA_1 and MB_1B are pairwise congruent.[27] Since the sides of a triangle are proportional to the opposite angles (see [13], Sec. 9), we conclude that

$$\frac{MA}{MA_1} = \frac{MB_1}{MB} \quad \text{or} \quad MA \cdot MB = MA_1 \cdot MB_1.$$

This proves that the product (24) is independent of the line l. This product is called *the power of the point M with respect to the cycle Z*. It is clear that the power of M with respect to Z is positive if M is outside Z (Fig. 107a; by taking l to be the tangent MT from M to Z we see that in this case the power of M with respect to Z is MT^2), zero if M is on Z (Fig. 107b), and negative if M is inside Z (Fig. 107c). [We say that a point M not on Z is inside Z if every line through M intersects Z; all other points M not on Z are said to be outside Z. If M is outside Z, then we can draw a tangent to Z through M. This cannot be done if M is inside Z.]

If M has coordinates (x_0, y_0), and S is a pair of (special) lines $x = x_1$ and $x = x_2$, then the segments MA and MB have Galilean lengths $x_1 - x_0$ and $x_2 - x_0$, and therefore the power of M with respect to S is

$$(x_1 - x_0)(x_2 - x_0) = (x_0 - x_1)(x_0 - x_2).$$

Since, in this case, the equation of S is

$$(x - x_1)(x - x_2) = 0, \tag{27}$$

or equivalently,

$$x^2 + 2bx + c = 0, \tag{27'}$$

where $2b = -(x_1 + x_2)$, $c = x_1 x_2$, it follows that *the power of a point M with respect to a circle S is the result of replacing x and y on the left-hand side of Eq.(27) [or (27')] of S by the coordinates (x_o, y_0) of M.*

---

[27] These triangles are "1-similar," i.e., they are related by a similitude of the first kind (see p. 53).

Now let Z be a cycle with equation

$$x^2 + 2b_1 x + 2b_2 y + c = 0 \tag{28}$$

[which is the same as Eq. (2), Sec. 7, with $a=1$], and let l be a line through $M(x_0, y_0)$ with equation

$$y - y_0 = k(x - x_0). \tag{29}$$

Then the coordinates (x_1, y_1) and (x_2, y_2) of the intersection points A, B of l and Z are determined from the system of simultaneous equations (28) and (29). From Eq. (29), we have

$$y = kx + (y_0 - kx_0).$$

Substitution of this value of y in (28) yields

$$x^2 + 2(b_1 + kb_2)x + [2b_2(y_0 - kx_0) + c] = 0,$$

so that

$$x_1 + x_2 = -2(b_1 + kb_2), \qquad x_1 x_2 = 2b_2(y_0 - kx_0) + c.$$

It follows that the power of M with respect to Z,

$$MA \cdot MB = (x_1 - x_0) \cdot (x_2 - x_0) = x_0^2 - (x_1 + x_2)x_0 + x_1 x_2,$$

is equal to

$$x_0^2 + 2(b_1 + kb_2)x_0 + 2b_2(y_0 - kx_0) + c = x_0^2 + 2b_1 x_0 + 2b_2 y_0 + c.$$

In other words, *the power of M with respect to a cycle Z is obtained by replacing x and y in the equation of Z [Eq. (28)] by the coordinates (x_0, y_0) of M.*

It follows immediately that *the equation of the set of points whose power with respect to the circle (27′) is k, is*

$$x^2 + 2bx + c = k, \qquad \text{i.e.,} \qquad x^2 + 2bx + (c - k) = 0,$$

so that *the set in question is a circle "concentric" with* (i.e., having the same line of centers as) *the given circle* [this, incidentally, follows directly from (25)]. Similarly, *the equation of the set of points whose power with respect to the cycle (28) is k, is*

$$x^2 + 2b_1 x + 2b_2 y + c = k, \qquad \text{i.e.,} \qquad x^2 + 2b_1 x + 2b_2 y + (c - k) = 0,$$

so that *the set in question is a cycle parallel to the given one*. Also, it is easy to see that *the set of points that have the same power with respect to two circles S and S_1* with Eqs. (27′) and

$$x^2 + 2b_1 x + c_1 = 0, \tag{27″}$$

or a circle S with equation (27′) and a cycle Z with equation (28), or two cycles Z and Z_1 with Eqs. (28) and

$$x^2 + 2b_1^{(1)} x + 2b_2^{(1)} y + c_1 = 0, \tag{28′}$$

is a (special or ordinary) line r; it is natural to call this line the *radical axis* of the relevant pair of curves (two circles, circle and cycle, or two cycles). Thus *the radical axis of two (nonconcentric) circles (27′) and (27″) is the*

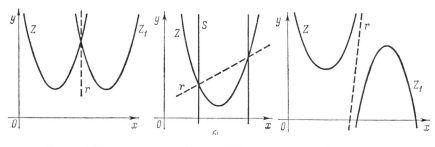

Figure 108a Figure 108b Figure 108c

special line

$$x^2 + 2bx + c = x^2 + 2b_1 x + c_1, \quad \text{i.e.,} \quad x = \frac{c_1 - c}{2(b - b_1)}.$$

(Two concentric circles have no radical axis.) *The radical axis of two (nonparallel) cycles Z and Z_1 of the same radius* [for such cycles we have $b_2 = b_2^{(1)}$ in Eqs. (28) and (28$'$)[28]] *is the special line*

$$x^2 + 2b_1 x + 2b_2 y + c = x^2 + 2b_1^{(1)} x + 2b_2 y + c_1, \quad \text{i.e.,} \quad x = \frac{c_1 - c}{2(b_1 - b_1^{(1)})}$$

(Fig. 108a; two parallel cycles have no radical axis). *In all remaining cases the radical axis* (if nonempty) *is an ordinary line* (Figs. 108b and 108c); for example, the equation of the radical axis of the cycles (28) and (28$'$) with unequal radii is

$$x^2 + 2b_1 x + 2b_2 y + c = x^2 + 2b_1^{(1)} x + 2b_2^{(1)} y + c_1,$$

or

$$y = \frac{b_1^{(1)} - b_1}{b_2 - b_2^{(1)}} x + \frac{c_1 - c}{2(b_2 - b_2^{(1)})}.$$

It is clear that if two curves Σ and Σ_1 (each either a cycle or a circle) intersect in two points P and Q, then the radical axis of Σ and Σ_1 is the line PQ (for P and Q have power 0 with respect to Σ and Σ_1, and thus belong to the radical axis of Σ and Σ_1; cf. Fig. 108b).

The analogy between the study of Euclidean circles and the study of Galilean circles and cycles can be carried further. Thus, for example, just as in Euclidean geometry, we can prove that *if each of the curves $\Sigma, \Sigma_1, \Sigma_2$ is a circle or cycle, and if two of the radical axes r_1, r_2, r_{12} of the three pairs of curves $(\Sigma_1, \Sigma_2), (\Sigma_2, \Sigma_1)$ and (Σ_1, Σ_1) intersect in a point R* (which obviously has the same power with respect to all three curves), *then the third radical axis passes through R* (Figs. 109a and 109b). It is natural to call R *the radical center* of the curves Σ, Σ_1 and Σ_2.

---

[28]Since Eq. (28) can be rewritten as $y = -(1/2b_2)x^2 - (b_1/b_2)x - (c/2b_2)$, the radius r of the cycle (28) is $-b_2$.

10. Power of a point with respect to a circle or a cycle; inversion

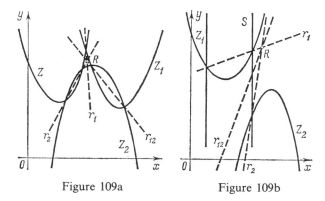

Figure 109a   Figure 109b

It is not difficult to establish the mechanical significance of the power of a point relative to a cycle. Thus consider the motion of a material point along a line o which moves with constant acceleration $w = 2a$ and is at $x = 0$ at time $t = 0$. The equation of the motion of this point is

$$x = at^2 + 2bt \tag{30}$$

[cf. (1'), Sec. 7]. An example of such a motion is the motion of a stone thrown vertically upward. The acceleration of the stone is the gravitational acceleration $w = -g$ with $g \approx 9.8$ m/sec² (cf. Figs. 110a and 110b). We select an inertial reference frame in such a way that the line l in Figure 110b represents a state of rest. Then this line determines a definite point $x = x_0$ of the (vertical) line o. Let M be the point of the Galilean plane which corresponds to the point $x = x_0$ of o and the moment $t = 0$ at which the motion begins. Then the segments MA and MB in Figure 110b represent the time intervals between the beginning of the motion and the moments when the moving point (stone) passes the point $x = x_0$. It is clear that these time intervals are determined by the equation

$$at^2 + 2bt = x_0 \quad \text{or} \quad at^2 + 2bt - x_0 = 0,$$

so that

$$t_1 t_2 = -\frac{x_0}{a}.$$

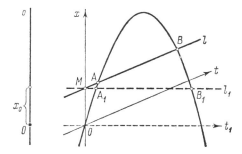

Figure 110a   Figure 110b

This shows that the product $MA \cdot MB = t_1 \cdot t_2$ depends only on the acceleration w $(=2a)$ of the moving point and the value of x_0, and is independent of the choice of the inertial coordinate system. This means that if we choose another coordinate system, for example, one in which the role of the "fixed point" $x = x_0$ in Figure 110a is played by the line l_1 in Figure 110b, then the product $MA_1 \cdot MB_1$ (see Fig. 110b) has the same value $-x_0/a$ as before.

We now turn to the study of inversions. Recall that in Euclidean geometry *an* **inversion** *with center Q and coefficient $k > 0$* (or the inversion in a circle S with center Q and radius $r = \sqrt{k}$) *is a map which associates to each point $A \neq Q$ the point A' on the ray QA such that*

$$QA \cdot QA' = k \tag{31}$$

(Fig. 111). The following is a slightly different description of this map. *The inversion leaves the points of the circle S pointwise fixed, it maps a point A exterior to S to the point A' of intersection of the diameter QA of S with the chord KL, where K and L are the points of contact of the tangents from A to S, and finally, it maps A' to A.* This follows from the fact that the right triangles QKA' and QAK are similar (since they share the acute angle Q) and, therefore,

$$\frac{QK}{QA'} = \frac{QA}{QK}, \quad \text{or} \quad QA \cdot QA' = QK^2 \ (= r^2 = k).$$

To determine the image A' of $A \neq Q$ under inversion in S we can also proceed as follows. *We draw all possible circles s (including the line $l = QA$) through A perpendicular to S*—in other words, all circles s such that the tangents to S and s at their intersection points K and L are perpendicular (i.e., s must be tangent to the radii QK and QL of S). *All such circles s and the line l intersect in a second point A', which is the image of A under inversion in S* (Fig. 112a). To justify this assertion (which can serve as another definition of inversion) we can use the concept of the power of a point with respect to a circle, as applied to Q and s. Specifically, if A' is the second intersection point of QA and a circle s, then

$$QA \cdot QA' = QK^2 \ (= r^2 = k)$$

(where QK is tangent to s). If A is a point of S, then all circles s (and the

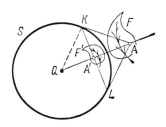

Figure 111

10. Power of a point with respect to a circle or a cycle; inversion

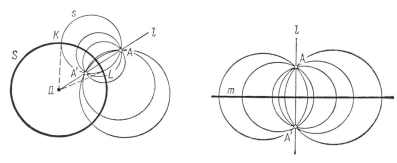

Figure 112a   Figure 112b

line *l*) are tangent at *A* and it is natural to make the convention that $A' = A$. This description of an inversion is very similar to that of a *reflection in a line m*, where we say that the image of a point *A* is the point *A'* of intersection of all circles *s* (and the line *l*) passing through *A* and perpendicular to *m* (Fig. 112b). This analogy justifies calling inversion in *S* a *reflection in S*.

The most important property of an inversion is that it *maps each circle in the plane* (including lines, which are regarded as "circles of infinite radius"), *to a circle* (or line). This assertion is most easily proved analytically. Thus the inversion with center at the origin $O(0,0)$ and coefficient k maps the point $A(x,y)$ to the point $A'(x',y') = A'(\lambda x, \lambda y)$ on the ray OA (Figure 113), such that

$$OA \cdot OA' = k, \quad \text{i.e.,} \quad \sqrt{x^2+y^2} \cdot \sqrt{(\lambda x)^2 + (\lambda y)^2} = k.$$

Hence

$$\lambda = \frac{k}{x^2+y^2},$$

and we have

$$x' = \frac{kx}{x^2+y^2}, \quad y' = \frac{ky}{x^2+y^2},$$

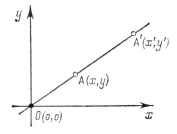

Figure 113

or

$$x = \frac{kx'}{x'^2+y'^2}, \quad y = \frac{ky'}{x'^2+y'^2}. \tag{32}$$

(The latter assertion follows from the fact that the inversion maps A' to A.) It follows that *the inversion* (32) *maps the circle* (*or line*) *s given by the equation*

$$a(x^2+y^2)+2b_1x+2b_2y+c=0 \tag{33}$$

[see (2'), Sec. 7] to the curve s' with equation

$$a\frac{k^2}{x'^2+y'^2}+2b_1\frac{kx'}{x'^2+y'^2}+2b_2\frac{ky'}{x'^2+y'^2}+c=0,$$

i.e., *to the circle* (*or line*)

$$a'(x^2+y^2)+2b_1'x+2b_2'y+c'=0 \tag{33'}$$

(here we write x and y in place of x' and y'), where

$$a'=c, \quad b_1'=kb_1, \quad b_2'=kb_2, \quad c'=k^2a. \tag{33a}$$

We note that if in (33) we have

(a) $a=0$, $c=0$ (i.e., s is a line through O);
(b) $a=0$, $c\neq 0$ (i.e., s is a line not through 0);
(c) $a\neq 0$, $c=0$ (i.e., s is a circle through 0);
(d) $a\neq 0$, $c\neq 0$ (i.e., s is a circle not through O);

then in (33') we have

(a) $a'=0$, $c'=0$ (i.e., s' is a line through O);
(b) $a'\neq 0$, $c'=0$ (i.e., s' is a circle through O);
(c) $a'=0$, $c'\neq 0$ (i.e., s' is a line not through O);
(d) $a'\neq 0$, $c'\neq 0$ (i.e., s' is a circle not through O).

This purely analytic argument can be replaced by a synthetic one. It is clear that *an inversion with center Q maps a line s through Q to itself.* Now let s be a line which does not pass through Q. Let M and A be two points on s with images M' and A' (Fig. 114a). The equality

$$QA \cdot QA' = QM \cdot QM' \ (=k)$$

implies that

$$\frac{QA}{QM} = \frac{QM'}{QA'}.$$

It follows that the triangles $QA'M'$ and QMA are similar and, therefore,

$$\angle QA'M' = \angle AMQ. \tag{34}$$

We now fix the points M and M', and let the point A range over s. The set s' of image points A' is determined by the condition (34). This means that the points of s' are characterized by the constancy of the angle $QA'M'$ subtended at A' by the segment QM'. Therefore, s' is a *circle* passing

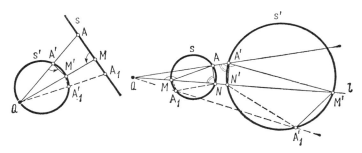

Figure 114a   Figure 114b

through Q and M'. Conversely, if A' in Figure 114a ranges over *a circle s' passing through the center of inversion* Q, then the set s of images A of A' is again characterized by the condition (34). This implies that $\angle AMQ = \angle QA'M' = $ const. Hence s is a *line* which forms the angle $\angle QMA$ with the line QM. This angle is congruent to the inscribed angle of the circle s' subtended by the chord QM'.[29]

Now *let s be a circle which does not pass through the center Q of inversion*. Let l be a line through Q intersecting s in points M and N, A another point of s, and M', N', A' the images of M, N, A under the inversion (Fig. 114b). The equalities

$$QA \cdot QA' = QM \cdot QM' \ (=k) \quad \text{and} \quad QA \cdot QA' = QN \cdot QN' \ (=k)$$

imply the proportionality of the sides of the triangles QAM and $QM'A'$, QAN and $QN'A'$:

$$\frac{QA}{QM} = \frac{QM'}{QA'} \quad \text{and} \quad \frac{QA}{QN} = \frac{QN'}{QA'},$$

and thus the congruence of their angles:

$$\angle QA'M' = \angle AMQ; \quad \angle QA'N' = \angle ANQ.$$

This and the theorem on the exterior angle of a triangle imply that

$$\angle N'A'M' = \angle QA'M' - \angle QA'N' = \angle AMQ - \angle ANQ = \angle MAN.$$

It follows that as A ranges over the circle s characterized by the constancy of the inscribed angle $\angle MAN$ subtended by the chord MN, A' ranges over the curve s' characterized by the constancy of the angle $\angle N'A'M'$, i.e., a circle.

Finally, we show that inversion is a *conformal mapping*, i.e., *it preserves angles between curves*. In fact, let Γ' be the image of a curve Γ under the inversion with center Q and coefficient k. Let A and A' be points on Γ and Γ' which correspond under the inversion, B a point on Γ close to A, and B' its image on Γ' (Fig. 115a). Since

$$QA \cdot QA' = QB \cdot QB' \ (=k), \quad \text{i.e.,} \quad \frac{QA}{QB} = \frac{QB'}{QA'},$$

---

[29] In this argument we must use directed angles, as discussed on p. 77.

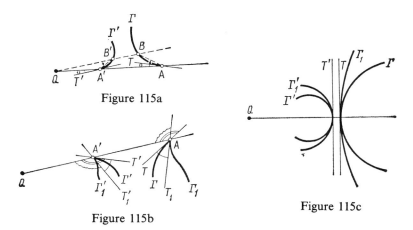

Figure 115a

Figure 115b

Figure 115c

it follows that the triangles QAB and $QB'A'$ are similar, i.e., $\angle BAQ = \angle QB'A'$. Now if B tends to A, then the line AB tends to the tangent AT to Γ at A, and the line $A'B'$ tends to the tangent $A'T'$ to Γ' at A' (see Fig. 115a). Hence in the limit the equality $\angle BAQ = \angle QB'A'$ becomes the equality

$$\angle TAQ = \angle QA'T',$$

i.e., the angles at A and A' between the line QA and the curves Γ and Γ' are congruent (but oppositely oriented; cf. Fig. 115a). This implies that if an inversion maps two curves Γ and Γ_1 intersecting at a point A to two curves Γ' and Γ'_1 intersecting at a point A' and if AT and AT_1 are the tangents to Γ and Γ_1 at A, while $A'T'$ and $A'T'_1$ are the tangents to Γ' and Γ'_1 at A' (Fig. 115b), then

$$\angle TAT_1 = \angle T'_1A'T'.$$

Since *the angle between two intersecting smooth curves Γ and Γ_1 is defined to be the angle between their tangents at their point of intersection*, the latter equality asserts that *the angle between Γ and Γ_1 is congruent to the angle between Γ' and Γ'_1*. In particular, the conformal nature of an inversion implies that *an inversion maps tangent curves Γ and Γ_1 (curves which form a zero angle; Fig. 115c) to tangent curves Γ' and Γ'_1* (for example, tangent circles to tangent circles).

After this considerable Euclidean detour, we return to Galilean geometry. In this geometry it is natural to define **an inversion of the first kind** *with center Q and coefficient $k > 0$* (*inversion with center Q and circle of inversion S of radius $r = \sqrt{k}$* [30]) *as a mapping which associates to each point A (at a*

---

[30]The center Q of inversion must coincide with the center of S, the circle of inversion. [In Euclidean geometry, the circle S fully determines an inversion. On the other hand, in order to determine an inversion of the first kind in Galilean geometry we must specify not only the circle S but also one of its (infinitely many) centers, which is to play the role of the center of inversion.]

nonzero distance from Q) the point A' on the ray QA such that

$$QA \cdot QA' = k \tag{31}$$

(Fig. 116). Using the circle S we can give the following equivalent description of an inversion of the first kind. *To determine the image A' of a point A* (at a nonzero distance from Q), *draw a cycle z through A intersecting S in points K and L such that the radii QK and QL of S are tangent to z at K and L, respectively. Then A' is the second point of intersection of all such cycles.* (Among the cycles z we include the line $l \equiv QA$ "tangent" to the rays QK_0 and QL_0, which join Q to the intersection points K_0 and L_0 of l and S.) To prove the equivalence of the two definitions of an inversion of the first kind, we use the concept of the power of a point with respect to a cycle (as applied to the point Q and any of the cycles z). From the definition of this concept we see that if A' is the second intersection point of the line QA and the cycle z, then

$$QA \cdot QA' = QK^2 \ (= r^2 = k),$$

since QK is tangent to z. *If A is on S, then* all the cycles z tangent to the ray QA at A are tangent to each other at A. In this case, *it is natural to make the convention that $A' = A$.* [We note that, just as in Euclidean geometry, the curves S and z are "perpendicular" at their intersection point K. However, in Galilean geometry this property has no special significance in view of the fact that S is perpendicular to every curve Γ which intersects S non-tangentially.]

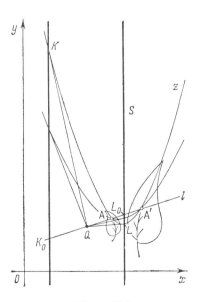

Figure 116

In Galilean geometry we sometimes call an inversion of the first kind with circle of inversion S and center Q *a reflection in the circle S with center Q*. Another analogue of Euclidean inversion in Galilean geometry is the so-called **inversion of the second kind** *with cycle of inversion Z*, also referred to as *a reflection in the cycle Z*. This transformation is defined as follows. *An inversion of the second kind with cycle of inversion Z leaves each point of Z fixed, maps each point A in the exterior of Z to the point A' where the diameter d of Z through A meets the chord KL joining the points of contact of the tangents to Z from A, and maps A' to A* (Fig. 117a). Equivalently, *an inversion of the second kind with cycle of inversion Z takes a point A to the point A' on the special line through A* (a diameter of Z) *such that Z bisects the segment AA'* (i.e., takes A to a point A' of the special line d such that

$$AP = PA',$$

or, in other words, such that

$$\delta_{AP} = \delta_{PA'},$$

where P is the intersection point of d and Z). For proof, note that $AK = AL$, i.e., d bisects the segment KL. Hence the line t tangent to Z at P is parallel to KL (cf. p. 96 above). Now if M is the intersection point of t and AK, then $PM = MK$, since the tangents to Z from M have the same length. In other words, the special line e through M is equidistant from the special line d and from the special line f through K. In turn, this implies that if N is the intersection point of e and KL, then

$$PA' = MN = \tfrac{1}{2} AA'$$

(since MN is a midline of triangle KAA'; here all the distances are special distances).

The last formulation of a reflection in a cycle relates it closely to a *reflection in an* (*ordinary*) *line l*, which we define as *a mapping of the plane which takes each point A to the point A' on the special line d through A such that l bisects the segment AA'* (Fig. 117b; cf. p. 43 above). In what follows,

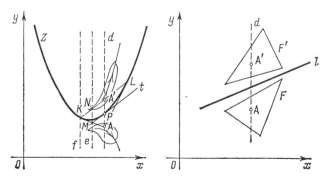

Figure 117a   Figure 117b

we shall view a reflection in a line as an inversion of the second kind, since an (ordinary) line may be viewed as a limiting case of a cycle (a "cycle of infinite radius").

It is clear that *an inversion of the first kind* (reflection in a circle) *as well as an inversion of the second kind* (a reflection in a cycle) *maps a special line p in the Galilean plane to a special line*; specifically, an inversion of the second kind maps p to itself, and an inversion of the first kind maps p to a special line p' such that the product of the distances from the center of inversion Q to p and p' is equal to the coefficient of inversion k. It follows that *an inversion* (of the first or second kind) *maps a circle in the Galilean plane* (i.e., a pair of special lines) *to a circle*. We shall now establish a deeper property of inversions (of either kind); *inversions map cycles* (including ordinary lines, which we regard as cycles of infinite radius) *to cycles*.

The simplest proof of this fact makes use of analytic geometry. Thus, an inversion of the first kind (a reflection in a circle) with center $O(0,0)$ and coefficient k maps the point $A(x,y)$ to the point $A'(x',y') = A'(\lambda x, \lambda y)$ of the ray OA such that

$$QA \cdot QA' = k \quad \text{or} \quad x \cdot \lambda x = k$$

(Fig. 118). It follows that $\lambda = k/x^2$, so that

$$x' = \frac{k}{x}, \quad y' = \frac{ky}{x^2},$$

or

$$x = \frac{k}{x'}, \quad y = \frac{ky'}{x'^2} \tag{35}$$

(since A' is mapped to A). Hence *the inversion of the first kind* (35) *maps the cycle* (*or line*) z with equation

$$y = ax^2 + 2bx + c \tag{1}$$

to the curve z' *with equation*

$$\frac{ky'}{x'^2} = a\frac{k^2}{x'^2} + 2b\frac{k}{x'} + c,$$

i.e., *to a cycle* (*or line*).

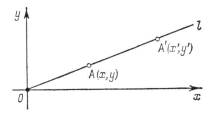

Figure 118

where

$$a' = \frac{c}{k}, \qquad b' = b, \qquad c' = ka. \tag{36a}$$

In particular, if in Eq. (1)

(a) $a=0$, $c=0$ (i.e., z is a line through O);
(b) $a=0$, $c\neq 0$ (i.e., z is a line not through O);
(c) $a\neq 0$, $c=0$ (i.e., z is a cycle through O);
(d) $a\neq 0$, $c\neq 0$ (i.e., z is a cycle not through O);

then in Eq. (36)

(a) $a'=0$, $c'=0$ (i.e., z' is a line through O);
(b) $a'\neq 0$, $c'=0$ (i.e., z' is a cycle through O);
(c) $a'=0$, $c'\neq 0$ (i.e., z' is a line not through O);
(d) $a'\neq 0$, $c'\neq 0$ (i.e., z' is a cycle not through O)

(see Figs. 119a and 119b; also, cf. p. 126 above).

On the other hand, if Z is a cycle with equation

$$y = \alpha x^2,$$

then a reflection in Z (an inversion of the second kind) maps the point $A(x,y)$ in the Galilean plane to the point $A'(x',y') = A'(x,y')$ of the special line l through A such that the midpoint of the segment AA' is the point $P(x,\alpha x^2)$, where l intersects Z (Fig. 120a). Hence

$$y' - \alpha x^2 = \alpha x^2 - y,$$

i.e.,

$$x' = x, \qquad y' = 2\alpha x^2 - y,$$

or

$$x = x', \qquad y = 2\alpha x'^2 - y' \tag{37}$$

(since A' is mapped to A). It follows that *the inversion of the second kind given by* (37) *maps the cycle (or line) z with Eq.* (1) *to the curve z' with*

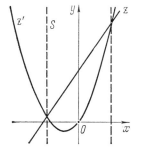
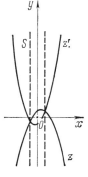

Figure 119a   Figure 119b

10. Power of a point with respect to a circle or a cycle; inversion

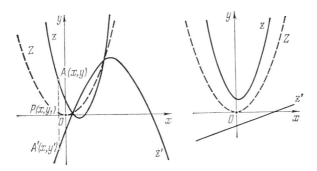

Figure 120a   Figure 120b

equation
$$2\alpha x'^2 - y' = \alpha x'^2 + 2bx' + c',$$
i.e., *to the cycle or line*
$$y = a'x^2 + 2b'x + c', \tag{36'}$$
where
$$a' = 2\alpha - a, \quad b' = -b, \quad c' = -c. \tag{36b}$$
Hence *the image of a cycle z with curvature* $\rho = 2a$ *under reflection in a cycle Z with curvature* $P = 2\alpha$ *is a cycle z' with curvature*
$$\rho' = 2P - \rho \tag{38}$$
(since $\rho' = 2a' = 4\alpha - 2a$). In particular, *reflection in Z maps cycles of curvature* $\rho = 2P$ *to lines* (cycles of infinite radius), *and lines to cycles of curvature* $2P$ (Fig. 120b).[31]

All of these results can also be deduced synthetically, without the use of analytic geometry. In the case of an inversion of the first kind (reflection in a circle), one can almost copy the arguments used in the Euclidean case (compare Figs. 121a and 121b with Figs. 114a and 114b). For example, in the Galilean case, just as in the Euclidean case, the equality
$$QA \cdot QA' = QM \cdot QM' \quad (= k)$$
implies that the sides of the triangles QAM and $QM'A'$ in Figure 121a as well as in Figure 121b are proportional:
$$\frac{QA}{QM} = \frac{QM'}{QA'}.$$
In view of Eqs. (13), Section 4, it follows that
$$\frac{\angle QMA}{\angle MAQ} = \frac{\angle QA'M'}{\angle A'M'Q}.$$

---

[31] We leave it to the reader to show that if reflection in a cycle Z maps a cycle z to a line z', then z' is the radical axis of the cycles Z and z.

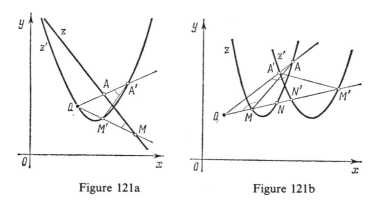

Figure 121a   Figure 121b

Since $\angle A'M'Q = -\angle QM'A'$ and $\angle QA'M' = -\angle M'A'Q$, we have $\angle MAQ/\angle QMA = \angle QM'A'/\angle M'A'Q$. Now $\angle MAQ + \angle QMA = \angle QM'A' + \angle M'A'Q$ ($=\angle MQA$). Hence, adding 1 to both sides of this last proportionality, we obtain $\angle MQA/\angle QMA = \angle MQA/\angle M'A'Q$, or finally

$$\angle M'A'Q = \angle QMA. \tag{34}$$

This implies that an inversion of the first kind maps the line z in Figure 121a to the cycle z' and conversely. Similarly, in Figure 121b the cycle z is mapped to a cycle z', and conversely. Finally, just as in the Euclidean case, we can prove that (with the notation of Fig. 122a) if an inversion of the first kind maps the (smooth) curve Γ to the curve Γ', and A and A' are corresponding points, then the angles between the line QA and the curves Γ and Γ' are congruent (but their directions are opposite), i.e.,

$$\angle QAT = \angle T'A'Q$$

(cf. Fig. 122a). It follows easily that an inversion of the first kind is a *conformal mapping*. This is a brief way of saying the following. *If an inversion of the first kind maps the curves Γ and Γ_1 intersecting in a point A to the curves Γ' and Γ'_1 intersecting in a point A', then the angle $\angle TAT_1$*

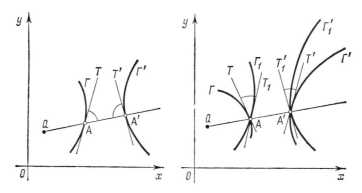

Figure 122a   Figure 122b

between the tangents AT and AT_1 to Γ and Γ_1 at A is congruent to the angle $\angle T_1'A'T'$ between the tangents $A'T'$ and $A'T_1'$ to Γ' and Γ_1' at A' (Fig. 122b). In particular, *an inversion of the first kind maps tangent curves* (for example, tangent cycles) *to tangent curves* (tangent cycles).

In the case of an inversion of the second kind (reflection in a cycle Z) we must argue in a completely different manner. Suppose reflection in a cycle Z maps points A, M, N to points A', M', N'. Let P, Q, R denote the intersection points of Z with the special lines AA', MM', NN', and let X, Y, U denote the intersection points of the special line NN' with the lines $AM, A'M', PQ$ (Fig. 123). Since the line PQ joins the midpoints of the bases of the trapezoid $AA'M'M$, it bisects the segment XY parallel to these bases. It follows that

$$N'Y = N'U + UY = (N'R - UR) + UY$$
$$= RN - UR + XU = (UN - UR) - UR + XU$$
$$= (XU + UN) - 2UR = XN - 2UR.$$

If we divide both sides of the equality $N'Y = XN - 2UR$ by the Galilean length of the segment $A'N' = AN = PR$) and use the fact (implied by the definition of angle in Galilean geometry) that

$$\frac{N'Y}{A'N'} = \angle N'A'M', \qquad \frac{XN}{AN} = \angle MAN$$

and

$$\frac{UR}{PR} = \angle QPR,$$

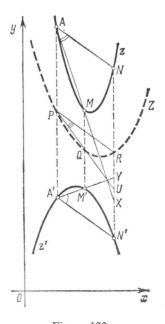

Figure 123

then we obtain the equalities
$$\angle N'A'M' = \angle MAN - 2\angle QPR,$$
$$\angle M'A'N' = 2\angle QPR - \angle MAN. \tag{39}$$

Now suppose M and N are fixed, and let A traverse a cycle (or line) z through M and N. Then we claim that A' traverses a cycle (or line) z'. For proof, note that $\angle MAN = \alpha = $ const. (inscribed angle of the cycle z subtended by the chord MN), and $\angle QPR = A = $ const. (inscribed angle of the cycle Z subtended by the chord QR), so that
$$\angle M'A'N' = \alpha' = 2A - \alpha = \text{const.}$$
Bearing in mind the definition of curvature given by (4′), Section 7, we see that the curvatures ρ, ρ' and P of the cycles z, z', and Z are $2\alpha/s$, $2\alpha'/s$, and $2A/s$, (where s is the length of the arcs MN, $M'N'$, and QR, i.e., the length of MN). This fact the last equation above yield the relation (38). To prove that an inversion of the second kind is a conformal, i.e., angle-preserving, mapping, we assume that the points A and N in (39) are fixed, and that M tends to A along some curve γ (Fig. 124a). Let γ' be the image of γ under the inversion. In the limit we obtain the relation
$$\angle t'A'N' = 2\angle TPR - \angle tAN,$$
where At is tangent to γ at A, $A't'$ is tangent to γ' at A', and PT is tangent to Z at P. It follows that *if two curves γ and γ_1 intersect at a point A, if their images γ' and γ_1' under an inversion of the second kind intersect at a point A', if t and t_1 are the tangents to γ and γ_1 at A, and t' and t_1' are the tangents to γ' and γ_1' at A' (see Fig. 124b), then*
$$\angle t'A't_1' = \angle t'A'N' - \angle t_1'A'N'$$
$$= (2\angle TPR - \angle tAN) - (2\angle TPR - \angle t_1AN)$$
$$= \angle t_1AN - \angle tAN$$
$$= \angle t_1At$$

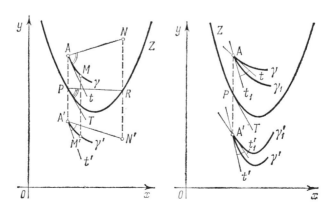

Figure 124a   Figure 124b

10. Power of a point with respect to a circle or a cycle; inversion 137

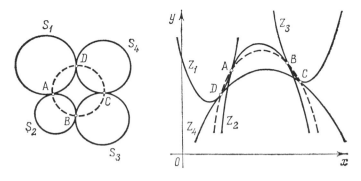

Figure 125a Figure 125b

(the lines AN, $A'N'$ and PR are not shown in Fig. 124b; cf. Fig. 124a). In particular, *an inversion of the second kind maps tangent curves to tangent curves*. One instance of the latter result is that an inversion of the second kind maps tangent cycles to tangent cycles.

We illustrate the many applications of inversion to Euclidean geometry (a large number of which can be found in [13]) and to Galilean geometry with an example. Specifically, we use inversion to prove the following Euclidean result and its Galilean analogue: *Let S_1, S_2, S_3, S_4 be four circles in the Euclidean plane (Z_1, Z_2, Z_3, Z_4 four cycles in the Galilean plane). Let each of S_1 and S_3 (Z_1 and Z_3) be tangent to each of S_2 and S_4 (Z_2 and Z_4).* In the case of the circles, we make the additional assumption that S_1 and S_3 as well as S_2 and S_4 are disjoint[32] (Figs. 125a and 125b). *Then the points A, B, C, D of contact of these four curves belong to a single circle (single cycle).*

We now give the proof of the Euclidean theorem. The proof of the corresponding Galilean result is entirely analogous, with the circles S_1, S_2, S_3, S_4 replaced throughout by the cycles Z_1, Z_2, Z_3, Z_4, and (Euclidean) inversion by inversion of the first kind.

We apply an inversion with center A, where A is the point of contact of S_1 and S_2, and arbitrary coefficient k. Since S_1 and S_2 pass through the

---

[32]The figures below show the need for the nonintersection requirement. [Actually, the theorem deals with oriented circles. While it complicates the statement of the theorem, the condition of nonintersection of the circles guarantees the possibility of orienting them in a manner compatible with their tangency. This is another illustration of the fact that Galilean geometry, for whose cycles there is a "natural" traversal direction, is simpler then Euclidean geometry.]

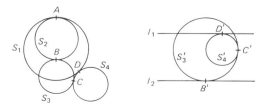

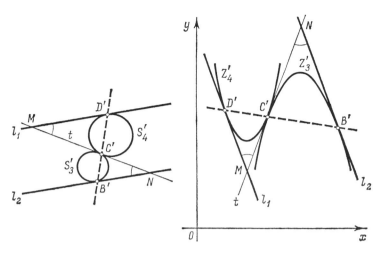

Figure 126a   Figure 126b

center of inversion, their images are lines l_1 and l_2. The images of the tangent circles S_3 and S_4 are tangent circles S_3' and S_4' with S_3' tangent to l_2 and S_4' tangent to l_1. (This follows from the fact that S_3 is tangent to S_2 and S_4 is tangent to S_1.) Since S_1 and S_2 have no common points other than A, and A is inverted to infinity, it follows that the lines l_1 and l_2 are parallel (Figs. 126a and 126b). We draw the tangent t to S_3' and S_4' at their point of intersection C', and denote the point of contact of S_3' and l_2 by B', and the point of contact of S_4' and l_1 by D'. Also, we denote by M and N the intersection points of t with l_1 and l_2. The triangles $MC'D'$ and $NC'B'$ are isosceles (since tangents from a point to a circle are congruent), and the alternate interior angles $\angle C'MD'$ and $\angle C'NB'$ formed by the parallel lines l_1, l_2 and the transversal t are congruent. It follows that $\angle D'C'M = \angle B'C'N$, which implies that $B'C'D'$ is a line. Since inversion with center A maps the points B, C, D to collinear points B', C', D', it follows that B, C, D are points of a circle through A, i.e., *the points A, B, C, D all lie on the same circle.*

    Here is another example of the use of inversion to prove a theorem. We wish to prove that the *six- (nine-) point circle of a triangle in the Euclidean plane is tangent to the incircle and three excircles of the triangle, and the six-point cycle of a triangle in the Galilean plane is tangent to the incycle of the triangle* (cf. p. 110 above). The proof is as follows. Let F be the point of contact of the side AB of a triangle ABC and its incircle s (incycle z). In the Euclidean case, let F_1 be the point of contact of the side AB and the excircle s_c. Let C_1 be the midpoint of the side AB (Figs. 127a and 127b; the notations are those used on pp. 109–116). It is easy to see that in the Euclidean case $C_1F = C_1F_1 = (b-a)/2$, where a, b, c are the lengths of the sides of triangle ABC (cf. p. 114), and that in the Galilean case we also have $FC_1 = |b-a|/2$ (cf. p. 114). We now apply an inversion (of the first kind) with center C_1 and coefficient $k = (C_1F)^2 = (C_1F_1)^2$. Suppose a line l through C_1 intersects the circle s (cycle z) in points M and N (and the circle s_c in points M_1 and N_1). Then, in view

10. Power of a point with respect to a circle or a cycle; inversion 139

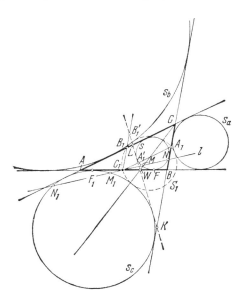

Figure 127a

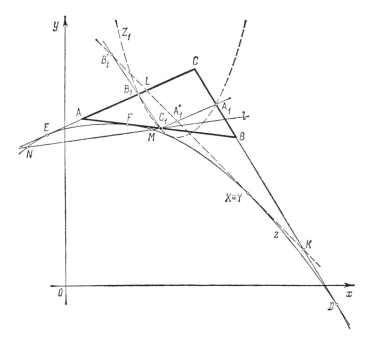

Figure 127b

of the definition of the power of a point with respect to a circle (cycle), we have
$$C_1 M \cdot C_1 N = (C_1 F)^2 \quad [\text{and} \quad C_1 M_1 \cdot C_1 N_1 = (C_1 F_1)^2],$$
which shows that our inversion interchanges the points M and N (M_1 and N_1). But then *our inversion maps each of the circles s and s_c (the cycle z) to itself.* On the other hand, the six-point circle S_1 (cycle Z_1) passes through the center of inversion C_1 and the midpoints B_1 and A_1 of the sides AC and BC. It follows *that our inversion maps the circle S_1 (cycle Z_1) to the line $A_1' B_1'$, where A_1' and B_1' are* points of the arcs $C_1 B_1$ and $C_1 A_1$ such that[33]

$$B_1' C_1 = \frac{k}{B_1 C_1} = \frac{[(b-a)/2]^2}{a/2} = \frac{(b-a)^2}{2a}$$

and

$$C_1 A_1' = \frac{k}{C_1 A_1} = \frac{[(b-a)/2]^2}{b/2} = \frac{(b-a)^2}{2b}$$

i.e., B_1' and A_1' are *the images of B_1 and A_1 under our inversion.* We shall now prove that *the line $A_1' B_1'$ is tangent to the circles s and s_c (cycle z).* In turn, this will prove that *the circle S_1 is tangent to the circles s and s_c (the cycle Z_1 is tangent to the cycle z).*

We denote by K and L the intersection points of the line $A_1' B_1'$ and the sides BC and AC of the triangle. The pairs of triangles $A_1 A_1' K$ and $C_1 A_1' B_1'$, $B_1 B_1' L$ and $C_1 B_1' A_1'$ yield the relations

$$\frac{A_1 K}{A_1' A_1} = \frac{B_1' C_1}{C_1 A_1'} \quad \text{and} \quad \frac{B_1' B_1}{B_1 L} = \frac{B_1' C_1}{C_1 A_1'}.$$

Now

$$\frac{B_1' C_1}{C_1 A_1'} = \frac{(b-a)^2 / 2a}{(b-a)^2 / 2b} = \frac{b}{a},$$

and (cf. Figs. 127a and 127b)

$$A_1' A_1 = C_1 A_1 - C_1 A_1' = \frac{b}{2} - \frac{(b-a)^2}{2b} = \frac{b^2 - (b-a)^2}{2b} = \frac{a(2b-a)}{2b},$$

$$B_1' B_1 = B_1' C_1 - B_1 C_1 = \frac{(b-a)^2}{2a} - \frac{a}{2} = \frac{(b-a)^2 - a^2}{2a} = \frac{b(b-2a)}{2a}.$$

In this way we obtain the relations

$$A_1 K = \frac{b}{a} \cdot \frac{a(2b-a)}{2b} = b - \frac{a}{2}, \quad B_1 L = \frac{a}{b} \cdot \frac{b(b-2a)}{2a} = \frac{b}{2} - a,$$

which imply that

$$CK = CA_1 + A_1 K = \frac{a}{2} + \left(b - \frac{a}{2}\right) = b \quad \text{and} \quad LC = B_1 C - B_1 L = \frac{b}{2} - \left(\frac{b}{2} - a\right) = a.$$

From this point it is convenient to argue differently in the two cases represented by Figures 127a and 127b. In the Euclidean case, the reflection in the angle bisector CW of the angle ACB interchanges the triangles CAB and CKL (where

---

[33]We assume that the lengths a, b, c of our triangle are positive. Also, the computations below reflect the situation in Figures 127a and 127b, where $b > 2a$. As a result, all the expressions below (lengths of various segments) are positive. (To insure independence of the argument from drawings we would have to use directed segments.)

10. Power of a point with respect to a circle or a cycle; inversion 141

$CK = CA = b$ and $CB = CL = a$). However, since CW passes through the centers of the circles s and s_c, this reflection maps each of these circles to itself. Since AB is tangent to s and s_c, this implies that KL is also tangent to s and s_c.

In the Galilean case, we can use the fact that the lines CB and CA are tangent to the cycle z at points D and E with $CD = EC = c$ [cf. (13), Sec. 9]. We now draw tangents KX and LY, different from KD and LE, from K and L to the cycle z. It is clear that

$$XK = KD = CD - CK = c - b = a \quad \text{and} \quad LY = EL = EC - LC = c - a = b.$$

Since

$$LK = CK + LC = a + b,$$

it follows that

$$LY + XK = LK,$$

which shows that the points X and Y coincide. (Actually, this equation merely shows that X and Y lie on the same special line; but, each special line is a diameter of the cycle and intersects it in just one point.) Our argument proves, moreover, that the line LK in Figure 127b is tangent at $X = Y$ to the cycle z, which is what we set out to prove.

We conclude this chapter with a discussion of some basic matters pertaining to inversive mappings. We know that a Euclidean inversion maps each point A of the plane other than the center Q of inversion to some point A', and that no point is the image of Q (for $QA = 0$, the equality

$$QA \cdot QA' = k \tag{31}$$

implies the meaningless "equality" $QA' = \infty$). Thus the domain of definition of an inversion with center Q is not the full Euclidean plane but rather *the plane "punctured" at the center Q of inversion*. This fact complicates arguments concerning inversive mappings. So far we have simply passed over these difficulties. Thus, for example, it is not quite correct to say that an inversion with center Q maps a line l not through Q onto a circle through Q, since Q is not the image of any point of l (cf. pp. 126–127, in particular, Fig. 114a). The correct statement is that *an inversion with center Q maps a line not through Q onto a circle punctured at Q* (Fig. 128a). Again, it is not quite correct to say that the image of a line through Q is the line iteself. The correct statement is that *the image of a line punctured at Q is the punctured line* (Fig. 128b); Q is not mapped to any point and is

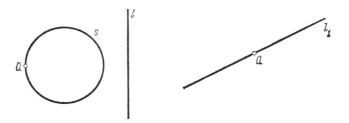

Figure 128a   Figure 128b

"nonexistent" from the point of view of an inversion with center Q. When a sequence of inversions $I_1, I_2, \ldots, I_n$ is applied (which is often the case when inversions are used to solve problems), it is necessary to puncture the plane at a number of points. For instance, if we apply three inversions I_1, I_2, and I_3, we must first remove their centers Q_1, Q_2, and Q_3. Then we must remove the point P_1 mapped onto Q_2 by I_1, the point P_2 mapped onto Q_3 by I_2, and finally the point N_1 mapped onto P_2 by I_1.

To avoid these difficulties it is best to proceed as follows. Rather than remove various points of the plane we supplement the plane by a single "point at infinity" Ω, and suppose that the inversion with center Q interchanges the points Q and Ω. This convention fits in with the equality (31) in the sense that if $QA = 0$ then $QA' = \infty$, and conversely. Now the domain of definition of our inversion is the "extended" plane (including the center of inversion Q and the point at infinity Ω), each point A has a definite image A' which, in turn, is mapped to A. An inversion of this so called (Euclidean) **inversive plane** maps circles (more precisely, ordinary circles and "circles of infinite radius," i.e., lines) to circles.[34] Since an inversion with center Q maps a line l not through Q to a circle s passing through Q and Q is the image of the point at infinity Ω, it is natural to suppose that every line "passes" through Ω. The image of l supplemented by the point Ω is a (complete) circle. Our convention also applies to lines through Q and surmounts the difficulties associated with such lines. Specifically, let l_1 be a line through Q extended by the addition of Ω. Our inversion maps l_1 onto itself, and interchanges Q and Ω.

To visualize the inversive plane we use a *stereographic projection*—i.e., a central projection of a plane π to a sphere σ from a center Q on σ. It is convenient to assume that σ is tangent to π at the point O antipodal to Q (Fig. 129). It is clear that the stereographic projection associates to each point A of π a unique point A' of σ. The inverse correspondence from σ to π is incomplete in the sense that it fails to associate any point of π to the point Q of σ. We note that if we move "to infinity" along a curve Γ in π, then the image A' of the moving point A on Γ tends to the center of projection Q. Thus, when we view our stereographic projection as a mapping of σ to π, it is natural to associate to Q the point at infinity Ω of π. In this way the stereographic projection becomes *a mapping of the inversive plane π onto the sphere σ* and σ becomes a natural model of the inversive plane.

A remarkable property of stereographic projection is that it *carries circles in the plane* (*including lines viewed as circles of infinite radius*) *to circles on the sphere* (i.e., to sections of the sphere by planes). Also, lines are carried to circles through Q, the image of the point at infinity Ω in π. The latter fact sheds additional light on our convention that all lines (circles of infinite radius) in π pass through Ω.

---

[34]Cf. [14], pp. 51–53.

10. Power of a point with respect to a circle or a cycle; inversion

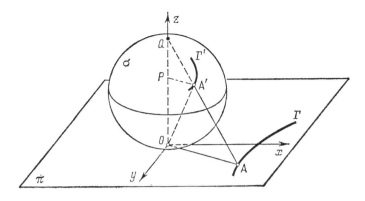

Figure 129

The simplest proof of these properties of stereographic projection uses coordinate geometry. Consider a coordinate system $Oxyz$ and let σ be the sphere given by the equation

$$x^2+y^2+\left(z-\frac{1}{2}\right)^2=\frac{1}{4} \quad \text{or} \quad x^2+y^2+z^2-z=0. \tag{40}$$

Then σ has radius $\frac{1}{2}$ and is tangent to the plane xOy (which we think of as the plane π) at the origin $O(0,0,0)$. We denote by $A'(X,Y,Z)$ the image of $A(x,y)$ under the stereographic projection of π to σ with center $Q(0,0,1)$ antipodal to O (see Fig. 129). Since the (right) triangles QOA and $QA'O$ with common angle Q are similar, the altitudes OA' and $A'P$ of these triangles are proportional to their hypotenuses QA and QO. Now

$$A'P=\sqrt{X^2+Y^2} \quad \text{and} \quad OA'=\sqrt{X^2+Y^2+Z^2}.$$

Also

$$QO=1 \quad \text{and} \quad QA=\sqrt{OA^2+QO^2}=\sqrt{x^2+y^2+1}.$$

Hence

$$\frac{\sqrt{X^2+Y^2+Z^2}}{\sqrt{X^2+Y^2}}=\frac{\sqrt{x^2+y^2+1}}{1}, \quad \text{or} \quad \frac{X^2+Y^2+Z^2}{X^2+Y^2}=x^2+y^2+1,$$

and therefore

$$\frac{Z^2}{X^2+Y^2}=x^2+y^2. \tag{41}$$

Since the projection of the ray PA' to π is the ray OA, we have $X=\lambda x$, $Y=\lambda y$. Substitution of these expressions in (41) yields the relation

$$Z=\sqrt{(X^2+Y^2)(x^2+y^2)}=\lambda(x^2+y^2).$$

The coordinates (X, Y, Z) of A' satisfy Eq. (40). Hence
$$\lambda^2 x^2 + \lambda^2 y^2 + \lambda^2 (x^2+y^2)^2 - \lambda(x^2+y^2) = 0, \quad \text{i.e.,} \quad \lambda = \frac{1}{x^2+y^2+1}.$$
In summary,
$$X = \frac{x}{x^2+y^2+1},$$
$$Y = \frac{y}{x^2+y^2+1}, \qquad (42)$$
$$Z = \frac{x^2+y^2}{x^2+y^2+1}.$$
But then, in particular,
$$1 - Z = \frac{1}{x^2+y^2+1}. \qquad (42a)$$
Equations (42) and (42a) imply that *the circle (or line) s in π* given by the equation
$$a(x^2+y^2) + 2b_1 x + 2b_2 y + c = 0 \qquad (33)$$
is mapped by the stereographic projection to the set s' of points $A'(X, Y, Z)$ of σ such that
$$aZ + 2b_1 X + 2b_2 Y + c(1-Z) = 0,$$
i.e., *to the circle of intersection of the sphere σ given by Eq. (40) with the plane*
$$az + 2b_1 x + 2b_2 y + c(1-z) = 0. \qquad (43)$$
If, in Eq. (33), $a = 0$, i.e., if our circle s is a line, then Eq. (43) reduces to
$$2b_1 x + 2b_2 y - cz + c = 0, \qquad (43a)$$
which shows that the corresponding circle s' contains the point $Q(0, 0, 1)$.

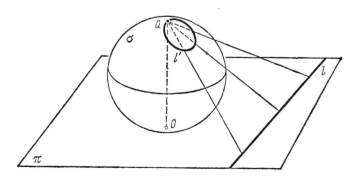

Figure 130

10. Power of a point with respect to a circle or a cycle; inversion

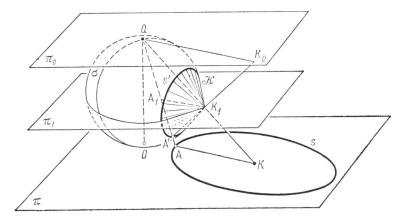

Figure 131a

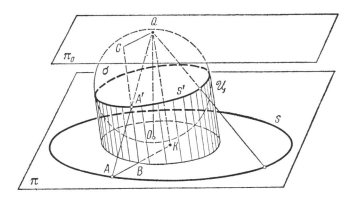

Figure 131b

The purely geometric counterpart of the above argument is somewhat more complicated. What is quite clear is that *if l is a line in the plane π, then its image on the sphere under stereographic projection is* the intersection of σ and the plane determined by l and the center of projection Q, i.e., *a circle l' on σ passing through Q* (Fig. 130). Conversely, the image of a circle l' on σ passing through Q, is the line l on π in which the plane of l' intersects π.

Now let s be a circle in π with center K, A an arbitrary point of s, and A' its image on σ (Fig. 131). Let K_1 denote the intersection point of the line QK and the plane τ tangent to σ at A' (Fig. 131a; we shall treat separately the case depicted in Fig. 131b when τ is parallel to QK). Let π_0 and π_1 be planes through Q and K_1 parallel to π. Finally, let K_0 and A_1 be the intersection points of these planes with the lines $A'K_1$ and QA. In view of the similarity of the triangles QKA and QK_1A_1, we have

$$\frac{K_1A_1}{QK_1} = \frac{KA}{QK}, \quad \text{i.e.,} \quad K_1A_1 = \frac{QK_1}{QK} \cdot KA = \frac{QK_1}{QK} r,$$

where r is the radius of s. On the other hand, the triangle $K_1A'A_1$ is similar to the isosceles triangle $K_0A'Q$ (K_0A' and K_0Q are tangents from K_0 to σ). This implies that the triangle $K_1A'A_1$ is also isosceles:

$$K_1A' = K_1A_1 = \frac{QK_1}{QK} r.$$

The last relation shows that the length of the tangent K_1A' from K_1 to σ is independent of the choice of A on s. But then K_1 must be the same for all A on s; indeed, if K_1 moves along the line QK, then the length of the tangent from K_1 to σ changes regardless of whether K_1 moves towards σ or away from σ. Thus the points A' on σ corresponding to the points of s are the points in which the lines through a fixed point K_1 touch σ. It is clear that the totality of such points is the *circle* of contact of σ and the cone $\mathcal{K}$ of tangents to σ from K_1. Conversely, suppose s' is any circle on σ except a great circle, and suppose Q does not lie on s'. Then s' is the circle of contact of σ and a cone $\mathcal{K}$ of tangents to σ from a point K_1. By reversing the above argument, we can show *that the image of s' under stereographic projection is a circle s in π*, whose center is the point K where the line QK_1 intersects π.

If the plane τ is parallel to the line QK then matters are even simpler. We denote by B and C the points where the planes π and π_0 (see above) intersect the line through A' parallel to QK. Since $A'B \parallel QK$, B is on the line AK (Fig. 131b). The similarity of the triangle BAA' and the isosceles triangle CQA' (CA' and CQ are tangents from C to σ) implies that $BA' = BA$, and the similarity of the triangles ABA' and AKQ implies that $KQ = KA$. Conversely, the equality $KQ = KA$ (which is obviously independent of the choice of A on s) implies that if BC is the line through A' parallel to KQ, then $CA' = CQ$ (Fig. 131b), i.e., CA' is tangent to σ. Hence, in this case (which is characterized by the equality $KQ = KA = r$), the set s' of points A' is the curve of contact of σ and the cylinder $\mathcal{X}$ with generators parallel to QK, i.e., *a circle*. Reversing our argument, we can show that *the image under stereographic projection of a great circle s' on σ not passing through Q is a circle s in π with center K such that QK is parallel to the generators of $\mathcal{X}$, and with radius $r = QK$*.

The property of a stereographic projection established above can be used to represent so-called *circular transformations* of the plane, i.e., transformations with a circle-preserving property analogous to that of inversions. Consider, for example, the reflection of a sphere σ in the equatorial plane δ, i.e., the transformation which associates to each point A_1 of σ the point A_1' of σ with $A_1A_1' \perp \delta$ (Fig. 132). It is clear that this transformation of σ maps circles to circles. But then the induced transformation of the plane π which associates to the point A corresponding to A_1

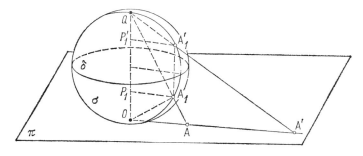

Figure 132

the point A' corresponding to A_1', is also circle-preserving. The nature of the induced transformation of π is clear: the point A' lies on the line OA of intersection of π with the plane $\varepsilon \perp \pi$ passing through QO and $A_1 A_1'$. If P_1 and P_1' are the projections of A_1 and A_1' onto QO, then the similarity of the triangles $QA_1 P_1$ and QAO and of the triangles $QA_1' P_1'$ and $QA'O$ implies that

$$OA = P_1 A_1 \cdot \frac{QO}{QP_1} = \frac{P_1 A_1}{QP_1}, \quad \text{and} \quad OA' = P_1' A_1' \cdot \frac{QO}{QP_1'} = \frac{P_1 A_1}{P_1 O}$$

(since $QO = 1$, $P_1' A_1' = P_1 A_1$ and $QP_1' = P_1 O$). On the other hand, $QA_1 O$ is a right triangle (since $\angle QA_1 O$ is subtended by the diameter OQ of σ). Hence $(A_1 P_1)^2 = QP_1 \cdot P_1 O$. But then

$$OA \cdot OA' = \frac{(P_1 A_1)^2}{QP_1 \cdot P_1 O} = 1.$$

In other words, *if A_1' is the image of A_1 under the reflection of the sphere σ in the plane δ, then A' (corresponding to A_1') is the image of A (corresponding to A_1) under the inversion of the plane π with center O and coefficient* 1.

Representation of an inversion of the plane π as a reflection of the sphere σ with π mapped to σ by means of a stereographic projection simplifies the study of the properties of the inversion. If, as is sometimes done, this representation is taken as the definition of an inversion, then the circle- and angle-preserving properties of an inversion can be deduced from the properties of stereographic projection. Stereographic projections are also very useful in the study of *circular transformations*, i.e., transformations of an (inversive) plane π that carry lines and circles to lines and circles. Using stereographic projection it is possible to prove the so-called fundamental theorem on circular transformations, which asserts that *every circular transformation of an* (inversive) *plane π is a similitude possibly followed by an inversion* (cf. Exercise **17** below).[35]

We now turn to Galilean geometry. Here matters are more involved than in the case of Euclidean geometry. An inversion of the first kind with center Q and coefficient k cannot be viewed as a transformation of the Galilean plane, since the inversion assigns no images to points at zero distance from Q, i.e., to all points on the special line q through Q. This complicates all considerations involving inversions of the first kind. For example, strictly speaking, such an inversion is a transformation of the Galilean plane with deleted line q which maps an (ordinary) line l not through Q with deleted point $L = q \cap l$ to a cycle z with deleted point $Q = q \cap z$ (Fig. 133a; cf. Fig. 119a). Similarly, our inversion maps a line m through Q with Q deleted to itself, and a cycle z not passing through Q with $N = q \cap z$ deleted to a cycle z' with $P = q \cap z'$ deleted (cf. Figs. 133b and 119b). In the case of a product of two or more inversions, the complications are horrendous.

---

[35]Cf. for example, Chap. 6 of [19] or Section 4 of [13].

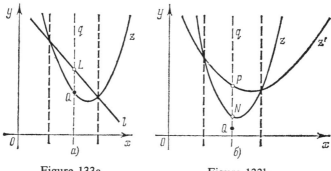

Figure 133a   Figure 133b

Given our experience with Euclidean inversions, it is not difficult to see how to surmount the present difficulties. We supplement the Galilean plane with "points at infinity," which we regard as the images of the points of q under our inversion of the first kind; the term "point at infinity" accords with the fact implied by the relation

$$QA \cdot QA' = k \tag{31}$$

that the smaller the distance from A to q the larger the distance from A' to Q. Since our inversion maps all cycles (and lines) through Q to lines, it is natural to say that all (ordinary) lines of the Galilean plane pass through the point Ω at infinity, which is the image of Q. Further, let N be the point of intersection of a cycle z with the line q, and let $QN = n$. If Q coincides with the origin O of the coordinate system xOy (Fig. 134), then our cycle is given by an equation

$$y = ax^2 + 2bx + c \tag{1}$$

with $c = n$ [since $N(0, n)$ is on z]. An inversion of the first kind with coefficient k maps z to a cycle z' given by the equation

$$y = a'x^2 + 2b'x + c', \tag{36}$$

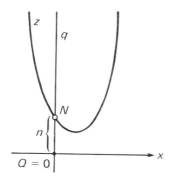

Figure 134

10. Power of a point with respect to a circle or a cycle; inversion   149

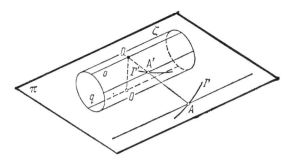

Figure 135

with

$$a' = \frac{c}{k} = \frac{n}{k}$$

[cf. Eqs. (36)–(36a)], i.e., to a cycle z' with curvature $\rho' = 2n/k$, and z' to z. It is therefore natural to say that *all cycles with curvature ρ pass through the same point at infinity Ω_ρ, and the inversion with coefficient k maps the point N on q with $\delta_{QN} = n$ to the point $\Omega_{2n/k}$*.

We see that it is convenient to supplement the Galilean plane with points at infinity, which are the images under an inversion of the first kind with center Q of the points on the special line q through Q. Since an inversion preserves special lines, it is natural to speak of the "*special line at infinity.*" Its points are denoted by Ω_ρ where ρ is any real number. All cycles with curvature ρ pass through the point Ω_ρ; in particular, "cycles with curvature 0," i.e., lines, pass through the point $\Omega_0 \equiv \Omega$. It is natural to call the Galilean plane supplemented with points at infinity Ω_ρ or, equivalently, with a special line at infinity, the **inversive Galilean plane**.[36] The inversive Galilean plane provides a convenient setting for the study of cycles. In this connection, we note that an inversion of the second kind, i.e., a reflection in a cycle Z with curvature P, maps a cycle with curvature ρ to a cycle with curvature $\rho' = 2P - \rho$ [cf. (38)]. Hence such an inversion maps Ω_ρ to $\Omega_{2P-\rho}$ (in particular, it maps $\Omega = \Omega_0$ to Ω_{2P} and Ω_{2P} to Ω), and so fixes each special line (including the special line at infinity).

To visualize an inversive Galilean plane, we map the Galilean plane π to a circular cylinder ζ by means of a *stereographic projection*. Specifically, we suppose ζ tangent to π along the special line q, choose O on q and Q on ζ

---

[36]It follows that the inversive Galilean plane, like the projective plane, (see the references listed in footnote 22 of Chap. 1), is obtained from the Galilean plane by the addition of a "line at infinity." (The two planes are nevertheless completely different. Thus, in contrast to the projective plane, all lines in the inversive Galilean plane pass through the same point.) On the other hand, the Euclidean inversive plane is obtained from the Euclidean plane by the addition of a single "point at infinity."

The concept of an infinite plane is a mathematical abstraction rather than a "physical" object. Depending on the problem, we may use any of the Euclidean, projective, inversive Euclidean or Galilean planes (all of which are different mathematical concepts), or indeed any other "plane" defined by suitable axioms.

diametrically opposite to O, and thus on the "upper generator" o of ζ (Fig. 135), and project ζ to π from Q. This projection establishes a one-to-one correspondence between the points of π and the points of ζ exclusive of the points of the generator o. As A' on ζ tends to some Ω' on o, the distance between q and the image A of A' increases beyond all bounds. This suggests that we regard the points at infinity of the Galilean plane as the images of the points of o. Since our projection maps generators of ζ to special lines of π, it is natural to say that it maps all of ζ to π supplemented by the special line at infinity, the image of the generator o of ζ. By now it is clear in what sense the cylinder ζ may be regarded as a good model of the inversive Galilean plane.

We claim that *the stereographic projection of the Galilean plane π to the cylinder ζ maps circles and cycles of π to plane sections of ζ*. It is easiest to prove this assertion analytically. We denote the coordinates of points of π by x and y, and the coordinates of points in space by x, y, and z. We take q as the y-axis, put the diameter of ζ equal to 1, and choose the point $Q(0,0,1)$ of ζ diametrically opposite to $O(0,0,0)$ as the center of the stereographic projection (Fig. 136). The equation of the circle in which the xz-plane intersects the cylinder ζ is

$$x^2 + \left(z - \tfrac{1}{2}\right)^2 = \tfrac{1}{4} \quad \text{or} \quad x^2 + z^2 - z = 0. \tag{40'}$$

The points of ζ are precisely the points (x,y,z) whose x and z coordinates satisfy (40'). Let $A(x,y)$ in π and $A'(X,Y,Z)$ on ζ denote points which correspond under our projection. Also, let $B'(X,0,Z)$ be the point of ζ in which the generator through A' meets the xOz plane, and let $B(x,0)$ in π correspond to B' under our projection (Fig. 136). In view of the similarity of the right triangles QOB and $QB'O$ with common angle Q and with altitudes OB' and $B'P$ dropped to their respective hypotenuses QB and

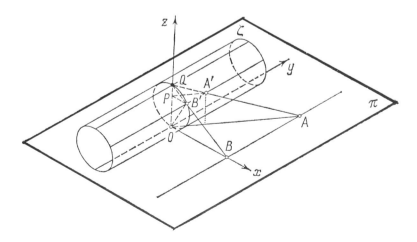

Figure 136

QO, we conclude that $OB'/PB' = QB/OQ$, or, since $OQ = 1$, $PB' = X$, $QB = \sqrt{OB^2 + OQ^2} = \sqrt{x^2 + 1}$, and $OB' = \sqrt{PB'^2 + OP^2} = \sqrt{X^2 + Z^2}$ (cf. p. 143 above), that

$$\frac{\sqrt{X^2 + Z^2}}{X} = \frac{\sqrt{x^2 + 1}}{1}, \quad \text{i.e.,} \quad \frac{X^2 + Z^2}{X^2} = x^2 + 1.$$

But then

$$\frac{Z}{X} = x \quad \text{and} \quad Z = xX. \tag{41'}$$

Since the coordinates (X, Y, Z) of A' satisfy Eq. (40'), we see readily that

$$X^2 + x^2 X^2 - xX = 0, \quad \text{and so} \quad X = \frac{x}{x^2 + 1}.$$

But then $Z = xX = x^2/(x^2 + 1)$. Finally, since A and A' lie in the same plane through Oz and therefore $X/Y = x/y$, we have $Y = (y/x)X = y/(x^2 + 1)$. In summary,

$$X = \frac{x}{x^2 + 1}, \quad Y = \frac{y}{x^2 + 1}, \quad Z = \frac{x^2}{x^2 + 1}, \tag{42'}$$

and in particular,

$$1 - Z = \frac{1}{x^2 + 1}. \tag{42'a}$$

It is now clear that the image of a circle or cycle

$$ax^2 + 2b_1 x + 2b_2 y + c = 0 \tag{2}$$

under the stereographic projection (42') is the section of the cylinder (40') by the plane

$$az + 2b_1 x + 2b_2 y + c(1 - z) = 0. \tag{43}$$

If in Eq. (2) $b_2 = 0$, i.e., if the curve (2) is a circle s (or a special line), then its image under the stereographic projection is the section of ζ by a plane (43) with $b_2 = 0$, i.e., by a plane λ parallel to the y-axis (parallel to the axis of ζ; see Fig. 137a). On the other hand, if the curve (2) is a cycle z (or an ordinary line; in other words, if $b_2 \neq 0$), then the stereographic projection maps this curve to the section of ζ by a plane μ not parallel to the axis of ζ (Fig. 137b).[37] The images of the cycles of fixed curvature ρ, i.e. the cycles (2) with $b_2 = -1$ and $a = \rho$, are the sections of ζ by the planes

$$\rho z + 2b_1 x - 2y + c(1 - z) = 0.$$

Since these planes intersect the line o given by the equations $x = 0$, $z = 1$ in the point $R(0, \rho/2, 1)$, it is natural to regard R as the image Ω'_ρ of the point at infinity Ω_ρ in the inversive Galilean plane π. In particular, $Q(0, 0, 1) = \Omega'_0$. This is a brief way of saying that the lines of the Galilean plane (cycles of curvature 0) are mapped to plane sections of ζ passing through $Q(0, 0, 1)$.

---

[37]The intersection of a cylinder and a plane not parallel to its axis is an ellipse (or a circle).

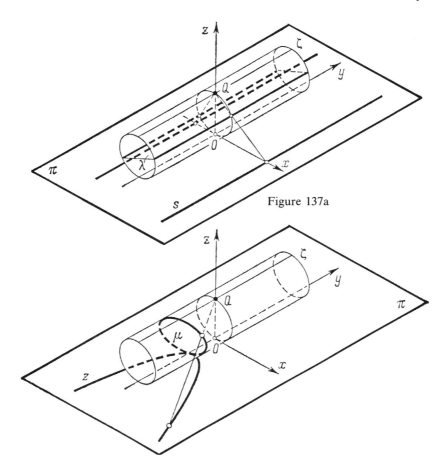

Figure 137a

Figure 137b

This property of stereographic projection from the plane π to the cylinder ζ makes this projection a very useful tool in the study of *cyclic transformations* of π, i.e., *transformations of the inversive Galilean plane which map cycles* (including lines that can be viewed as cycles of infinite radius or of zero curvature) *to cycles*. As an example we consider the transformation of π induced by the reflection of ζ in its (horizontal) equatorial plane δ, so that each point A_1 and its image A'_1 lie on a line perpendicular to δ (Fig. 138a). It is clear that this transformation of ζ maps plane sections to plane sections. But then the induced transformation of π, i.e., the transformation of π which associates to the stereographic image A of A_1 the stereographic image A' of A'_1, maps cycles to cycles. Since the orthogonal projections of the rays QA_1 and QA'_1 to π coincide with the ray OA, it follows that A' is on the ray OA. Figure 138b represents a section of the configuration in Figure 138a by means of the xOz plane (B_1 and B'_1 in Fig. 138b are the points in which the generators of ζ through A_1 and A'_1

10. Power of a point with respect to a circle or a cycle; inversion

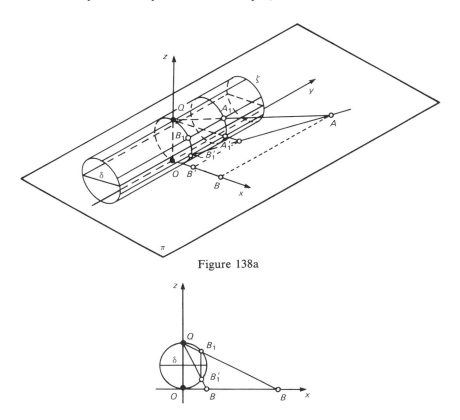

Figure 138a

Figure 138b

meet the xOz plane, and B and B' are their stereographic images in π). Comparison of Figure 138b with Figure 132 shows that

$$OB \cdot OB' = 1,$$

or, in view of the fact that $AB \| Oy$ and $A'B' \| Oy$ (bear in mind the definition of length of a segment in Galilean geometry!), that

$$OA \cdot OA' = 1 \qquad (\text{or } d_{OA} \cdot d_{OA'} = 1).$$

Hence the transformation of π which associates A to A' is an inversion of the first kind with center O and coefficient 1.

Next we consider the transformation of π induced by a reflection of the cylinder ζ in a plane ε parallel to the xz-plane. It is obvious that if A_1 and A'_1 are points of ζ symmetric with respect to the xz-plane, then the points A and A' in π, their images under the stereographic projection with center Q, are symmetric with respect to the x-axis (Fig. 139a). Thus a reflection of ζ in the plane xOz induces a reflection of π in the line Ox. In the more general case, when A_1 and A'_1 are points of ζ symmetric with respect to a plane ε parallel to the xz-plane and given by the equation $y = \rho$ (Fig. 139b), then Eqs. (43) and (2) (p. 151) show that the stereographic image in π of

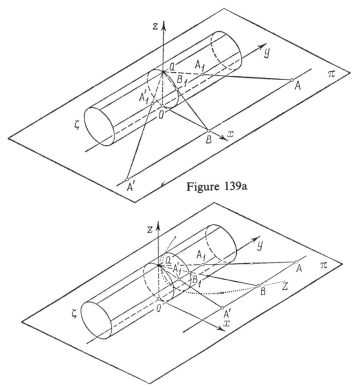

Figure 139a

Figure 139b

the section of ζ by the plane

$$y - \rho = 0$$

is the curve

$$\rho x^2 - y + \rho = 0 \quad \text{or} \quad y = \rho x^2 + \rho, \tag{44}$$

i.e., a cycle with curvature 2ρ. Further, if B_1 is the point of intersection of the line $A_1 A_1'$ with the plane ε, then the points A_1, B_1, A_1' of the generator $A_1 A_1'$ of ζ are projected from Q to collinear points A, B, A' of π such that $AA' \| Oy$, and (in view of the equality $A_1 B_1 = B_1 A_1'$) $AB = BA'$. In other words, the midpoint B of the segment AA' of a special line in π is its intersection point with the cycle (44). Thus the reflection of ζ in the plane ε induces an inversion of the second kind, i.e., a reflection in a cycle Z of π.

The connection between inversions of the first and second kind of the Galilean plane and reflections of a cylinder ζ in planes δ and ε is very helpful in the study of inversions. Using such representations of inversions and properties of the stereographic projection of the Galilean plane to a cylinder, it is possible to prove what might well be called the fundamental theorem of the theory of cyclic transformations: *Every cyclic transformation of the (inversive) Galilean plane is a similitude, possibly followed by an inversion of the first or second kind*. In this connection, we note that a similitude of the Galilean plane with first similarity coefficient k and

10. Power of a point with respect to a circle or a cycle; inversion   155

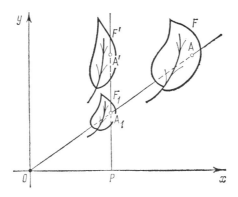

Figure 140

second similarity coefficient κ is defined as a transformation of the plane which carries points to points and lines to lines, and multiplies distances by k and angles by κ; in other words, if A and B are mapped to A' and B', and a and b are mapped to a' and b', then

$$\frac{A'B'}{AB} = \frac{d_{A'B'}}{d_{AB}} = k; \qquad \frac{\angle(a',b')}{\angle(a,b)} = \frac{\delta_{a'b'}}{\delta_{ab}} = \kappa;$$

here k and κ can be positive or negative. [An example of such a transformation is the product of a central dilatation with center O and coefficient k and a compression with axis x and coefficient κ (Fig. 140).]

We conclude with the observation that the principle of duality (cf. Sec. 5, Chap. I) enables us to obtain many new results dual to the results obtained in this chapter. We leave it to the interested reader to carry this out. However, we will at least suggest the duals of some of the concepts underlying our development. Thus the dual of the power of a point M with respect to a cycle Z is *the power of a line m with respect to a cycle Z*, i.e., the product

$$\angle aLm \cdot \angle bLm, \tag{24a}$$

where a and b are tangents from a point L on m to Z [Fig. 141a; the product (24a) is independent of the choice of L on m]; the dual of an inversion of the first kind is *an axial inversion of the first kind*, i.e., a transformation which associates to each

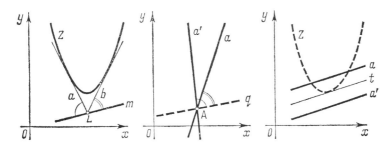

Figure 141a   Figure 141b   Figure 141c

line a of π the line a' passing through the point A, where a intersects the fixed "axis of inversion" q, and such that

$$\angle qAa \cdot \angle qAa' = k, \tag{31a}$$

where k is the fixed "coefficient of the axial inversion" (Fig. 141b); the dual of an inversion of the second kind is *an axial inversion of the second kind*, or *an axial reflection in a cycle Z*. It associates to each line a the line $a' \| a$ such that the tangent t to the cycle Z parallel to a and a' is equidistant from a and a' (Fig. 141c). The suggested theory is far simpler than the theory we are led to in Euclidean geometry when, say, we view a circle not as a set of points but rather as a set of lines.[38]

PROBLEMS AND EXERCISES

13   Let A, B be two points on a Euclidean circle (Galilean cycle) Σ. Draw all possible pairs of circles (cycles) σ_1, σ_2 tangent to Σ at A and B respectively, and tangent to each other at a point M. Describe the set of points M.

14   Let $\Sigma_1, \Sigma_2, \Sigma_3, \Sigma_4$ be four Euclidean circles (four Galilean cycles) such that $\Sigma_1 \cap \Sigma_2 = \{A_1, B_1\}$, $\Sigma_2 \cap \Sigma_3 = \{A_2, B_2\}$, $\Sigma_3 \cap \Sigma_4 = \{A_3, B_3\}$, $\Sigma_4 \cap \Sigma_1 = \{A_4, B_4\}$. Prove that if A_1, A_2, A_3, A_4 lie on a circle (cycle), then the same is true of B_1, B_2, B_3, B_4.

15   Give examples of the use of inversion to prove (Euclidean and Galilean) geometric theorems.

16   State the duals of Exercises **13–15** and give their direct proofs (i.e., proofs independent of the principle of duality). In the proofs, make use of axial inversions of the Galilean plane; cf. pp. 155–156.

17   Prove the fundamental theorem (**a**) on circular transformations of the Euclidean plane of (cf. p. 147); (**b**) on cyclic transformations of the Galilean plane (cf. p. 154).

XII (**a**) "Transplant" the theory of pencils and bundles of circles in the Euclidean plane (discussed in Chap. 6 of [19], Chap. 3 of [31], Chap. 3 of [33], Section 3 of [13], and in part A of [29]) to the Galilean plane. (**b**) Dualize the theory developed in (**a**).

XIII (**a**) Discuss "linear cyclic transformations" of the Galilean plane; such transformtions map lines to lines and cycles (including points, which may be viewed as cycles with zero radius) to cycles. This topic is hinted at in pp. 155–156. (**b**) Outline the theory of the "most general" cyclic transformations of the Galilean plane which map cycles, including lines (viewed as cycles of zero curvature) and points (viewed as cycles of zero radius) to cycles, and preserve contact of cycles (cf. Part C of [31] dealing with plane Euclidean geometry; [31] contains references to the literature).

XIV Discuss the concepts and results of this section in the setting of three-dimensional Galilean geometry (cf. Exercise **5** in the Introduction).

XV   Discuss the issue of Problem **XIV** in the setting of three-dimensional semi-Galilean geometry (cf. Exercise **5** in the Introduction).

[38]Cf., for example, Part B of [29]; Problems **VIII** and **X** in Chap. I; Chap. II, Section 5 of [13]; or Appendix 6.I in [32]. For the suggested constructions in Galilean geometry, see [35].

10. Power of a point with respect to a circle or a cycle; inversion

XVI Outline the theory of quadric curves in the Galilean plane (i.e., curves given by equations of the form $F(x,y)=0$, where $F(x,y)$ is a quadratic polynomial). List the various types of such curves and describe them geometrically.

XVII Outline the theory of quadric surfaces (**a**) in three-dimensional Galilean space; (**b**) in three-dimensional semi-Galilean space (cf. Exercise **5** in the Introduction).

Conclusion

11. Einstein's principle of relativity and Lorentz transformations

In Section 2 of the Introduction we formulated the Galilean principle of relativity as follows. *No mechanical experiments conducted within a physical system can disclose a uniform motion of this system* (cf. p. 18). This implies that *no experimentally observable property of the system is changed when the velocity of the system is changed by the addition of a uniform velocity*. When we deduced the formulas describing a Galilean transformation we used this principle and, implicitly, another fundamental condition which we propose to discuss in detail.

The Galilean principle of relativity put an end to the concept of "absolute" space, fixed by its very nature, relative to which it would be possible to determine the position and velocity of material bodies. Galileo declared an equivalence of all reference frames moving relative to each other with uniform velocity which rendered meaningless the concept of "absolute" rest. Isaac Newton wrote in his immortal "Principia" that "Absolute space, in its own nature, without relation to anything external, remains always similar and immovable." But he then went on to say, in what is virtually a quotation from Galileo, that "...because the parts of space cannot be seen, or distinguished from one another by our senses, therefore in their stead we use sensible measures of them," i.e., measurements carried out in the "relative," moving space. Newton then deals a final blow to the concept of absolute space: "For it may be that there is no body really at rest to which the places and motions of others may be referred."[1] Thus Newton's absolute space is a figment of the imagination; it cannot be apprehended by observation or experiment and is superfluous for a precise formulation of the laws of nature.

[1] See Newton [16].

Figure 142

Matters are different when it comes to time, the second basic concept connected with our perception of the universe. True, Newton deliberately links the two concepts: "Absolute, true, and mathematical time of itself and from its own nature flows equably without relation to anything external, and by another name is called duration..." (compare this formulation with his formulation of the concept of absolute space); but he adds almost immediately that "It may be that there is not such thing in nature as an equable motion, whereby time may be accurately measured."[2] Nevertheless, both Galileo and Newton always asumed the existence of absolute time not linked to any frame of reference (we are ignoring the freedom of choice of the initial moment for which $t=0$) and flowing the same way at all points of the universe. Following A. EINSTEIN (1879–1955), we shall now analyze the assumptions underlying the belief in the absolute nature of the concept of time.

Let A and B be two points on a line o which are at rest with respect to each other (Fig. 142). Relative to a given reference frame, A and B may either be at rest or in motion. However, this is of no consequence, for the very choice of a reference frame is arbitrary and cannot be motivated by physical considerations. Observers at A and B have clocks which enable them to measure time at these points. Our problem is to establish the absolute nature of time, independent of A or B after "absolute synchronization" of the clocks of our observers.

There are two ways of synchronizing the clocks at A and B. One is to synchronize the clocks at A, say, and to transfer one of them to B; this imitates the procedure followed by ships, which are equipped with chronometers whose movements are regulated before leaving port by comparison with a land-based "standard" clock. This method of synchronization of clocks is basically unsatisfactory, since it merely establishes the fact that the clocks were synchronous when they were both at A but not that they are synchronous when they are at A and B, respectively (after all, it is conceivable that mechanical displacement affects the movement of a clock and that while the clock is transferred from A to B it slows down or speeds up). The other way is to set a clock at B by means of signals from A (just as the time indicated by a clock on a ship is checked against "standard time" readings radioed from land). But it takes time for a signal from A to reach B. Accurate determination of that time requires knowledge of the distance from A to B and of the speed of propagation of the signal. For accurate determination of the speed of propagation of the signal we must have synchronized clocks which give the time at different points. Thus our

---

[2]See Newton [16].

second method of synchronization of clocks also leads to logical circularity and is just as unsatisfactory as the first.

We could, of course, agree to transmit a signal from the midpoint S of the segment AB and set the clocks when the signal reached A and B. However, this procedure assumes that A and B are at rest. If both points moved in the direction indicated, say, by the arrow in Figure 142, then the signal from the "midpoint" S would reach A before reaching B. Now our earlier discussion stressed the fact that the assertions "A and B are at rest" or "A and B move uniformly" have no physical meaning, i.e., the validity of one or another of these statements cannot be checked by means of a physical experiment. This being the case, the problem of whether the clocks at A and B are synchronous or not remains unresolved.

A related difficulty is involved in the determination of the order of events which take place at different points A and B of the line o. Earlier we associated to each point of o a "time axis t." We represented these axes as parallel lines (cf. Fig. 143). We assumed that we could associate to a moment $t_0^{(A)}$ at A a number $t_0^{(B)}$ on the time axis $t^{(B)}$. It is then natural to regard moments at B preceding $t_0^{(B)}$ as preceding $t_0^{(A)}$ and, similarly, moments at B following $t_0^{(B)}$ as following $t_0^{(A)}$. Now we might ask, how well founded is the assumption implicit in the above procedure about the possibility of "absolute ordering" in time of two events which take place at different points of o.

Let us consider the issue in some detail. If a signal (or letter or telegram) is sent from B to A at time $t_1^{(B)}$ and arrives at A at $t_0^{(A)}$, then any event at B preceding $t_1^{(B)}$ must be thought of as preceding $t_0^{(A)}$ (otherwise an observer at A might receive information about an event at B before its occurrence, and this would violate the principle of causality which asserts that a consequence of an event cannot precede the event). Similarly, if a signal is sent from A to B at $t_0^{(A)}$ and arrives at B at $t_2^{(B)}$, then an event at B following $t_2^{(B)}$ must be thought of as following $t_0^{(A)}$. This leaves undetermined the time order of events at B between $t_1^{(B)}$ and $t_2^{(B)}$ relative to the

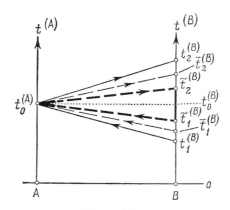

Figure 143

event which occurs at A at $t_0^{(A)}$. If we had at our disposal a faster signal, then we could shorten the interval of indeterminacy $(t_1^{(B)}, t_2^{(B)})$ at B associated with the moment $t_0^{(A)}$ at A (the broken lines $\bar{t}_1^{(B)} t_0^{(A)}$ and $t_0^{(A)} \bar{t}_2^{(B)}$ in Fig. 143 correspond to the faster signal). Given signals or arbitrary speed, we could order in time any two events occurring at different points A and B of o and reduce the interval of indeterminacy to the single moment $t_0^{(B)}$. In this way, we arrive at the concept of simultaneity of events implicit in the preceding part of the book (see, in particular, Sec. 2 of the Introduction). On the other hand, if there is a signal of maximal speed, then the corresponding interval of indeterminacy $(\tilde{t}_1^{(B)}, \tilde{t}_2^{(B)})$ (see Fig. 143, where the broken heavy lines $\tilde{t}_1^{(B)} t_0^{(A)}$ and $t_0^{(A)} \tilde{t}_2^{(B)}$ correspond to the signal of maximal speed) cannot be shortened, i.e., Newton's absolute (place-independent) time is just as much a figment of the imagination as his absolute space.

A ray of light is the fastest known signal. But is it also the fastest possible signal? This question was answered by a physical experiment.

It is clear that in the mechanics of Galileo and Newton there can be no signals of maximal speed. In fact, the classical law of composition of velocities (cf., in particular, Sec. 4, pp. 48–49) shows that if a signal moves from A along o with speed c and a signal receiver B moves along o towards the signal with speed v (cf. Fig. 144), then the speed of the signal relative to B is $c + v$, which is greater than c. In particular, the speed of light would vary depending on whether the light receiver moved towards or away from the signal or was at rest.

In view of these considerations, the results of refined experiments carried out in 1881 by the eminent American physicist A. A. MICHELSON (1852–1931), [and repeated in even more conclusive form in 1887 by Michelson and his colleague, the chemist E. W. MORLEY (1838–1923)] were truly astonishing. These experiments showed that *the speed of light is the same in all inertial systems* and is independent of the direction of the motion of the (moving) reference frame relative to the ray of light.[3] The Michelson–Morley experiment undermined the view of the universe based on the principles underlying the mechanics of Galileo and Newton.

In 1905, Einstein explained the results of the Michelson–Morley experiment by postulating a new foundation for mechanics. Specifically, Einstein accepted the Galilean principle of relativity in the form in which we restated it in the beginning of this section, but completely rejected the existence of absolute (i.e., space-independent) time and replaced it with the

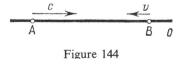

Figure 144

---

[3] A description of the Michelson–Morley experiment is found in virtually every book dealing with the special theory of relativity (cf. the Bibliography).

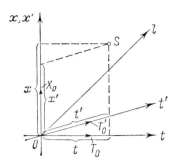

Figure 145a

assumption of constancy of the speed of light (i.e., its independence of the selected inertial reference frame). Thus Einstein's principle of relativity consists of two postulates, one of which asserts *the equivalence of all inertial reference frames* (a postulate first advanced by Galileo), while the other asserts *the constancy of the speed of light in all inertial reference frames*.[4] The latter postulate is inconsistent with the mechanics of Galileo and Newton based on the Galilean transformations (13), p. 23. The transformations associated with the mechanics of Einstein are the so-called **Lorentz transformations**, first obtained (in another connection) by the eminent Dutch physicist H. A. LORENTZ (1853–1928).

We shall now derive the Lorentz transformations. We shall retain the geometric setting used in deriving the Galilean transformations of classical mechanics, and thereby introducing Galilean geometry. Specifically, we choose a reference frame on the line o and associate to each (space–time) event of one-dimensional kinematics (kinematics on the line o) the point of the plane with coordinates x (the abscissa of the appropriate point on o) and t (time); as before, the axes Ot and Ox are assumed to be horizontal and vertical, respectively (Fig. 145). We choose units OX_0 and OT_0 of length and time so that the speed of light c is 1. This means that if the unit of time is one second, then the unit of length is about 300,000 kilometers (since the approximate speed of light is 300,000 km/sec). The line l in Figure 145 whose equation is

$$x = t \quad \text{or} \quad x - t = 0$$

represents the trajectory (or "world line") of a ray of light propagating with speed $c = 1$ in the direction of increasing x from the event $O(0,0)$ corresponding to the initial moment $t = 0$ and the origin $x = 0$ on o.

In addition to the inertial coordinate system $\{x,t\}$, we consider another inertial coordinate system $\{x',t'\}$ whose origin coincides with the old origin at $t = 0$ but moves along o with velocity v. Thus the origin $x' = 0$ of the new coordinate system is given in the old coordinate system by the line Ot', whose equation is

$$x = vt.$$

---

[4]Similarly, Galileo's principle of relativity consists of two postulates, of which one asserts *the equivalence of all inertial reference frames* and the other asserts *the absolute nature of time*.

11. Einstein's principle of relativity and Lorentz transformations

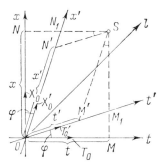

Figure 145b

In the mechanics of Galileo and Newton, the transition from the reference frame $\{x,t\}$ to $\{x',t'\}$ is characterized by the transition from the axes Ot, Ox to $Ot', Ox' = Ox$ (Fig. 145a). The transition from the coordinates (x,t) to (x',t') is given by the Galilean formulas[5]

$$x' = x - vt,$$
$$t' = t \qquad (1)$$

[here we ignore the constants a and b in Eqs. (13) of Sec. 2, which reflect the possibility of changing the initial point $x=0$ on o and the initial moment $t=0$]. However, in the new reference frame the speed of light (the Galilean angle between the line l in Fig. 145a and the t'-axis) is no longer 1 but $1-v$, and this contradicts the outcome of the Michelson–Morley experiment.

If our transformations are not to contradict the postulate of the constancy of the speed of light, then we must eliminate the concept of absolute time, which is expressed geometrically by the invariance of the x-axis (the axis $t=0$). In other words, a Lorentz transformation must involve a transition from a coordinate system $\{x,t\}$ to a coordinate system $\{x',t'\}$ where both new axes Ot' and Ox' have directions differing from those of the old axes Ot and Ox (Fig. 145b). The invariance of the speed of light means that if the unit segments OX'_0 and OT'_0 on the axes Ox' and Ot' have, as before, the same (Euclidean) length, then the world line of a ray of light is given by the same line l. This implies that the axes Ot' and Ox' must form equal (Euclidean) angles with the axes Ot and Ox, i.e., $\angle tOt' = \angle x'Ox = \varphi$. The latter conclusion follows from the equivalent requirements: the line l must bisect the angle $t'Ox'$; the equation of l must be of the form

$$x' = t' \quad \text{or} \quad x' - t' = 0;$$

---

[5] Here v denotes the velocity of O' relative to O. On p. 23, v denotes the velocity of O relative to O'. (Translator's note.)

the speed of light in the $\{x',t'\}$ reference frame must have the value 1.

We shall now derive the connection between the old coordinates (x,t) of an event S and its new coordinates (x',t') (Fig. 145b). The coordinate x is equal to the length $d(MS)$ of the segment MS measured in terms of the unit of length OX_0 on Ox. The coordinate x' is equal to the new length $d'(M'S)$ of the segment $M'S$ measured in terms of the unit of length OX_0' on Ox'. If we recall that the equation of the line Ot' in the coordinate system $\{x,t\}$ is $x=vt$, then it is clear that $d(M_1S)$ (the length of M_1S measured in terms of the unit OX_0) is $x-vt$. Further, the ratio $d'(M'S)/d(M_1S)$ depends on the angles $\angle M'SM_1 = \angle x'Ox$ ($=\varphi$) and $\angle M_1M'S = \angle t'Ox'$,[6] and on the units OX_0 and OX_0' on Ox and Ox' but is independent of the event S. Hence if we denote by k the quotient $d'(M'S)/d(M_1S)$ (i.e., $x'/x-vt$), then we have[7]

$$x' = k(x - vt). \tag{2a}$$

The connection between the new time coordinate t' of an event S and its old coordinates (x,t) is found in much the same way: Figure 145b shows that

$$t = d(NS), \quad t' = d'(N'S),$$

where the lengths $d(NS)$ and $d'(N'S)$ are measured in terms of the units OT_0 ($=OX_0$) and OT_0' ($=OX_0'$). The similarity of the triangles ONN_1 and OMM_1 (these are right triangles in Fig. 145b) implies the equality

$$\frac{d(NN_1)}{d(ON)} = \frac{d(MM_1)}{d(OM)},$$

where all lengths are measured in terms of the same unit $OX_0 = OT_0$. This equality can be written as

$$\frac{d(NN_1)}{x} = \frac{vt}{t}.$$

Hence

$$d(NN_1) = \frac{vt}{t} x = vx,$$

and therefore

$$d(N_1S) = d(NS) - d(NN_1) = t - vx.$$

In view of the similarity of the triangles $SN'N_1$ and $SM'M_1$ (and the equalities $OT_0' = OX_0'$ and $OT_0 = OX_0$) we have

$$\frac{d'(N'S)}{d(N_1S)} = \frac{d'(M'S)}{d(M_1S)} = k, \quad \text{or} \quad \frac{t'}{t-vx} = k.$$

---

[6] Of course, in Figure 145b, we have $\angle t'Ox' = \pi/2 - 2\varphi$, so it is redundant to say that $d'(M'S)/d(M_1S)$ depends on both $\angle x'Ox$ and $\angle t'Ox'$. Presumably, the author is anticipating the general case, where Ox is no longer perpendicular to Ot. (Translator's note.)

[7] We note that Eq. (2a) differs from the first of the formulas (1) which give the classical Galilean transformation by the factor k. (In general, this factor depends on the velocity v of the new reference frame with respect to the old. In turn v determines the "angle" φ between Ot and Ot'.)

This yields the second equation of the Lorentz transformation:
$$t' = k(t - vx). \tag{2b}$$

Before going further, we note that Eqs. (2a) and (2b) imply a surprising fact. Consider a rod of length δ at rest in the reference frame $\{x, t\}$. Suppose one end of the rod is at the point $x = 0$ of the line o and its other end at the point $x = \delta$ of o. In Figure 146, the world lines of these two points are represented by the lines Ot and e, where $e \| Ot$ and e intersects Ox at $E(\delta, 0)$. Let us now measure the length of our rod in the reference frame $\{x', t'\}$ relative to which it moves with speed v (with velocity $-v$ if we take into consideration the direction of motion). To determine the new length of the rod we must determine the x'-coordinates of both its ends at the same moment t', say, at $t' = 0$. This amounts to finding the x'-coordinates of the points O and E' in Fig. 146. The coordinates (x', t') of O are $(0, 0)$. To determine the coordinates of the event E', we put $t' = 0$ in (2b) and obtain $t = vx$. Further, for E' we have $x = \delta$ ($x = \delta$ is the equation of the line e), and therefore $t = v\delta$. Substituting these values for x and t in (2a), we obtain
$$x' = k(x - vt) = k(\delta - v^2\delta) = k(1 - v^2)\delta.$$
Thus at $t' = 0$ the space coordinate of our rod is $x' = k(1 - v^2)\delta$. Therefore, its length δ' in the reference frame $\{x', t'\}$ is
$$\delta' = k(1 - v^2)\delta. \tag{3}$$

The above argument shows that *the length of a rod depends on the reference frame in which it is measured.*

Suppose that $\{x', t'\}$ is the reference frame of a physicist's laboratory, while $\{x, t\}$ is a reference frame fixed in the rod R. Then R is at rest with respect to $\{x, t\}$, but is moving with speed v with respect to $\{x', t'\}$. We have just seen that if R has length δ with respect to $\{x, t\}$, then it has length $k(1 - v^2)\delta$ with respect to $\{x', t'\}$. In other words, if δ is the length of a rod R when at rest, and if R moves with speed v relative to an observer in a laboratory, R will appear to him to have length $k(1 - v^2)\delta$. Here k is a

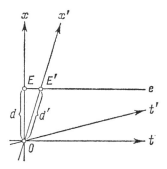

Figure 146

constant which depends solely on v. We will determine the precise value of k in a moment, and in particular we will show that $k(1-v^2)<1$. This means that a moving rod appears to be shorter than the same rod at rest. The possibility of contraction of moving objects was first envisaged by the Irish physicist G. F. FITZGERALD (1851–1901), who thought of it as a formal way of reconciling the results of the Michelson–Morley experiment with Galilean mechanics. Accordingly, the change by the factor $k(1-v^2)$ of the length of an object moving with speed v is known as the *Fitzgerald contraction*.

We now invert Eqs. (2a) and (2b), i.e., we express the old coordinates (x,t) of an event in terms of its new coordinates (x',t'). Leaving (2a) essentially unchanged and multiplying both sides of (2b) by v, we obtain the equations

$$x' = kx - kvt, \qquad vt' = kvt - kv^2 x.$$

Addition of these equations yields

$$x' + vt' = k(1-v^2)x.$$

Hence

$$x = \frac{1}{k(1-v^2)}(x' + vt'). \tag{2'a}$$

Multiplying both sides of (2a) by v and leaving (2b) essentially unchanged, we obtain the equations

$$vx' = kvx - kv^2 t, \qquad t' = kt - kvx.$$

Addition of these equations yields

$$vx' + t' = k(1-v^2)t.$$

Hence

$$t = \frac{1}{k(1-v^2)}(t' + vx'). \tag{2'b}$$

We compare Equations (2a) and (2b) with their inverses, (2'a) and (2'b). If we write Eqs. (2') in the form

$$x = k'(x' - v't'), \qquad t = k'(t' - v'x'),$$

then v', the velocity of $\{x,t\}$ relative to $\{x',t'\}$, is the negative of v, the velocity of $\{x',t'\}$ relative to $\{x,t\}$. On the other hand,

$$k' = \frac{1}{k(1-v^2)}. \tag{4}$$

The length of a rod R at rest with respect to $\{x',t'\}$ is multiplied by

$$k'(1-v^2) \tag{3'}$$

with respect to $\{x,t\}$ (relative to which R moves with speed v). But we saw earlier that when a rod moves with speed v, its "rest length" is multiplied by $k(1-v^2)$. Hence $k(1-v^2) = k'(1-v^2)$, so that $k=k'$. Substituting this

11. Einstein's principle of relativity and Lorentz transformations

into (4), we obtain

$$k^2 = \frac{1}{1-v^2}, \quad k = \frac{1}{\sqrt{1-v^2}}. \tag{5}$$

In particular, the Fitzgerald contraction coefficient in (3) is less than one, since

$$k(1-v^2) = \frac{1}{k} = \sqrt{1-v^2}.$$

Thus *the length of a moving rod is actually less than its rest length*:

$$\delta' = \sqrt{1-v^2} \cdot \delta. \tag{6a}$$

Substitution of the value (5) of k in (2) yields the Lorentz transformation formulas

$$\boxed{\begin{aligned} x' &= \frac{x-vt}{\sqrt{1-v^2}}, \\ t' &= \frac{-vx+t}{\sqrt{1-v^2}}, \end{aligned}} \tag{7}$$

In 1905, Einstein adopted Eqs. (7) as the foundation of mechanics. The change from the Galilean transformation (1) to the Lorentz transformation (7) implied many baffling physical phenomena which were a serious obstacle to the general acceptance of the theory of relativity. One such phenomenon was the Fitzgerald contraction (6a) of moving objects. Another was the stretching (or contraction) of time as a result of motion. The latter hypothesis, which we are about to discuss, was due to Lorentz, who thought of it as a complement to the Fitzgerald contraction hypothesis. Both hypotheses antedated the theory of relativity and thoroughly confused physicists and philosophers. Coming before Einstein's rejection of the concept of absolute time, the formal constructs of Fitzgerald and Lorentz were entirely baffling.

To clarify the effect of stretching (or contraction) of time we consider two observers H and H' whose frames of reference are $\{x,t\}$ and $\{x',t'\}$, respectively. Each observer has a clock indicating time in his own frame of reference. Let $(0,\tau)$ be a time interval of length τ in $\{x,t\}$. The time $t=0$ is given by the world line Ox (Fig. 147a) and the time $t=\tau$ by the world line $f\|Ox$ which intersects Ot at $F(0,\tau)$. Suppose that at $t'=0$ the observer H' is at the point $x=0$ of o. This means that the space–time position of H' is given by the point O in Figure 147a. In $\{x',t'\}$, the time $t'=0$ is represented by the world line Ox' whose points correspond to all events that are simultaneous with O from the point of view of H'. Similarly, from the point of view of H', the events simultaneous with F are represented by the points of the line f' passing through F and parallel to Ox'. The coordinates of F in $\{x,t\}$ are $(0,\tau)$ and its coordinates in $\{x',t'\}$ are

$$x' = \frac{-v\tau}{\sqrt{1-v^2}} \quad \text{and} \quad t' = \frac{\tau}{\sqrt{1-v^2}}$$

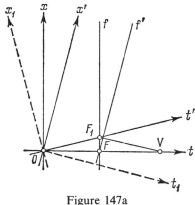

Figure 147a

[cf. (7) above]. Thus from the point of view of the observer H', F occurs at time $\tau/\sqrt{1-v^2} > \tau$, i.e., later than from the point of view of the observer H. This means that *the time interval separating the events O and F, which has length τ for H, has the greater length*

$$\tau' = \frac{1}{\sqrt{1-v^2}}\tau, \tag{6b}$$

for the observer H' moving with velocity v relative to H.

The seemingly paradoxical relation (6b) can be deduced in another way. Consider a thin rod AB of length τ moving with velocity v in the direction of the line o perpendicular to AB (cf. Fig. 147b representing the plane xOy of the motion of the rod; our figure does not include a time axis). For an observer H at rest relative to the rod, a light signal will traverse the rod from A to B in time τ (the speed of light is taken as 1). From the point of view of an observer H' at rest with respect to the reference frame relative to which the rod is in motion, the path of the light signal from A to B (or rather to B', the position of B when it is reached by the light signal from A) is

$$AB' = \sqrt{(AB)^2 + (BB')^2} = \sqrt{\tau^2 + (v\tau')^2} \; ;$$

here $BB' = vt'$ is the path traversed by the point B moving with velocity v during the time τ' which elapses, in view of the observer H', between the moment the light

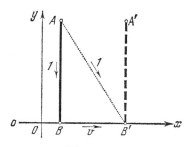

Figure 147b

11. Einstein's principle of relativity and Lorentz transformations

signal leaves A and the moment it reaches B. [It should be noted that since the rod does not move in the direction AB, its length is the same for both observers; cf. (10), p. 173.] Since the speed of light has the same value 1 for the observers H and H', the time τ' (as measured by the clock of H') taken by the ray of light to traverse the path

$$AB' = \sqrt{\tau^2 + v^2\tau'^2}$$

is numerically equal to that path. Thus

$$\tau' = \sqrt{\tau^2 + v^2\tau'^2}, \quad \text{i.e.,} \quad \tau'^2 = \tau^2 + v^2\tau'^2$$

or

$$\tau' = \frac{1}{\sqrt{1-v^2}} \tau. \tag{6b}$$

The above argument is particularly surprising when presented in the form of the so-called "twin paradox" which even today, after a detailed explanation given by Einstein more than half a century ago, is sometimes subject to false interpretations. Thus suppose that our observers H and H' are twins and that H stays at the point $x=0$ of o associated with the origin of the reference frame $\{x,t\}$, while H' moves with respect to H with "cosmic" velocity v. Let Ot' in Figure 147a be the world line of H', i.e., the line $x=vt$. At time τ by the clock of H, i.e., beginning with the point $F_1 = Ot' \cap f$, H' changes the direction of his motion and begins to approach the (fixed) observer H with velocity v. Equivalently, H' now moves relative to H with velocity $-v$. Now the world line of H' is F_1V, which is parallel to Ot_1, the mirror image of Ot' with respect to Ot. The twins meet at the point $x=0$ of o. The corresponding space–time moment is $V = Ot \cap F_1V$ in Figure 147a.

It is clear that by the clock of H the time elapsed between the events O and V is

$$\tau + \tau = 2\tau \text{ units of time.}$$

If we go by the clock of H', the corresponding time interval is different. Specifically, the time coordinate of F_1 in $\{x,t\}$ is $t=\tau$ and, since it lies on Ot', its space coordinate is

$$x = v\tau.$$

But then the second equation in (7) implies that the time coordinate of F_1 in $\{x',t'\}$ is

$$\tau' = \frac{-v^2\tau + \tau}{\sqrt{1-v^2}} = \frac{\tau(1-v^2)}{\sqrt{1-v^2}} = \tau\sqrt{1-v^2},$$

i.e., the time coordinate of F_1 from the point of view of H' is $\tau\sqrt{1-v^2}$. The time of passage from the world point F_1 to the world point V must "fit in" with the frame of reference $\{x_1,t_1\}$ (Fig. 147a) relative to which H' is now at rest. Considerations of symmetry show that in passing from the world point F_1 to the world point V the observer H' ages by

$$\tau\sqrt{1-v^2} \text{ units of time}$$

(check by a direct computation). Thus the trip of H' from the world point O to the world point V takes

$$\tau\sqrt{1-v^2} + \tau\sqrt{1-v^2} = 2\tau\sqrt{1-v^2} \text{ units of time.}$$

Since $2\tau\sqrt{1-v^2} < 2\tau$, it follows that after his return H' is younger than his twin brother.

It is clear that actual experimentation along these lines would require spaceships moving with speeds close to that of light and physiques capable of sustaining the stresses caused by the tremendous accelerations involved in launching, stopping, and almost instantaneously reversing direction at the world point F_1. However, a far more important point is that the above result cannot in principle be used to lengthen the human lifespan. This is due to the fact that all physical and physiological processes in a "moving"[8] medium transpire in accordance with the "indigenous time" of the medium; the trip takes precisely $2\tau' = 2\tau\sqrt{1-v^2}$ time units of the life of H'. One faulty interpretation of our result invokes the equivalence of all inertial systems. Careless opponents of the theory of relativity tend to argue that "H also moves with respect to H', and when they meet it is equally correct to argue that it is H who is younger than H'." This argument overlooks the fact that the frame of reference of H' is *not* an inertial reference frame, as witness his instant change of direction at the world point F_1 (see the above discussion of the resulting acceleration and the associated inertial stress). It is thus simply incorrect to speak of the equivalence of the reference frame $\{x,t\}$ of H and the "reference frame" $\{x',t'\}-\{x_1,t_1\}$ of H'.[8a]

---

[8]The quotation marks are intended to emphasize the relative nature of the term which is meaningless without involving a reference frame; here we use the term to reflect the view of the twin H.

[8a]It seems difficult to find two people who agree as to the correct explanation of the twin paradox. So it is perhaps not surprising that the editor finds himself uneasy over the remark about physiological process in a "moving" medium transpiring in accordance with the "indigenous time" of the medium. With considerable trepidation, he offers the following alternative explanation of the paradox. It at least has the merit of remaining within the framework of the special theory of relativity, which deals only with unaccelerated motions.

Let us imagine *three* observers H_1, H_2, and H_3. For convenience we will carry out our calculations in a reference frame in which H_1 is stationary and situated at the origin. Suppose that in this frame H_2 moves according to the equation $x = vt$ ($v > 0$), while H_3 moves according to the equation $x = -v(t - 2T)$, where T is a constant. Then at time $t = 0$, H_2 is at the origin, while H_3 is at a point $2vT$ units to the right of the origin. Moreover H_2 moves to the right with speed v, while H_3 moves to the left with speed v. Suppose that as H_2 moves past H_1 at time $t = 0$, they synchronize their watches. From the viewpoint of H_1, observers H_2 and H_3 meet at time $t = T$. (This is found by setting $vt = -v(t - 2T)$ and solving for t.) At this moment the reading on H_2's watch is $T\sqrt{1-v^2}$. Now suppose that H_3 synchronizes his watch with H_2's as they pass one another; thus H_3 sets his watch at a reading of $T\sqrt{1-v^2}$. Then H_3 continues moving to the left, and after the elapse of an additional time of T units in H_1's frame, he passes H_1. The reading on his watch at this moment is $T\sqrt{1-v^2} + T\sqrt{1-v^2} = 2T\sqrt{1-v^2}$.

This formulation does not really seem very paradoxical. For one thing the observer who "comes back" is not the same as the observer who "went away." Also, there is no problem with symmetry since we are dealing with three observers, not two. (Editor's note.)

11. Einstein's principle of relativity and Lorentz transformations

We shall now discuss the Einsteinian law of composition of velocities. First, however, we remind the reader of the corresponding relation in Newtonian mechanics. Let $\{x,t\}$ be a fixed reference frame and let $\{x',t'\}$ move relative to $\{x,t\}$ with velocity u. Let A be an object moving with velocity v in $\{x',t'\}$. We wish to determine the velocity w of A relative to $\{x,t\}$. The quantities v,u,w are, respectively, the "relative velocity," "transport velocity," and "absolute velocity," of A. In the mechanics of Galileo and Newton these velocities are connected by the relation (cf. pp. 48–49)

$$w = u + v. \tag{8}$$

To determine the corresponding law in the theory of relativity we argue as follows. Let a (Fig. 148) be the world line of A. Let O and S be the positions of A whose coordinates in $\{x',t'\}$ are $(0,0)$ and (δ',τ'). Since the velocity of A in $\{x',t'\}$ is v, we have $\delta'/\tau' = v$, and thus $\delta' = v\tau'$. The coordinates of the point (event) O in $\{x,t\}$ are also $(0,0)$ and the coordinates of S in $\{x,t\}$ can be determined from (7) by setting $x' = \delta'$, $t' = \tau'$, $v = u$, and solving for x,y. This yields

$$x = \frac{\delta' + u\tau'}{\sqrt{1-u^2}} \quad \text{and} \quad t = \frac{u\delta' + \tau'}{\sqrt{1-u^2}}.$$

These relations imply that in time $\tau = (u\delta' + \tau')/\sqrt{1-u^2}$, A traverses the distance $\delta = (\delta' + u\tau')/\sqrt{1-u^2}$. It follows that the velocity of A in $\{x,t\}$ is

$$w = \frac{\delta}{\tau} = \frac{(\delta' + u\tau')/\sqrt{1-u^2}}{(u\delta' + \tau')/\sqrt{1-u^2}} = \frac{\delta' + u\tau'}{u\delta' + \tau'} = \frac{v\tau' + u\tau'}{uv\tau' + \tau'} = \frac{u+v}{uv+1}.$$

Thus the relativistic counterpart of Eq. (8) is the equation

$$w = \frac{u+v}{uv+1}. \tag{9}$$

In particular, if A is a ray of light for which $v = 1$, then

$$w = \frac{u+1}{u \cdot 1 + 1} = 1.$$

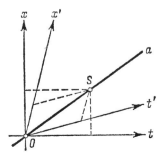

Figure 148

In other words, *the speed of light is invariant in all inertial reference frames.* We note that the presence of the term $\sqrt{1-v^2}$ in the Eqs. (7) of a Lorentz transformation shows that there can be no signals whose speed of propagation exceeds the speed of light (i.e., signals for which $v > 1$).

We conclude this section with a further exploration of the topic of simultaneity and of the concepts of "before" and "after" in the theory of relativity. Since the velocity of an inertial frame $\{x',t'\}$ relative to an inertial frame $\{x,t\}$ is subject to just one restriction (namely, that its absolute value must be less than the speed of light, 1), the axis Ot' of the "moving" frame can occupy any position in the angle mOn (Fig. 149). Two events S and S_1 in some reference frame $\{x',t'\}$ are simultaneous if they lie on a line parallel to Ox'. Hence an event R represented by a point belonging to one of the two shaded angles nOm_1 and n_1Om is simultaneous with O relative to a suitably selected (inertial) reference frame $\{x',t'\}$. For a different choice of reference frame, R may either precede O (this applies to the frame $\{x_1,t_1\}$ in Fig. 149) or follow it (this applies to the frame $\{x_2,t_2\}$ in Fig. 149). On the other hand, mOn and m_1On_1 represent the "absolute future" and "absolute past" of O, respectively. Two events R_1 and R_2 can be causally connected only if one of them (the "cause" R_1) absolutely precedes the other (the "effect" R_2; cf. Fig. 149). In particular, if Ot and $p_1p_2 \| Ot$ are the time axes corresponding in some frame of reference $\{x,t\}$ to the origin and to some point M of the line o, then, relative to the event O, all events at M belong to one of three classes: the class of points of the ray P_1p_1 representing the events preceding O, the class of points of the ray P_2p_2 representing the events following O, and the class of points of the segment P_1P_2 representing the events that are "time-neutral" relative to O. Depending on the choice of reference frame, each of these last events may occur before O, at the same time as O, or after O (compare Figs. 149 and 143). It is clear that *the causes of O are just*

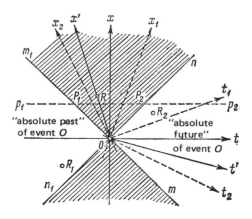

Figure 149

the events represented by the points of the ray P_1p_1 and the effects of O are just the events represented by the points of the ray P_2p_2.

Experimental verification of relativistic effects in the range of commonly encountered speeds is difficult because of the fact that for such speeds the coefficient $k = 1/\sqrt{1-v^2}$ in Equations (2a), (2b) or in (7), and the Fitzgerald contraction coefficient $\sqrt{1-v^2}$ are very close to 1 (the value we have assigned to the speed of light). Thus, for example, if $v \sim 1000$ km/hr $\sim$ .3 km/sec $\sim$ .000001 of the speed of light (which is the speed of some jets), then $k = 1/\sqrt{1-v^2} \sim 1.0000000000005$ and $\sqrt{1-v^2} \sim$ .9999999999995. But in nuclear and particle physics, as well as in astrophysics, we routinely encounter speeds close to the speed of light, and in such cases ignoring Einstein's theory of relativity may lead to serious errors. We shall return to this issue in Section 13.

It should be noted that in classical physics the Galilean principle of relativity applies solely to mechanical phenomena. This is connected with the fact that the fundamental laws of mechanics (Newton's famous laws of motion) are invariant (i.e., retain their form) under Galilean transformations (1). On the other hand, the fundamental laws of electromagnetics, i.e., the laws (or equations) of Maxwell, first stated in 1864 by the great English physicist J. C. MAXWELL (1831–1879) and forming the cornerstone of his "Treatise on Electicity and Magnetism" (1873), are invariant with respect to Lorentz transformations (7) but not with respect to Galilean transformations (1). It follows that the Galilean principle of relativity is not applicable to electromagnetic phenomena (including the propagation of light, which Maxwell's theory linked to electromagnetism). Such considerations were the basis of the thinking of the great French mathematician and physicist H. POINCARÉ (1854–1912), who arrived at some of the ideas of the (special) theory of relativity independently of, and simultaneously with, Einstein (i.e., in 1905).

A final observation. So far we have discussed the Einsteinian principle of relativity and Lorentz transformations only in the context of "one-dimensional" kinematics, i.e., kinematics of motions on a fixed line o. Clearly, the Einsteinian principle of relativity as formulated on p. 162 is applicable to plane-parallel (two-dimensional) as well as to arbitrary (three-dimensional) motions. The transition from one inertial frame of reference to another is governed by more complex Lorentz transformations of a three- and four-dimensional space of events, respectively. In the special case when the "moving reference frame" $\{x',y',z',t'\}$ and the "fixed reference frame" $\{x,y,z,t\}$ are at all times related in the manner illustrated in Figure 150 ($Ox \equiv O'x'$, $O'y' \| Oy$, $O'z' \| Oz$), the corresponding Lorentz transformations take the simple form

$$\begin{aligned} x' &= \frac{1}{\sqrt{1-v^2}} x - \frac{v}{\sqrt{1-v^2}} t, \\ y' &= y, \\ z' &= z, \\ t' &= -\frac{v}{\sqrt{1-v^2}} x + \frac{1}{\sqrt{1-v^2}} t, \end{aligned} \qquad (10)$$

where v is the velocity of the moving reference frame relative to the fixed reference frame.

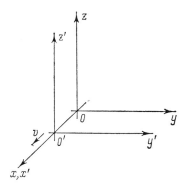

Figure 150

Problems and Exercises

1 A square is moving in the direction of one of its sides with velocity 0.9 (in terms of the units adopted in this section). Compute the angle between the diagonals of the (moving) square.

2 A solid moves in space with uniform velocity v. What is the effect of this motion on the volume of the solid?

3 In the theory of relativity, a "distance" between two events $A(x,t)$ and $B(x_1,t_1)$ is defined as the quantity $\nu_{AB} = \sqrt{\delta^2 - \tau^2}$, where $\delta = x_1 - x$ is the "space distance" and $\tau = |t_1 - t|$, the "time distance" between A and B. Show that ν is independent of the inertial frame in which A and B are considered.

4 A scientist aged 50 left for an interstellar voyage. After 41 years (relative to a reference frame tied to the fixed stars) he landed on a planet 40 light years away from the earth. Compute the age of the scientist at the time of his arrival on the planet. [*Hint*: use the result of Exercise 3.]

I Consider the three-dimensional Galilean geometry defined in Exercise 5 of the Introduction. Its motions are given by Equations (12) or (12') (pp. 20 and 30, respectively). Compare this geometry with three-dimensional Minkowsian geometry (See Supplement A; in particular, pp. 221 and 227). Now discuss the geometry of plane-parallel motions in classical Newtonian mechanics and in the (special) Einsteinian theory of relativity. (Newtonian plane-parallel motions are discussed in Section 2 of the Introduction.)

12. Minkowskian geometry

We saw above that the transition from the Galilean to the Einsteinian principle of relativity amounts to replacing the elementary Galilean transformations of classical mechanics

$$x' = x - vt,$$
$$t' = t, \qquad (1)$$

which govern the transition from one inertial frame of reference to

12. Minkowskian geometry

another, by the somewhat more complicated Lorentz transformations

$$x' = \frac{x - vt}{\sqrt{1-v^2}},$$
$$t' = \frac{-vx + t}{\sqrt{1-v^2}}. \tag{7}$$

We devoted the first ten sections of this book largely to the study of the geometric system consisting of the plane $\{x,t\}$ and the motions (1). We called this system "Galilean geometry" and pointed out that each of its theorems could be interpreted as a fact about (Newtonian) kinematics on a line o. Given the significance of the Lorentz transformations (7), it is quite natural for us to study the geometric system consisting of the plane $\{x,t\}$ and the motions (7). This geometric system, usually referred to as (pseudo-Euclidean) **Minkowskian geometry**,[9] was first investigated by the eminent German mathematician and physicist H. MINKOWSKI (1864–1909) who in the years 1907–1908 suggested its use for the description of the phenomena of relativistic mechanics.

The motions of the Galilean plane are given by the transformations

$$x' = x \quad\quad + a.$$
$$y' = vx + y + b \tag{1'}$$

The differences between (1) and (1') are a change of notation (t,x are replaced by x,y with x playing the role of time t and y playing the role of the abscissa x of a point on the line o) and also a difference in scope, in that the transformations (1') include the translations

$$x' = x + a,$$
$$y' = y + b, \tag{1a}$$

which correspond to changes in the "time origin" and "space origin" on o. Similarly, *Minkowskian geometry is defined as the study of the properties of figures in the x,y plane which are invariant under the motions*

$$\boxed{\begin{aligned} x' &= \frac{1}{\sqrt{1-v^2}} x - \frac{v}{\sqrt{1-v^2}} y + a, \\ y' &= -\frac{v}{\sqrt{1-v^2}} x + \frac{1}{\sqrt{1-v^2}} y + b, \end{aligned}} \tag{11}$$

where v, a, and b are arbitrary parameters of the motion. Again, the differences between (7) and (11) involve notation (t,x are replaced by x,y with x playing the role of time t and y playing the role of the abscissa x of a point on the line o) and scope, in that the transformations (11) include the translations (1a), which correspond to changes in the time origin and

---

[9] The term "pseudo-Euclidean" reflects the closeness of Minkowskian and Euclidean geometry. We shall see below, however, that the assertions of Euclidean geometry are sometimes altered in Minkowskian geometry in a singular manner.

space origin on o. The rest of this section is devoted to a sketch of Minkowskian geometry with hardly any mechanical interpretations of this remarkable geometry in terms of special relativity.

In Minkowskian geometry we use two types of point coordinates, (x,y) and (X,Y), connected by the relations

$$X = \frac{1}{\sqrt{2}}(x+y), \qquad Y = \frac{1}{\sqrt{2}}(-x+y),$$

or equivalently,

$$x = \frac{1}{\sqrt{2}}(X-Y), \qquad y = \frac{1}{\sqrt{2}}(X+Y) \tag{12}$$

(Fig. 151). Addition and substraction of the equations in (11) enables us to express the motions of the Minkowskian plane in terms of the coordinates X and Y. Specifically,

$$\sqrt{2}\, X' = \frac{1-v}{\sqrt{1-v^2}} \frac{X-Y}{\sqrt{2}} + \frac{1-v}{\sqrt{1-v^2}} \frac{X+Y}{\sqrt{2}} + (a+b),$$

$$\sqrt{2}\, Y' = -\frac{1+v}{\sqrt{1-v^2}} \frac{X-Y}{\sqrt{2}} + \frac{1+v}{\sqrt{1-v^2}} \frac{X+Y}{\sqrt{2}} + (-a+b),$$

or

$$\boxed{\begin{aligned} X' &= \lambda X &&+ A, \\ Y' &= &\frac{1}{\lambda} Y &+ B, \end{aligned}} \tag{11a}$$

where $\lambda = \sqrt{1-v}/\sqrt{1+v} = (1-v)/\sqrt{1-v^2}$ (so that $1/\lambda = \sqrt{1+v}/\sqrt{1-v} = (1+v)/\sqrt{1-v^2}$), $A = (a+b)/\sqrt{2}$, and $B = (-a+b)/\sqrt{2}$. The motion (11a) can be regarded as the transformation

$$\begin{aligned} X_1 &= \lambda X, \\ Y_1 &= \frac{1}{\lambda} Y, \end{aligned} \tag{13a}$$

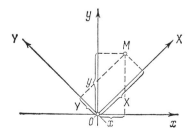

Figure 151

12. Minkowskian geometry

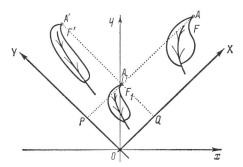

Figure 152

which represents a product of two compressions[10] [a compression with axis OY and coefficient λ and a compression with axis OX and coefficient $1/\lambda$ (cf. Fig. 152, where $\lambda < 1 < 1/\lambda$)] followed by the translation

$$X' = X_1 + A,$$
$$Y' = Y_1 + B. \tag{13b}$$

The transformation (13a) plays the role of a rotation in Minkowskian geometry. For reasons given below, when viewed as a transformation of the Euclidean coordinate plane $\{X, Y\}$, (13a) is called a *hyperbolic rotation*. In terms of the $\{x,y\}$ coordinates, it takes the form

$$\begin{aligned} x' &= \frac{1}{\sqrt{1-v^2}} x - \frac{v}{\sqrt{1-v^2}} y, \\ y' &= -\frac{v}{\sqrt{1-v^2}} x + \frac{1}{\sqrt{1-v^2}} y \end{aligned} \tag{13}$$

[compare (13) with Eqs. (7) of a Lorentz transformation].

By an argument similar to that employed in Section 3, Chapter 1 in connection with shears (cf. p. 54), one finds that *a compression, and* thus also *the transformation (13a), maps lines to lines and parallel lines to parallel lines, and preserves the ratio of collinear (or parallel) segments*. Thus *the concepts of a line, of parallel lines, and of the ratio of collinear (or parallel) segments are meaningful concepts of Minkowskian geometry*. Also, Equations (11a) imply that *a motion* (11) *or* (11a) *maps a line parallel to the OY or OX axis*, i.e., a line

$$X = \text{const.} \quad \text{or} \quad Y = \text{const.},$$

to a line with the same direction. This means that in Minkowskian geometry we have two families of "special lines"; the image of a special line l under

---

[10] A compression with axis OY and coefficient λ maps the point $A(X, Y)$ whose distance from OY is X, to the point $A_1(\lambda X, Y)$ on the perpendicular AP from A to OY whose distance from OY is λX. If $\lambda > 1$, it would be more appropriate to speak of a stretching rather than a compression (cf. text on p. 53 and Fig. 47b).

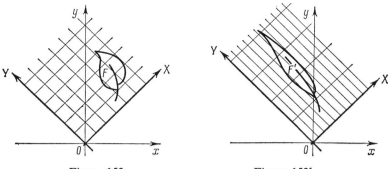

Figure 153a Figure 153b

any motion is parallel to l. Through each point of the plane there passes exactly one line of each of the two families. Finally, a motion (11a) maps a square whose sides are parallel to the OX and OY axes to a rectangle of the same area (if the sides of the square are ε, ε, then the sides of its image are $\lambda\varepsilon, (1/\lambda)\varepsilon$). But then a net of squares of the same size is mapped to a net of rectangles of the same area. This suggests the result that *a motion (11) or (11a) maps a figure F to a figure F' of the same area* (Figs. 153a and 153b; cf. Sec. 3, Chap. 1, in particular Fig. 28). It follows that *the area of a figure is* also *a meaningful concept of Minkowskian geometry*.

We now consider an arbitrary segment which we first assume is not parallel to one of the axes OX and OY. Let $AKBL$ be the rectangle whose sides are parallel to the axes OX and OY and whose diagonal is the segment AB (Fig. 154a). A motion (11) or (11a) maps this rectangle to a rectangle $A'K'B'L'$ (Fig. 154b) whose sides are also parallel to the axes OX and OY and *whose area is the same as that of $AKBL$*. If A and B are close to each other, then the area of $AKBL$ is small. Hence it is not unreasonable to regard $S(AKBL)$ (i.e., the area of $AKBL$) as a measure of separation of the points A, B which tends to zero as A tends to B and has the same value for pairs of points A, B and A', B' which are congruent in the sense of Minkowskian geometry. This allows us to think of $S(AKBL)$ as the "Minkowskian distance" between A and B. Since area is measured

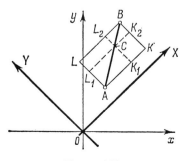

Figure 154a

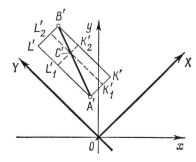

Figure 154b

12. Minkowskian geometry

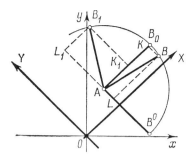

Figure 155

in square units rather than linear units,[11] it is convenient to suppose that in Minkowskian geometry the *distance* d_{AB} between A and B is proportional to the square root of the area of the rectangle $AKBL$, i.e., equal to $k\sqrt{S(AKBL)}$. The choice of the coefficient k is dictated by the choice of units of length and area. We shall find it convenient to put

$$d_{AB} = \sqrt{2S(AKBL)}. \qquad (14)$$

It is clear that if C is a point of the segment AB, then

$$d_{AB} = d_{AC} + d_{CB}$$

(for the areas of the similar rectangles $AKBL$, AK_1CL_1, and CK_2BL_2 in Fig. 154a are proportional to the squares of the diagonals AB, AC, and BC of these rectangles).

Let us now rotate the segment AB about A while preserving its Euclidean length (Fig. 155). It is clear that as AB approaches the position $AB_0 \| OX$ (or $AB^0 \| OY$), the area of $AKBL$ decreases, and when AB coincides with $AB_0 \| OX$ (or $AB^0 \| OY$), the corresponding "rectangle" reduces to a segment, so that its area vanishes. That is why *the length of a segment on a special line is assigned the value zero*. When in the process of rotation about A the segment AB passes the position $AB_0 \| OX$ (or $AB^0 \| OY$), the area of the rectangle $AKBL$ begins to increase again, but its orientation (determined by the ordering $A \to K \to B \to L$; here $AK \| LB \| OX$ and $AL \| KB \| OY$) is reversed (cf. the rectangles $AKBL$ and $AK_1B_1L_1$ in Fig. 155). In view of this fact, it is convenient to regard segments in the plane as being of two different kinds; two segments AB and CD of different kinds are viewed as not comparable in length. A segment or line is said to be *of the first* (*second*) *kind* if it is parallel to a line through the origin located in the pair of vertical angles formed by the axes OX and OY and containing the axis Ox (Oy). Special lines and segments of such lines are called *null lines* and *null segments*, respectively. A motion (11) or (11a) maps each segment to a segment of the same kind. *Two segments are*

---

[11] In other words, if the linear dimensions of a figure are multiplied by r, then its area is multiplied by r^2.

congruent [i.e., are related by a motion (11) or (11a)] *if and only if they have the same length and are of the same kind (if AB and A'B' are null segments then,* in addition, *they must belong to the same family of special lines*, i.e., they must be parallel).

Now let A and B have coordinates (X_1, Y_1) and (X_2, Y_2) in the coordinate system $\{X, Y\}$ and coordinates (x_1, y_1) and (x_2, y_2) in the coordinate system $\{x, y\}$. It is easy to see that the sides of the rectangle $AKBL$ are $|X_2 - X_1|$ and $|Y_2 - Y_1|$ (Fig. 156). Hence $S(AKBL) = |(X_2 - X_1)(Y_2 - Y_1)|$, and thus

$$d_{AB} = \sqrt{2|(X_2 - X_1)(Y_2 - Y_1)|} . \tag{15}$$

Using Eqs. (12) to go from the coordinates (X, Y) to the coordinates (x, y) we see that the distance between the points $A(x_1, y_1)$ and $B(x_2, y_2)$ is equal to

$$\sqrt{2|\tfrac{1}{2}[(x_2 - x_1) + (y_2 - y_1)][(x_2 - x_1) - (y_2 - y_1)]|}$$
$$= \sqrt{|(x_2 - x_1)^2 - (y_2 - y_1)^2|} ,$$

i.e.,

$$d_{AB} = \sqrt{|(x_2 - x_1)^2 - (y_2 - y_1)^2|} \tag{15a}$$

[compare this with the corresponding Euclidean formula (1) in the Introduction]. It is clear that if we ignore absolute value signs, then the radicands on the right-hand sides of (15) and (15a) are positive if AB is of the first kind and negative if AB is of the second kind. With this fact in mind, we often put

$$d_{AB} = \sqrt{2(X_2 - X_1)(Y_1 - Y_2)} , \tag{15'}$$

or equivalently,

$$d_{AB} = \sqrt{(x_2 - x_1)^2 - (y_2 - y_1)^2} . \tag{15'a}$$

Then we see that the lengths of segments of the first and second kind are

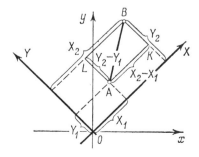

Figure 156

12. Minkowskian geometry

measured in different (and unrelated) units. Specifically, the length of a segment of the first kind is given by a real number and the length of a segment of the second kind by a complex number.

Fix an inertial reference frame, and let Ox be its time axis and Oy its space axis (i.e., the line o at time $x=0$). Then the endpoints of a segment of the first kind are two events *which occur at the same point of o* relative to a suitably chosen inertial (in the sense of Einstein's theory of relativity) reference frame, and the length of this segment is simply the *time interval* between the two events relative to that frame. Similarly, given a segment of the second kind, we can choose an inertial reference frame so that the endpoints of the segment correspond to *simultaneous events*. Then the length of the segment, measured in the sense of Minkowskian geometry, is the distance between the corresponding points of o. We may, therefore, assume that the length of segments of the first kind is measured in units of time (sec), while the length of segments of the second kind is measured in units of distance (cm). That is why, in Minkowskian geometry, lines of the first kind are called *timelike* and lines of the second kind are called *spacelike*.

In Minkowskian geometry, it is natural to define a **circle** S with center Q and radius CD (where CD is a segment of the first or second kind) as *the set of points M such that the segment QM is congruent to CD*. We shall denote the coordinates of the center Q of S relative to the coordinate systems $\{X,Y\}$ and $\{x,y\}$ by (A,B) and (m,n), respectively, and the coordinates of the endpoints C and D of CD by (X_1,Y_1) and (X_2,Y_2) and by (x_1,y_1) and (x_2,y_2), respectively. We shall call the quantity

$$2(X_2-X_1)(Y_1-Y_2)=(x_2-x_1)^2-(y_2-y_1)^2$$

(which can be positive or negative) *the square of the radius* of S, and denote it by $\pm r^2$ (where $r>0$). It is clear that the equation of the circle S with center Q and radius of square $\pm r^2$ is

$$2(X-A)(Y-B)=\pm r^2, \tag{16}$$

or

$$(x-m)^2-(y-n)^2=\pm r^2 \tag{16a}$$

[cf. Eq. (2), Sec. 1, of a Euclidean circle]. The latter equation can be rewritten in the form

$$x^2-y^2+2px+2qy+f=0 \tag{16b}$$

(where $p=-m, q=n, f=m^2-n^2\pm r^2$), or in the form

$$a(x^2-y^2)+2b_1 x+2b_2 y+c=0, \tag{16'}$$

which includes circles (if $a\neq 0$) as well as lines (if $a=0$). In particular, the equation of a circle with center at the origin $O(0,0)$ is

$$XY=\text{const.} \tag{17}$$

or

$$x^2-y^2=\text{const.} \tag{17a}$$

It is clear that from the point of view of Euclidean geometry the circle given by (16) or (16a) represents a *hyperbola* whose asymptotes are the

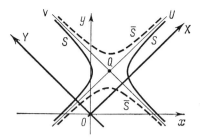

Figure 157

special lines passing through its center (Fig. 157). We say that the circle S is of the first or second kind according as its radius is of the first or second kind. The circles S and $\bar{S}$ in Figure 157 (of the first and second kind, respectively) with common center Q and radii of square r^2 and $-r^2$ are said to be *conjugate*. The "cross" consisting of the special lines QU and QV is the locus of points at zero distance from Q. This set of points is sometimes referred to as *a circle of radius zero* (or *a null circle*). It is easy to see that *three points of the Minkowskian plane determine a unique line* (of the first or second kind or a null line) *or a unique circle* (which may be a circle of the first or second kind or a null circle).

The transformations (13) and (13a) are called hyperbolic rotations for they map the Minkowskian circles (17) or (17a), i.e., the Euclidean hyperbolas with center A, onto themselves (cf. Figure 158; the transformation $X_1 = \lambda X$, $Y_1 = Y$ maps the hyperbolas S and $\bar{S}$ given by the equations $XY = \pm c$ to hyperbolas S_1 and $\bar{S}_1$, and the transformation $X' = X_1$, $Y' = (1/\lambda) Y_1$ maps the hyperbolas S_1 and $\bar{S}_1$ back to the initial hyperbolas S and $\bar{S}$).

We now consider the concept of angle between lines. If l and l_1 are two lines in the Minkowskian plane which intersect in a point Q, then it is natural to define *the angle* δ_{ll_1} *between* l *and* l_1 as the Minkowskian length of the arc NN_1 between l and l_1 belonging to the "unit circle" S with center

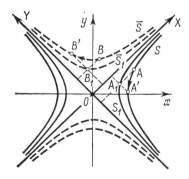

Figure 158

12. Minkowskian geometry

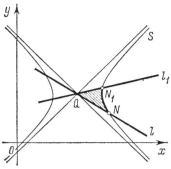

Figure 159a

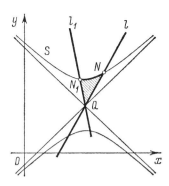

Figure 159b

Q; here the term unit circle refers to a circle within radius of square $+1$ or -1. We note that our definition makes sense only if l and l_1 are both of the first kind (Fig. 159a) or both of the second kind (Fig. 159b), for only in those cases do the lines intersect the same unit circle and thus determine an arc NN_1 of that circle. The Minkowskian length of the arc NN_1 can be defined as *the limit of the lengths of polygonal lines $NM_1M_2...M_nN_1$ inscribed in NN_1 whose longest edges approach zero.* In this connection, it is of interest to point out that all edges of a polygonal line inscribed in a circle of the first kind are segments of the second kind, and conversely. It is convenient to replace this relatively complicated definition of an angle (involving Minkowskian arc length and thus a new limiting process) by the following simpler (and equivalent[12]) definition: *the magnitude δ_{ll_1} of the angle between lines l and l_1 intersecting at Q is equal to twice the (Euclidean) area of the sector NQN_1 of the unit circle with center Q* (cf. Figs. 159a and 159b). We can also speak of the directed angle δ_{ll_1} between the (ordered) lines l and l_1 by regarding δ_{ll_1} as positive or negative according as the

---

[12]In defining the arc length of a curve $y=f(x)$ in Euclidean geometry, we start with the relation $\Delta s^2 = \Delta x^2 + \Delta y^2 = [1+(\Delta y/\Delta x)^2]\Delta x^2$ and are led to an integral of the form $\int_a^b \sqrt{1+y'^2}\, dx$. In Minkowskian geometry, we start with the relation $\Delta s^2 = |\Delta x^2 - \Delta y^2| = |1-(\Delta y/\Delta x)^2|\Delta x^2$ and are led to an integral of the form $\int_a^b \sqrt{|1-y'^2|}\, dx$. In particular, in Figure 166a, the (positive) Minkowskian length of the arc AM of the hyperbola $S: x^2-y^2=1$, is

$$L_{AM} = \int_1^x \sqrt{\left(\frac{x}{\sqrt{x^2-1}}\right)^2 - 1}\, dx = \int_1^x \frac{dx}{\sqrt{x^2-1}} = \log(x+\sqrt{x^2-1}),$$

and twice the shaded area ("area" has the same meaning in Minkowskian geometry and in Euclidean geometry; cf. p. 178) is

$$E_{AM} = 2\left(\frac{xy}{2} - \int_1^x y\, dx\right) = 2\left(\frac{x\sqrt{x^2-1}}{2} - \int_1^x \sqrt{x^2-1}\, dx\right) = \log(x+\sqrt{x^2-1}),$$

which shows the asserted equality $\varphi = L_{AM} = E_{AM}$. (Translator's note.)

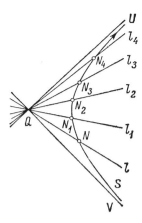

Figure 160

rotation which carries the ray QN to QN_1 is counterclockwise or clockwise.[13] Two angles APB and CQD, each formed by lines of the same kind, are congruent in the sense of Minkowskian geometry [i.e., related by a motion (11) or (11a)] if and only *if the segments PA, PB, QC and QD, are all of the same kind and* $\delta_{PA,PB} = \delta_{QA,QC}$. The question of the magnitude of the angle between lines of different kinds and the congruence of such angles will be discussed below.

Now let l and l_1 be two lines of, say, the first kind intersecting at Q, and let NN_1 be the arc of the unit circle S with center Q cut off by l and l_1 (Fig. 160). Consider the hyperbolic rotation which maps the ray QN to QN_1. This rotation maps the ray QN_1 to a ray QN_2, QN_2 to a ray QN_3, QN_3 to a ray QN_4, and so on; here $N, N_1, N_2, N_3, N_4, \ldots$ are points of S. Since the lines QN ($\equiv l$), QN_1 ($\equiv l_1$), $QN_2, QN_3, QN_4, \ldots$ are of the first kind, it follows that the rays QN, QN_k ($k=1,2,3,\ldots$) belong to the interior of the angle UQV bounded by the special lines QU and QV passing through Q and separating lines of the first kind from lines of the second kind. It follows that in Minkowskian geometry we can rotate a ray about Q through an arbitrarily large angle (since $\delta_{QN,QN_k} = k \cdot \delta_{QN,QN_1}$) without ever reaching the line QU, which forms an "infinitely large" angle with QN. Similarly, a line of the second kind can be rotated through an arbitrarily large angle, and the angle it forms with special lines is infinite.

Next we define the distance from a point to a line and the distance between parallel lines. For this we must first define perpendicularity of lines in the Minkowskian plane. A line l through the center of a Minkowskian circle S either does not intersect S at all or intersects it in

---

[13]For a different approach start with footnote 12 and define the magnitude of the negative angle $-\varphi$ in Figure 166a by the equation $-\varphi = L_{AM'} = -L_{AM}$, or by the equation $-\varphi = -E_{AM}$, twice the signed area of the image of the shaded area in Figure 166a under reflection in the x-axis. Obviously $-\varphi = -L_{AM} = -E_{AM}$.

We leave it to the reader to establish the asserted equality for a general angle φ (see Fig. 187c). (Translators note.)

12. Minkowskian geometry

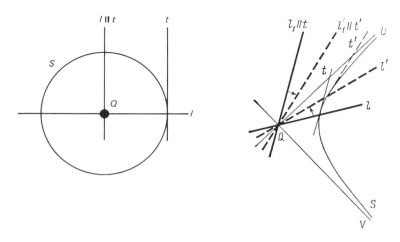

Figure 161a   Figure 161b

exactly two points P and Q. In the latter case, we call the segment $d = PQ$ a *diameter of S*. Following the example of Euclidean geometry, we shall say that *each diameter d of a circle S is perpendicular to the tangent t at the endd points of d on S*; here by a tangent t to S we mean a line of the first or second kind (but not a special line!) which has just one point in common with S.[14] It is, therefore, natural to say that l is perpendicular to l_1 if there is a circle S such that l is parallel to a diameter of S, while l_1 is parallel to the tangent to S at an end point of d (see Fig. 161b; compare this figure and Fig. 161a, which refers to Euclidean geometry). It is easy to see that *a line perpendicular to a line of the first kind is of the second kind, and conversely*. [It is a simple exercise in calculus to show that if we can establish the perpendicularity of two lines l and l_1 (more exactly, of two directions) using *some* circle S, then we can do so using *any* circle of the same kind as S. (This shows that our definition of perpendicularity is sensible.) In particular, we may choose S so that its center coincides with the point $l \cap l_1$ (see Figs. 161a and 161b).]

If a line l rotates about one of its points Q and approaches a special line QU then, in distinction to Euclidean geometry, the line l_1 through Q such that $l \perp l_1$ (we use the same perpendicularity symbol in Minkowskian and in Euclidean geometry) rotates in a direction opposite to that of l (Fig. 161b), and l and l_1 simultaneously approach the line QU. [The latter assertion also follows from the fact that *two lines l and l_1 perpendicular in the sense of Minkowski are symmetric in the sense of Euclid relative to the special lines QU and QV passing through their point of intersection Q* (cf. Exercise **5** below)]. It is therefore natural to say that *every special line in the Minkowskian plane is perpendicular to itself*.

---

[14]It is clear that a special line passing through an endpoint of a diameter d also has just one point in common with S. Naturally, we will not refer to such lines as tangents.

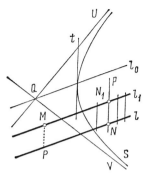 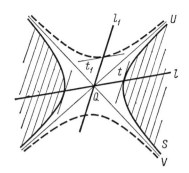

Figure 162   Figure 163

We note that our definition does not automatically imply that the relation of perpendicularity of two lines is symmetric. This is nevertheless true: *If l is perpendicular to l_1, then l_1 is perpendicular to l* (Fig. 162; cf. Exercise 6). Also, it is not difficult to prove that *if $l \perp l_1$, then the diameter parallel to l of a circle S bisects the chords of S parallel to l_1*; this is illustrated in Figure 162.

It is now natural to define *the distance d_{Ml} from a point M to a line l of the first or second kind as the length d_{MP} of the perpendicular MP from M to l* (Fig. 163). There exists one and only one such perpendicular. *The distance $d_{l_1 l}$ between two parallel lines* can be defined as the distance d_{Ml} from an arbitrary point M on l_1 to l (the latter distance is easily seen to be independent of the choice of M; cf. Fig. 163). Equivalently, $d_{l_1 l}$ can be defined as *the length $d_{N_1 N}$ of the segment $N_1 N$ cut off by the lines l_1 and l on any line p perpendicular to them*. We do not define the distance between parallel special lines k and k_1 or the distance from a point M to a special line k.[15]

We note that in Minkowskian geometry two perpendicular lines are always of different kinds. This fact enables us to associate to every (ordered) pair l and l_1 of lines of different kinds a number $\tilde{\delta}_{ll_1}$ such that two such pairs (l,l_1) and (m,m_1) are congruent [i.e., there exists a motion (11) or (11a) which takes l to m and l_1 to m_1] if and only if l and m are of the same kind and $\tilde{\delta}_{ll_1} = \tilde{\delta}_{mm_1}$. The number in question is the directed angle $\delta_{l'l_1}$, where $l' \perp l$ (Fig. 164). [In particular, the equality $\tilde{\delta}_{ll_1} = 0$ signifies that $l \perp l_1$.] The quantity just defined is not comparable with the magnitude of the angle between two lines of the same kind.

The following result is of interest. *If A, B, M_0 are three points in the Minkowskian plane such that none of the lines AB, AM_0, BM_0 is special, then the set S of points M for which the (directed) angle between MA and MB is*

---

[15]This is due to the fact that given two parallel special lines k and k_1 and two points M and M_1 (with $M \in k$, $M_1 \in k_1$ or $M \notin k$, $M_1 \notin k_1$) there is always a motion (11) or (11a) which takes k to k_1 and M to M_1.

12. Minkowskian geometry

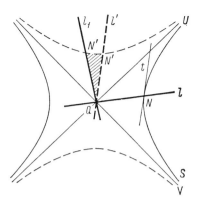

Figure 164

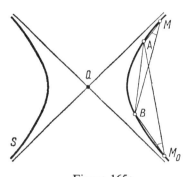

Figure 165a

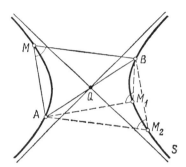

Figure 165b

congruent to the angle between M_0A and M_0B,

$$\delta_{MA,MB} = \delta_{M_0A, M_0B} \quad \text{or} \quad \tilde{\delta}_{MA,MB} = \tilde{\delta}_{M_0A, M_0B},$$

is a circle passing through A, B, and M_0 (Fig. 165a). *In particular, if A and B are fixed, then the set of points M for which $MA \perp MB$ is a circle with diameter AB* (Fig. 165b).

We shall require some results pertaining to the "trigonometry of hyperbolas." Thus let S be the unit circle

$$x^2 - y^2 = 1 \tag{18}$$

in the Minkowskian plane with center $O(0,0)$, OA the radius of S joining O to the vertex $A(1,0)$ of the hyperbola (18), and OM a variable radius of S (cf. Figs. 166a and 166b; the latter represents the Euclidean unit circle

$$x^2 + y^2 = 1 \tag{18a}$$

with radii OA and OM). The directed angle $\delta_{OA,OM} = \varphi$ (regarded as an angle in Minkowskian geometry, and thus equal to twice the signed area of the shaded sector AOM of the hyperbola S in Fig. 166a) is often called the

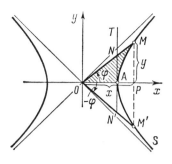
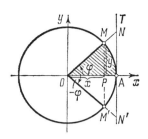

Figure 166a   Figure 166b

hyperbolic angle between OA and OM; the coordinates $x = OP > 0$ and $y = PM$ of the variable point M are called the *hyperbolic cosine* and the *hyperbolic sine* of φ and are denoted by $\cosh \varphi$ and $\sinh \varphi$, respectively. The ratio $PM/OP = \sinh \varphi / \cosh \varphi$ is called the *hyperbolic tangent* of φ and is denoted by $\tanh \varphi$.[16] The absolute value of $\tanh \varphi$ is equal to the (Euclidean) length of the segment AN cut off by the variable radius OM on the tangent AT to the hyperbola S at its vertex A.

Figure 166a shows that if φ increases indefinitely, then $\cosh \varphi$ and $\sinh \varphi$ increae indefinitely and $\tanh \varphi$ tends to 1. Also, if $\varphi = 0$, then $\cosh \varphi = 1$, $\sinh \varphi = 0$, and $\tanh \varphi = 0$. Finally, Figure 166a implies that

$$\cosh(-\varphi) = \cosh \varphi \tag{19a}$$

[just as $\cos(-\varphi) = \cos \varphi$],

$$\sinh(-\varphi) = -\sinh \varphi \tag{19b}$$

[just as $\sin(-\varphi) = -\sin \varphi$], and therefore

$$\tanh(-\varphi) = -\tanh \varphi \tag{19c}$$

[just as $\tan(-\varphi) = -\tan \varphi$]. The graphs of the functions $u = \cosh \varphi$, $u = \sinh \varphi$, and $u = \tanh \varphi$ are shown in Figure 167.[17]

Hyperbolic functions have many properties analogous to those of the trigonometric functions $\sin \varphi$, $\cos \varphi$, and $\tan \varphi$. These properties can be easily proved using the hyperbolic rotations (13a) or (13).[18] Thus, for example, Eq. (18) of the hyperbola S implies that *for every choice of the angle* φ,

$$\cosh^2 \varphi - \sinh^2 \varphi = 1 \tag{20}$$

---

[16] We note for later use that the mechanical sense of $\tanh \varphi$ is that of the velocity of the uniform motion represented by the line OM. (Translator's note.)

[17] We note for later use that if k is any given real number, there is one and only one hyperbolic angle φ such that $\sinh \varphi = k$. We also note that $\cosh \varphi \geq 1$ for all φ. (Translator's note.)

[18] See, for example, [58], Chapter IX of [65], or Section 27 of [57].

12. Minkowskian geometry

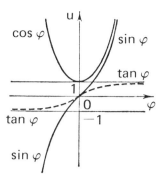

Figure 167

Again, *for every choice of the angles φ and ψ, we have*

$$\sinh(\varphi \pm \psi) = \sinh\varphi \cosh\psi \pm \cosh\varphi \sinh\psi \tag{21a}$$

and

$$\cosh(\varphi \pm \psi) = \cosh\varphi \cosh\psi \pm \sinh\varphi \sinh\psi, \tag{21b}$$

so that

$$\tanh(\varphi \pm \psi) = \frac{\tanh\varphi \pm \tanh\psi}{1 \pm \tanh\varphi \tanh\psi}. \tag{21c}$$

From Eqs. (21a)–(21c) we can easily deduce that

$$\left. \begin{array}{l} \sinh 2\varphi = 2\sinh\varphi \cosh\varphi, \\ \cosh 2\varphi = \cosh^2\varphi + \sinh^2\varphi, \\ \tanh 2\varphi = \dfrac{2\tanh\varphi}{1+\tanh^2\varphi}; \end{array} \right\} \tag{22a}$$

$$\left. \begin{array}{ll} & \sinh\varphi = \dfrac{2\tanh\varphi/2}{1-\tanh^2\varphi/2}, \\ \cosh\varphi = \dfrac{1+\tanh^2\varphi/2}{1-\tanh^2\varphi/2}, & \tanh\varphi = \dfrac{2\tanh\varphi/2}{1+\tanh^2\varphi/2}; \end{array} \right\} \tag{22b}$$

and so on.

Note that the coefficients $1/\sqrt{1-v^2}$ and $-v/\sqrt{1-v^2}$ in the formulas (11) are connected by the relation

$$\left(\frac{1}{\sqrt{1-v^2}}\right)^2 - \left(\frac{-v}{\sqrt{1-v^2}}\right)^2 = 1,$$

which coincides with the equality (20) connecting the hyperbolic functions $\cosh\varphi$ and $\sinh\varphi$. It follows[19] that if α is the unique hyperbolic angle such

---

[19] See the remarks in footnote 17. (Translator's note.)

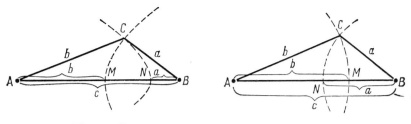

Figure 168a   Figure 168b

that $\sinh\alpha = -v/\sqrt{1-v^2}$, then $\cosh\alpha = 1/\sqrt{1-v^2}$, and (11) can be rewritten in the form

$$\boxed{\begin{aligned} x' &= x\cosh\alpha + y\sinh\alpha + a, \\ y' &= x\sinh\alpha + y\cosh\alpha + b \end{aligned}} \qquad (23)$$

[compare (23) with the Euclidean motion (6) in the Introduction].

We shall now consider **triangles** in the Minkowskian plane. Assume that the sides $BC \equiv a$, $CA \equiv b$, and $AB \equiv c$ of the triangle ABC are nonspecial. If all the sides are of the same kind, then the lengths of the segments BC, CA, and AB can be compared. (The lengths of these segments will be denoted by the same letters a,b,c as the segments themselves.) Then if A, B, and C are not collinear, *the largest side of the triangle ABC is larger than the sum of the other two sides* (see Fig. 168a, where $AM = AC = b$ and $BN = BC = a$, so that $a + b = BN + AM < AB = c$; compare this with Fig. 168b illustrating the inequality $a + b > c$ which holds for the sides of a Euclidean triangle).

The inequality

$$a + b < c$$

(where c is the largest side of $\triangle ABC$) follows from the "law of cosines" of Minkowskian geometry, which states that *if all the sides of $\triangle ABC$ are of the same kind, then*

$$a^2 = b^2 + c^2 - 2bc\cosh A, \qquad (24a)$$

where $A = \delta_{AB,AC}$ is the undirected (i.e., positive) Minkowskian[20] angle between the sides AB and AC of $\triangle ABC$. It turns out that the area $S(ABC)$ of $\triangle ABC$ can be defined by the formula

$$S(ABC) = \tfrac{1}{2}ab\sinh C. \qquad (25)$$

(Remember that the Minkowskian area of any figure in the Minkowskian plane is by definition the same as its Euclidean area.) The proofs of (24a) and (25) are straightforward and are left to the reader.

---

[20] So far we have defined the functions $\cosh\varphi$, $\sinh\varphi$ and $\tanh\varphi$ only when φ is the angle between two lines of the first kind. If φ is the angle between two lines m and m_1 of the second kind, then we define $\cosh\varphi = \cosh\psi$, $\sinh\varphi = \sinh\psi$, and $\tanh\varphi = \tanh\psi$, where ψ is the angle between lines $l \perp m$ and $l_1 \perp m_1$.

12. Minkowskian geometry

Formula (25) implies that

$$\tfrac{1}{2} ab \sinh C = \tfrac{1}{2} ac \sinh B = \tfrac{1}{2} bc \sinh A,$$

from which we readily obtain the Minkowskian "law of sines"

$$\frac{a}{\sinh A} = \frac{b}{\sinh B} = \frac{c}{\sinh C}. \tag{24b}$$

It turns out that the relations (24a), (24b), and (25) also hold for triangles with sides of different kinds.

We shall say a few words about isosceles and right triangles in the Minkowskian plane (the inequality $c > a + b$ rules out the existence of equilateral triangles). If the sides AC and BC of $\triangle ABC$ are equal (so that AC and BC are lines of the same kind), then the vertices A and B lie on the circle S with center C and radius $CA = CB$ (Fig. 169a). Let C be the origin of a rectangular coordinate system. Then there exists a rotation (13a) or (13) which carries the median CM of $\triangle ABC$ (M is the midpoint of the side AB) to some point on the x-axis. If $\triangle A'B'C$ is the image of $\triangle ABC$ under this rotation, then the sides CA' and CB' of $\triangle A'B'C$ are symmetric (in the Euclidean sense) with respect to the line Ox (Fig. 169b), which shows that the *median CM of an isosceles triangle is also an angle bisector and an altitude*. [The fact that $CM \perp AB$ implies that the base AB of an isosceles triangle is a segment of the opposite kind from its sides.] Figure 169b also implies the equality of the base angles of an isosceles triangle.

Now let $\triangle ABC$ be a right triangle with $AC \perp BC$ (Fig. 170a). Then the sides AC and BC are necessarily of different kinds. We assume that the hypotenuse AB is of the same kind as the side BC. Then it is easy to show (compare Figs. 170a and 170b, where C' is the vertex of the hyperbola S) that

$$a \equiv c \cosh B, \qquad b \equiv c \sinh B, \qquad b = a \tanh B. \tag{26}$$

The first two relations in (26) and the relation (20) imply the Minkowskian

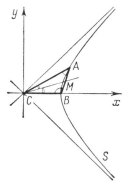 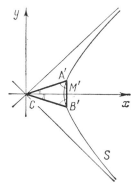

Figure 169a	Figure 169b

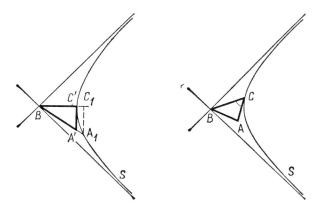

Figure 170a   Figure 170b

version of Pythagoras' theorem:
$$a^2 - b^2 = c^2. \tag{27}$$
We note that the second of the relations (26) and the relation (25) imply for the area of a triangle ABC the alternative formula
$$S(ABC) = \tfrac{1}{2} a \cdot h_a, \tag{25'}$$
where $h_a = b \sinh C$ is the length d_{AP} of the altitude AP of $\triangle ABC$ dropped from A to the line BC (Fig. 171). *The theorem on the concurrence of the medians of a triangle* is valid in Minkowskian geometry, and is proved just as in Euclidean geometry (cf. Sec. 4, Chap. 1, in particular p. 51). *The perpendicular bisectors of the sides of $\triangle ABC$ are also concurrent*; they meet at the center O of its circumcircle S (Fig. 172a). [We observe that if the sides of $\triangle ABC$ are all of the same kind and the radius of the circumcircle is R, then
$$\frac{a}{\sinh A} = \frac{b}{\sinh B} = \frac{c}{\sinh C} = 2R$$
(cf. Exercise **10**).] Just as in Euclidean geometry, this fact implies that *the altitudes of $\triangle ABC$ are concurrent* (Fig. 172b).

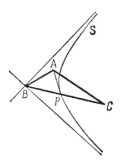

Figure 171

12. Minkowskian geometry

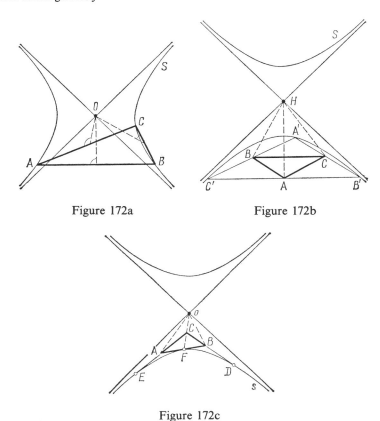

Figure 172a

Figure 172b

Figure 172c

A triangle in the Minkowskian plane has three angle bisectors only if all of its sides are of the same kind; for obviously an angle formed by lines l and l_1 of different kinds has no bisector (i.e., a line forming equal angles with l and l_1).[21] However, *if the triangle has three angle bisectors* (we know that this is not the case for, say, an isosceles or a right triangle) *then they are concurrent*; in fact, they meet at the center o of the incircle s tangent to all sides of the triangle (Fig. 172c). (It is clear that if the sides of a triangle are not all of the same kind, then no circle is tangent to all of them. This is due to the fact that all the tangents to a circle of one kind are lines of the other kind.) It is also true that *the midpoints of the sides of a triangle ABC and the feet of its altitudes* (*as well as the midpoints of the segments joining the orthocenter of $\triangle ABC$ to its vertices*) *lie on a circle S_1* whose radius is half the radius of the circumcircle of the triangle. It is natural to refer to S_1

---

[21]The bisector of the angle formed by lines l and l_1 can be described as *the set of points equidistant from l and l_1* (we note that this description implies the theorem, formulated below, on the concurrence of the angle bisectors of a triangle). If l and l_1 are of different kinds, then there are no points equidistant from l and l_1, for we cannot compare the distance from a point to a line of the first kind with its distance from a line of the second kind. The first of these distances is measured in complex units (cf. p. 181) and the second in real units.

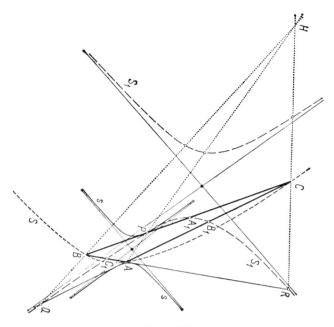

Figure 173

as the six- (nine-) point circle of the (Minkowskian) triangle ABC; if a triangle ABC has an incircle s, then *the six- (nine-) point circle S_1 of $\triangle ABC$ touches its incircle s* (Fig. 173).

Next we consider a circle S in the Minkowskian plane and a point M (Fig. 174). It is not difficult to show that *the product*

$$MA \cdot MB = d_{MA} \cdot d_{MB}, \qquad (28)$$

where A and B are the points of intersection of S and a line l through M, depends on S and M but not on l. We shall regard this product as positive if (a) the directions of the segments MA and MB (from M to A and from M to B) are the same and l is a line of the first kind or (b) the directions of

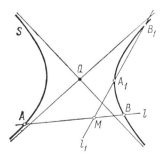

Figure 174

the segments MA and MB are opposite and l is of the second kind; and as negative in all other cases. [At this point it is convenient to regard the lengths of segments of the second kind as imaginary. The sign of the product (28) is the same as in Euclidean or Galilean geometry.] We shall refer to this signed product as *the power of the point M with respect to the circle S*.

It can be shown that *the power of a point $M(x_0,y_0)$ with respect to a circle S given by Eq. (16b) is*

$$x_0^2 - y_0^2 + 2px_0 + 2qy_0 + f,$$

i.e., *is equal to the result of substituting the coordinates (x_0,y_0) of M in the equation (16'b) of S* (see Exercise **10** below). From this it readily follows that *the set of points in the Minkowskian plane whose power relative to a circle S has the value k is a circle concentric with S* (Fig. 175a) and that *the set of points which have the same power relative to two circles S and S_1 given respectively by the Eqs. (16b) and*

$$x^2 - y^2 + 2p_1 x + 2q_1 y + f = 0, \tag{16'b}$$

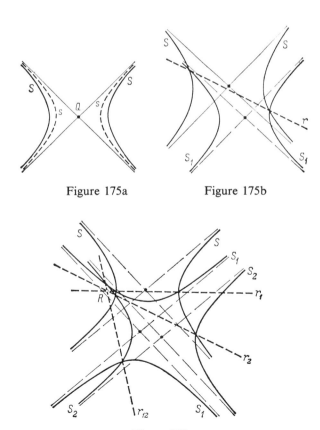

Figure 175a   Figure 175b

Figure 175c

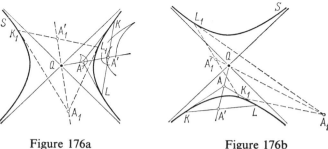

Figure 176a Figure 176b

is the line

$$2(p-p_1)x + 2(q-q_1)y + (f-f_1) = 0$$

(Fig. 175b). It is natural to call this line (the line r in Fig. 175b; r passes through the points of intersection of S and S_1, if any) the *radical axis* of S and S_1. Just as in Euclidean and Galilean geometry, we can show that *the radical axes of pairs of three Minkowskian circles S, S_1, and S_2 are all parallel or intersect in a point R* (Fig. 175c), the *radical center* of the three circles, and so on.

The study of inversions can be pursued in the context of Minkowskian geometry. An **inversion** with center Q and coefficient k, (also called an inversion in the circle S with center Q and radius r, where $k = \pm r^2$), is *a mapping of the Minkowskian plane which maps a point A to a point A' on the line QA such that*

$$QA \cdot QA' = d_{QA} \cdot d_{QA'} = k. \tag{28a}$$

Further, for $k > 0$, the segments QA and QA' have the same or opposite directions (from Q to A and from Q to A') according as they are of the first or second kind. If $k < 0$, then QA and QA' have the same direction if they are of the second kind and opposite directions if they are of the first kind (see Figs. 176a and 176b)[22]. It can be shown that if A is a point from which two tangents can be drawn to S, then inversion in S maps A to the point A' in which the line QA intersects the line KL which joins the points of tangency (see Figs. 176a and 176b; see also Exercise **11** below). Moreover, inversion in S leaves each point of S fixed. (A point A is said to be exterior to S, on S, or interior to S, according as the number of tangents from A to S is *two, one,* or *zero*.)

The following is an alternative definition of inversion in a circle S: the image A' of a point A is the point (other than A) of intersection of all circles s (and the line m) passing through A and perpendicular to S (i.e., such that the tangents to S and s at the points of $S \cap s$ are perpendicular in the sense of Minkowskian geometry; Fig. 177a). (See Exercise **11** below.) This approach to inversions permits us to regard a reflection in a (nonspecial) line l (defined as the mapping which takes each point A to the point

---

[22]The exposition can be simplified by using the convention on real and imaginary lengths of segments (see pp. 181 and 195).

12. Minkowskian geometry

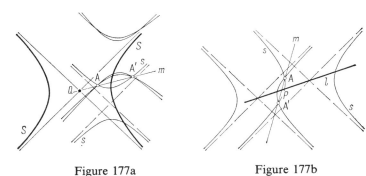

Figure 177a   Figure 177b

A' on $AA' \perp l$ such that $P = AA' \cap l$ is the midpoint of the segment AA') as a special case of an inversion. That this is reasonable follows from the fact that a reflection in a line l can also be described as a mapping which takes a point A to the point $A' \neq A$ of intersection of all circles s (including the line m) passing through A and perpendicular to l (i.e., such that the tangent to s at $s \cap l$ is perpendicular to l; see Fig. 177b). In turn, this explains why in Minkowskian geometry, just as in some other geometries, inversion in a circle S is often called *reflection in S*.

The definition (28a) readily implies that the inversion with center $O(0,0)$ and coefficient k maps $A(x,y)$ or $A(X,Y)$ to the point $A'(x',y')$ or $A'(X',Y')$ such that

$$x' = \frac{kx}{x^2 - y^2}, \qquad y' = \frac{ky}{x^2 - y^2},$$

or

$$X' = \frac{k}{2Y}, \qquad Y' = \frac{k}{2X}. \qquad (29)$$

Using Eq. (16') of a circle in the Minkowskian plane, we can now easily show that *an inversion maps a line of the first or second kind passing* (*not passing*) *through the center of inversion Q to itself* (*to a circle passing through the center of inversion*; Fig. 178a), *a circle passing* (*not passing*) *through the center of inversion to a line passing through the center of inversion* (*to a circle*

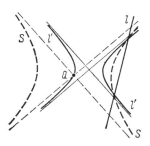 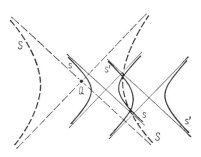

Figure 178a   Figure 178b

not passing through the center of inversion; Fig. 178b), *and special lines to special lines*. It can also be shown that inversions are *conformal*, i.e., angle preserving, *mappings*: *an inversion takes two curves* Γ *and* Γ_1 *intersecting at a point A to two curves* Γ' *and* Γ'_1 *intersecting at a point A' such that the (Minkowskian) angle between the tangents t and* t_1 *at A to* Γ *and* Γ_1 *is equal to the (Minkowskian) angle between the tangents at A' to* Γ' *and* Γ'_1. All these properties of inversions can be used to prove many theorems of Minkowskian geometry. We leave it to the reader to find examples of such theorems (cf. pp. 137–141).

We conclude this section with a discussion of certain important general considerations pertaining to inversions. It is clear that the definition (28a) assigns no image to a point A whose distance d_{QA} from the center of inversion is zero. Yet the set of points M of the Minkowskian plane defined by the equality

$$QM = d_{QM} = 0 \tag{30}$$

consists of the pair of special lines q_1 and q_2 passing through Q.[23] Strictly speaking, the domain of definition of an inversion is not the whole Minkowskian plane but rather *the set obtained by removing from the plane the lines* q_1 *and* q_2; if Q is the origin of the coordinate system $\{x,y\}$ (or $\{X,Y\}$), then the lines q_1 and q_2 are given by the equations $x=y$ and $x=-y$ (or $X=0, Y=0$). It is clear that the complicated character of the domain of definition of an inversion in Minkowskian geometry complicates its study and applications.

It is possible to surmount these difficulties in much the same way as in Euclidean and Galilean geometry. We stipulate that the domain of definition of an inversion with center $O(0,0)$ is the Minkowskian plane π supplemented with "infinitely distant points" which are the images of the points on the lines q_1 and q_2 passing through the origin. Specifically, we assume that the inversion with center O and coefficient 1 takes the point $M(m,m)$ to the point at infinity $\Omega_m^{(1)}$, and the point $N(n,-n)$ to the point at infinity $\Omega_n^{(2)}$, and the center of inversion O to the point $\Omega \equiv \Omega_0^{(1)} \equiv \Omega_0^{(2)}$; here m and n range over the reals. The Minkowskian plane supplemented with the points at infinity $\Omega_m^{(1)}$ and $\Omega_n^{(2)}$ is called the **inversive Minkowskian plane**,[24] and is the domain of definition of inversive and other circular transformations, i.e., *transformations that carry every circle* (16′) *of the Minkowskian plane* [*including lines*, obtained by putting $a=0$ in (16′)] *to a circle (or line)*.

We can visualize the inversive Minkowskian plane by means of a sterographic projection (cf. Chap. II, Sec. 10, pp. 142–155). We introduce a coordinate system $\{x,y,z\}$ in three-dimensional space, and assume that the Minkowskian plane is the plane $z=0$. We consider the one-sheeted hyperboloid γ given by the equation

$$x^2 - y^2 + \left(z - \tfrac{1}{2}\right)^2 = \tfrac{1}{4} \quad \text{or} \quad x^2 - y^2 + z^2 - z = 0, \tag{31}$$

which touches the plane π (the plane $z=0$) at $O(0,0,0)$ and intersects it in the (special) lines q_1 and q_2 with equations $y=x$ and $y=-x$ (Fig. 179). Let Q denote the point $(0,0,1)$ of γ diametrically opposite to O (i.e., symmetric to O with respect to the center of symmetry of γ). Associate to each point $A(x,y)$ of π the point

---

[23] We recall that the term we applied earlier to the set (30) was "a null circle" or a "circle with center Q and radius zero" (see p. 182).

[24] For reasons which we cannot go into here, it is necessary to add to the inversive Minkowskian plane two more fictitious points at infinity σ_1 and σ_2, which are left fixed by inversions with center O (see p. 277 below and, say, Sec. 5 of [80]).

12. Minkowskian geometry

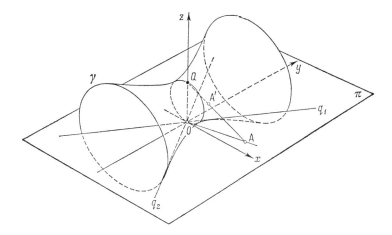

Figure 179

$A'(\bar{x},\bar{y},\bar{z}) \ne Q$ in which γ intersects the line QA. It is easy to see that

$$\bar{x} = \frac{x}{x^2-y^2+1}, \qquad \bar{y} = \frac{y}{x^2-y^2+1}, \qquad \bar{z} = \frac{x^2-y^2}{x^2-y^2+1}, \qquad (32)$$

[compare (32) with (42) and (42′) on pp. 144 and 151, respectively]. The sterographic projection (32) associates to every point of π a unique point of γ, but the correspondence is not "onto". Specifically, no points of π are mapped to the points on the lines o_1 and o_2 passing through the center of projection Q and parallel to π; these are the lines with equations $y = x$, $z = 1$ and $y = -x$, $z = 1$. It is therefore convenient to assume that γ is the image of π supplemented by two fictitious lines which are projected onto o_1 and o_2.[25] If we associate with the points of o_1 and o_2 (other than Q) the points at infinity $\Omega_m^{(1)}$ and $\Omega_n^{(2)}$ of π, and with the center of projection Q the point at infinity Ω, then we can say that the sterographic projection (32) is a one-to-one map of the inversive Minkowskian plane π onto γ. This enables us to view a (one-sheeted) hyperboloid as a model of the inversive Minkowskian plane.

Equations (32) imply that *stereographic projection maps the points of a circle (or line) (16′) of the Minkowskian plane to the points of intersection of the hyperboloid γ and the plane*

$$az + 2b_1 x + 2b_2 y + c(1-z) = 0, \qquad (16'')$$

and that, conversely, to every plane section of γ there corresponds a circle (or line) of π. In particular, lines (in the plane) are mapped to plane sections of γ passing through Q, special lines to plane sections of γ passing through Q and containing a rectilinear generator of γ (see footnote 25), and null circles to plane sections of γ

---

[25]It is easy to see that the stereographic projection (32) associates to the rectilinear generators $x - y = c(1 - z)$, $x + y = (1/c)z$ and $x + y = d(1 - z)$, $x - y = (1/d)z$ of γ the special lines

$$y = x - c \quad \text{and} \quad y = -x + d$$

of π. It is therefore natural to associate to the rectilinear generators o_1 and o_2 the "special lines at infinity" of π. This terminology reflects the fact that an inversion of the Minkowskian plane maps special lines to special lines. Thus it is reasonable to suppose that an inversion maps the special lines q_1 and q_2 to the special lines at infinity of the inversive Minkowskian plane.

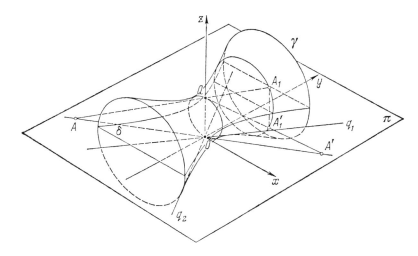

Figure 180

containing a pair of rectilinear generators. Indeed, the sterographic projection maps the points of the lines obtained, say, by putting $a=0$ in (16′) to the section of γ by the plane

$$2b_1 x + 2b_2 y + c(1-z) = 0,$$

which, obviously, passes through $Q(0,0,1)$. Since our convention associates $Q(0,0,1)$ to the point at infinity Ω of the inversive Minkowskian plane, we must suppose that all the lines pass through Ω. We must also suppose that every circle (16′) contains a point at infinity $\Omega_m^{(1)}$ and a point at infinity $\Omega_n^{(2)}$; these points are represented by the points of intersection of the plane (16″) and the lines o_1 and o_2.

The fact that the stereographic projection (32) maps circles (and lines) in the Minkowskian plane to plane sections of the hyperboloid enables us to use this projection to represent circular transformations of the (inversive) Minkowskian plane π. For example, consider *the reflection of γ in the* (horizontal) *plane δ* given by the equation $z = \frac{1}{2}$, i.e., the mapping which takes each point A_1 of γ to the point A_1' of γ such that $A_1 A_1' \perp \delta$ (Fig. 180). It is clear that this reflection takes a plane section of γ to a plane section. It follows that the induced transformation of π, i.e., the mapping which takes a point A of π (corresponding via stereographic projection to the point A_1 of γ) to the point A' of π (corresponding to the point A_1' of γ), must be *circular*. It is easy to show that this mapping of the Minkowskian plane is just the inversion with center O and coefficient 1 (the inversion with circle of inversion S corresponding via stereographic projection to the section of γ by the plane δ). This fact can be taken as the definition of inversion, and enables us to deduce all of its properties from those of stereographic projection. Using stereographic projection, one can prove the fundamental theorem that *every circular transformation is a similitude* (i.e., a mapping which takes each pair of points A, B of π to a pair of points A', B' such that $A'B'/AB = d_{A'B'}/d_{AB} = k$, k fixed) *or the product of an inversion and a similitude* (cf. Problem **III** below).

Problems and Exercises

5   Let the Ox and Oy axes and the units on them be chosen as in the text of this section. Prove that the relation $l \perp l_1$ (in the sense of Minkowskian geometry) amounts to saying that the special lines QU and QV passing through $Q = l \cap l_1$ are Euclidean bisectors of the angles between l and l_1 (cf. Fig. 161).

6   (a) Show that the relation of perpendicularity in Minkowskian geometry is symmetric (cf. p. 322). (b) Show that if AB is a chord of a circle in the Minkowskian plane and QP is a diameter of S, then the statements "$QP \perp AB$" and "QP bisects AB" are equivalent.

7   (a) Prove the properties of Minkowskian circles illustrated in Figures 165a and 165b. (b) Show that these properties of Minkowskian circles can be used to define such circles.

8   Prove the formulas (a) (24a); (b) (24b); (c) (25).

9   Prove the concurrence of (a) the perpendicular bisectors of the sides of a triangle; (b) the altitudes of a triangle; (c) the angle bisectors of a triangle in the Minkowskian plane.

10 Show that in Minkowskian geometry (a) The tangents PA and PB from a point P to a circle S with center Q are congruent; the line QP bisects the segment AB; $QA \perp AB$ (in the Minkowskian sense). (b) The midpoints of all chords of a circle S which pass through a point M in the Minkowskian plane lie on a circle s.

11 Prove the theorems stated on pp. 193–196 concerning the power of a point with respect to a circle in the Minkowskian plane, the radical axis of two circles, and the radical center of three circles (see pp. 195–196). Show that the tangents to two circles S_1 and S_2 from a point on their radical axis external to S_1 and S_2 are congruent, and establish other properties of the radical axis.

12 Prove the properties of inversion in the Minkowskian plane stated on pp. 196–198. Give examples of the use of inversion in proving theorems of Minkowskian geometry.

13 In Minkowskian geometry, prove (a) the existence of the nine-point circle of a triangle; (b) the theorem that the nine-point circle and the incircle of a triangle are tangent (cf. Fig. 173).

14 Formulate and prove other theorems of Minkowskian geometry.

II Develop a theory of pencils and bundles of circles in the Minkowskian plane (cf. Problem **XII**, Chap. II).

III Develop a theory of circular transformations in the Minkowskian plane. In particular, prove the fundamental theorem on circular transformations formulated on p. 200. [*Hint:* Use stereographic projection of a hyperboloid to the Minkowskian plane; see pp. 198–200.]

IV Outline the main features of a theory of quadric curves in the Minkowskian plane (compare this problem and Problem **XVI**, Chap. II).

V   Investigate three-dimensional Minkowskian geometry (cf. p. 221).

13. Galilean geometry as a limiting case of Euclidean and Minkowskian geometry

By now we know of three different geometries in the ordinary (affine) plane, viz. Euclidean, Galilean (whose exposition forms the main part of this book), and Minkowskian (sketched in the last section). Throughout the book we have noted similarities between these geometries. Many theorems have identical formulations in all three geometries, for example, the concurrence of the medians of a triangle, the existence of the six-point circle (cycle), the theorem on the power of a point with respect to a circle (cycle), the definition and properties of an inversion (of the first kind), and so on. Other theorems may differ in the three geometries but nevertheless retain certain common features—for example, it is difficult to deny the analogies between the Euclidean law of sines

$$\frac{a}{\sin A} = \frac{b}{\sin B} = \frac{c}{\sin C}, \tag{33a}$$

the Galilean "law of sines" [cf. (13), Sec. 4]

$$\frac{a}{A} = \frac{b}{B} = \frac{c}{C}, \tag{33}$$

and the Minkowskian theorem [cf. (24b), Sec. 12]

$$\frac{a}{\sinh A} = \frac{b}{\sinh B} = \frac{c}{\sinh C}. \tag{33b}$$

In the case of such related but nonidentical results, Galilean geometry is often "intermediate" between Euclidean and Minkowskian geometry. For example, if $a \leq b \leq c$ are the sides of a (nondegenerate) triangle, then in Euclidean geometry

$$a + b > c, \tag{34a}$$

in Galilean geometry

$$a + b = c \tag{34}$$

[cf. (11), Sec. 4], and in Minkowskian geometry

$$a + b < c \tag{34b}$$

(cf. p. 190)[26] Again, in Euclidean geometry the equation of a circle (in rectangular coordinates) is

$$a(x^2 + y^2) + 2b_1 x + 2b_2 y + c = 0 \tag{35a}$$

(cf. p. 79), in Galilean geometry the equation of a cycle is

$$ax^2 + 2b_1 x + 2b_2 y + c = 0 \tag{35}$$

[cf. (2), Sec. 7], and in Minkowskian geometry the equation of a circle is

$$a(x^2 - y^2) + 2b_1 x + 2b_2 y + c = 0 \tag{35b}$$

[cf. (16′), Sec. 12], and so on.

Another example: To visualize the inversive Euclidean plane, we used the central projection of the plane $z = 0$ in three-dimensional space $\{x, y, z\}$ from the point $(0, 0, 1)$ to the sphere σ given by

$$x^2 + y^2 + \left(z - \tfrac{1}{2}\right)^2 = \tfrac{1}{4} \quad \text{or} \quad x^2 + y^2 + z^2 - z = 0 \tag{36a}$$

---

[26]The assumed inequality $a \leq b \leq c$ implies that the sides of the Minkowskian triangle are comparable, i.e., of the same kind.

13. Galilean geometry as a limiting case of Euclidean and Minkowskian geometry

[see (40), Sec. 10 and Fig. 129). In Galilean geometry, a similar role is played by the projection of the plane $z=0$ to the cylinder ζ given by

$$x^2+\left(z-\tfrac{1}{2}\right)^2=\tfrac{1}{4} \quad \text{or} \quad x^2+z^2-z=0 \tag{36}$$

[see (40′), Sec. 10 and Fig. 135), and in Minkowskian geometry by the projection of the same plane to the (one-sheeted) hyperboloid γ given by

$$x^2-y^2+\left(z-\tfrac{1}{2}\right)^2=\tfrac{1}{4} \quad \text{or} \quad x^2-y^2+z^2-z=0 \tag{36b}$$

[see (31), Sec. 12 and Fig. 179). Also, the very concept of an inversive plane is connected with the fact that in each case there is a set of "special" points ("circle of zero radius" with center Q) with no images under inversion. In Euclidean geometry, this set consists of the point Q, while in Galilean geometry it is a line through Q, and in Minkowskian geometry a pair of lines through Q.

The connection between Euclidean (and Minkowskian) and Galilean geometry can be explained as follows. Earlier (see p. 173) we saw that for speeds which are very small in comparison with the speed of light the principles of relativity of Galileo and Einstein are practically the same. The "geometrized" version of this statement is that if the unit of length on the y-axis in the Minkowskian plane (the axis associated with the position of a point on the line o in one-dimensional kinematics) is very large relative to the unit of length on the x-axis [the time axis; in our expositon above we chose the unit of length on the line o so that the speed of light was 1, i.e., we chose the unit of length to be equal to the distance (300,000 km) travelled by light during a unit of time (1 sec)], then Minkowskian geometry turns out to be very close to Galilean geometry. In physics, this fact is linked to the possibility of applying the classical mechanics of Galileo and Newton in all physical problems not involving speeds close to that of light (see the end of this section). In geometry, it sheds additional light on the proximity of Minkowskian and Galilean geometry.

The "intermediate" character of Galilean geometry vis-a-vis Euclidean and Minkowskian geometry explains the analogy between the expressions for the distance d between points $A(x,y)$ and $A_1(x_1,y_1)$ in the Euclidean plane,

$$d^2=(x_1-x)^2+(y-y)^2 \tag{37a}$$

[cf. (1), Sec. 1], in the Galilean plane,

$$d^2=(x_1-x)^2 \tag{37}$$

[cf. (5), Sec. 3], and in the Minkowskian plane,

$$d^2=(x_1-x)^2-(y_1-y)^2 \tag{37b}$$

[cf. (15′a), Sec. 12]. Now suppose that in Euclidean and Minkowskian geometry the unit of length OE_2 on the y-axis is replaced by a new unit $OE_2'=(1/c)OE_2$, while the unit of length on the x-axis remains unchanged: $OE_1=OE_1'$ (Fig. 181). Denoting by (x,y) the old coordinates, and by (x',y') the new coordinates resulting from the change of unit on the y-axis, we see that

$$x'=x \quad \text{and} \quad y'=cy$$

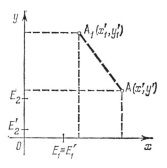

Figure 181

or

$$x = x' \quad \text{and} \quad y = \frac{1}{c}y'. \tag{38}$$

It follows that the old Euclidean and Minkowskian distances (37a) and (37b) between A and A_1 change to the distances

$$d^2 = (x'_1 - x')^2 + \frac{1}{c^2}(y'_1 - y')^2$$

in the Euclidean plane, and

$$d^2 = (x'_1 - x')^2 - \frac{1}{c^2}(y'_1 - y')^2$$

in the Minkowskian plane. Ignoring primes, i.e., denoting the new coordinates by (x,y), we can rewrite the latter expressions as

$$d^2 = (x_1 - x)^2 + \frac{1}{c^2}(y_1 - y)^2 \tag{37'a}$$

and

$$d^2 = (x_1 - x)^2 - \frac{1}{c^2}(y_1 - y)^2, \tag{37'b}$$

respectively. Letting c tend to infinity, i.e., viewing the unit OE_1 on the x-axis as being very small relative to the unit OE_2 on the y-axis, we obtain (in the limit) the familiar expression

$$d^2 = (x_1 - x)^2 \tag{37}$$

for the distance between two points in the Galilean plane.

We shall continue to use the new coordinates, in which the distance between the points A and A_1 is given by the formulas (37'a) and (37'b), but designate coordinates with the letters x and y. As before, we shall represent the units of length OE'_1 and OE'_2 in diagrams as equal and denote them by OE_1 and OE_2, but bear in mind that the unit $O\overline{E}_2$ of length on the y-axis "equivalent" to OE_1 is c times the unit OE_2 (see Figs. 182a and 182b, where $c = 5$). Then the lengths of the segment AA_1 computed by means of (37'a) and (37'b) will be the same as the Euclidean and Minkowskian

13. Galilean geometry as a limiting case of Euclidean and Minkowskian geometry

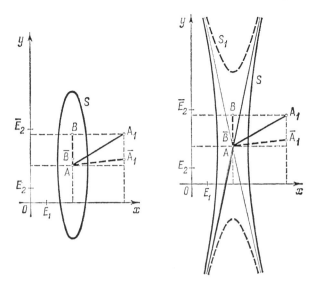

Figure 182a   Figure 182b

lengths of the segment $A\overline{A}_1$ [computed by means of (37a) and (37b)] whose projection to the x-axis is the same as that of the segment AA_1, and whose projection to the y-axis is c times smaller than that of AA_1. As c tends to infinity, we approach in the limit Galilean geometry, where the length of a segment is expressed solely in terms of its projection to the x-axis.

We now replace the segment AA_1 by its projection AB parallel to the y-axis. Similarly, we replace $A\overline{A}_1$ by the vertical projection $A\overline{B}$. Thus the segment $A\overline{B}$ is $(1/c)$th of the segment AB (see Figs. 182a and 182b). As c tends to infinity, $A\overline{B}$ tends to an infinitesimal segment in the sense of both Euclidean and Minkowskian geometry (the units are the same on both axes). This shows that as $c \to \infty$ the length d of such a segment tends to zero, $d \to 0$. The latter fact is easiest to perceive when we view Galilean geometry as a limiting case of Minkowskian geometry. We consider the lines

$$y = \pm cx. \qquad (39)$$

In Minkowskian geometry, any segment on these lines has length zero; on p. 177 such lines (as well as lines parallel to them) were referred to as special lines of Minkowskian geometry. From Figure 182b, where two special lines intersect at A, we infer that as $c \to \infty$ the special lines tend to a vertical line (i.e., a line parallel to the y-axis), and in the limit coincide with it. Thus the vertical line through A becomes the line of segments of zero length.

There is another way of illustrating the latter state of affairs. Figures 182a and 182b show, among other things, the "unit circle" S with center A (whose coordinates we now denote by a and b). The expressions (37'a) and

(37′b) imply that its equation is, in the Euclidean case,

$$(x-a)^2 + \frac{(y-b)^2}{c^2} = 1, \tag{40a}$$

and in the Minkowskian case

$$(x-a)^2 - \frac{(y-b)^2}{c^2} = 1, \tag{40b}$$

i.e., that the curve S represents in the first case an ellipse and in the second a hyperbola. In both cases the semiaxes are 1 and c. As c increases, S is stretched increasingly in the direction of the y-axis. Thus as $c \to \infty$, S tends to *the pair of* (parallel) *lines*

$$(x-a)^2 = 1 \tag{40}$$

—a Galilean unit circle. [Fig. 182b also shows the curve S_1 given by the equation

$$(x-a)^2 - \frac{(y-b)^2}{c^2} = -1, \tag{40c}$$

whose points are likewise at a distance 1 from A, but this distance 1 is qualitatively different, in that it corresponds to lengths of segments on the y-axis. As $c \to \infty$ this "circle" S_1 becomes ever narrower, following the narrowing of the angle between the special lines (through A) which contains it. It also recedes indefinitely from A and thus eventually disappears from any drawing, however large. This corresponds to the disappearance of segments of nonzero length whose nature is different from that of the segment OE_1.]

Next we consider the transformation of (Euclidean) angles resulting from the change of units of length. The (Euclidean) angle δ between the lines

$$y = kx + s \quad \text{and} \quad y = k_1 x + s_1 \tag{41}$$

is determined by the formula

$$\tan \delta = \frac{k_1 - k}{kk_1 + 1} \tag{42a}$$

[see (3), Sec. 3]. As a result of the change of variables (38), the equations (41) change to

$$y' = ckx' + cs \quad \text{and} \quad y' = ck_1 x' + cs_1$$

or

$$y' = k'x' + s' \quad \text{and} \quad y' = k_1'x' + s_1', \tag{41′}$$

where $k' = ck$ and $k_1' = ck_1$, i.e.,

$$k = \frac{1}{c} k' \quad \text{and} \quad k_1 = \frac{1}{c} k_1'.$$

13. Galilean geometry as a limiting case of Euclidean and Minkowskian geometry

Substituting these values of the slopes k and k_1 in (42a) and dropping all primes (i.e., replacing x',y' by x,y, and k',k_1' by k,k_1), we see that as a result of the change of coordinates (38) the angle δ between the lines (41) is now determined by the formula

$$\tan \delta = \frac{(1/c)k_1 - (1/c)k}{(1/c^2)kk_1 + 1}. \tag{42'}$$

As $c \to \infty$, the right-hand side of (42') tends to zero. In other words, $\tan \delta \to 0$ and therefore also $\delta \to 0$—an outcome both unexpected and unwelcome.

However, our difficulty is not very serious. Equation (42') shows that for large c the quantity

$$\frac{1}{c} \frac{k_1 - k}{1 + (kk_1/c^2)},$$

i.e., the tangent of the angle δ between the lines (41), is very small—of the order of $1/c$.[27] If angles are measured in radians, then

$$\lim_{\delta \to 0} \frac{\sin \delta}{\delta} = 1, \qquad \lim_{\delta \to 0} \frac{\cos \delta}{\delta} = 1,$$

and therefore

$$\lim_{\delta \to 0} \frac{\tan \delta}{\delta} = 1. \tag{43}$$

It follows that for large c the magnitude of the angle δ [computed by means of (42')] is very small: it is of the order of $1/c$.

We now change the angular unit in the new coordinate system $\{x',y'\}$ [where the connection between x',y' and x,y is given by (38)] from the old unit ε (radians) to a new unit ε' such that $\varepsilon' = (1/c)\varepsilon$. Then the new magnitude δ' of an angle is c times larger than its old magnitude δ:

$$\delta' = c\delta \quad \text{or} \quad \delta = \frac{1}{c}\delta'. \tag{44}$$

Substituting this value of δ in (42') and (as usual) dropping primes, we obtain the equality

$$\tan \frac{\delta}{c} = \frac{1}{c} \frac{k_1 - k}{1 + (1/c^2)kk_1},$$

or

$$c \tan \frac{\delta}{c} = \frac{k_1 - k}{1 + (1/c^2)kk_1}. \tag{42''}$$

Finally, letting $c \to \infty$ and bearing in mind the fact that

$$\lim_{c \to \infty} c \tan \frac{\delta}{c} = \delta \lim_{\delta/c \to 0} \frac{\tan(\delta/c)}{\delta/c} = \delta \cdot 1 = \delta,$$

---

[27] More precisely, as $c \to \infty$, $\tan \delta \to 0$, $1/c \to 0$, but $\tan \delta/(1/c) \to k_1 - k \neq 0$.

we see that
$$\delta = k_1 - k, \qquad (42)$$
i.e., we obtain the formula for the angle δ between the lines (41) in the Galilean plane [cf. (8), Sec. 3]. In much the same way, we use the substitution (38) to obtain (42) from the formula
$$\tanh \delta = \frac{k_1 - k}{1 - k_1 k} \qquad (42b)$$
for the magnitude of the angle between two lines (41) (of the same kind) in the Minkowskian plane [Eq. (42b) can be easily deduced from Eq. (21c), Sec. 12].

Our argument shows that the procedure (just described) for changing the units of length, which in conjunction with the limiting process $c \to \infty$ reduces Euclidean (or Minkowskian) geometry to Galilean geometry, is accompanied by a decrease in all angles. Indeed, in order to obtain finite expressions for the magnitude of angles we must resort to the change (44) of angular unit. This implies that in order to go from Euclidean to Galilean geometry, we must make the replacement
$$\delta \to \frac{\delta}{c} \qquad (44')$$
in all relations involving angles [see the second relation in (44)], and the replacements
$$\sin \delta \to \sin \frac{\delta}{c} \sim \frac{\delta}{c}, \quad \cos \delta \to \cos \frac{\delta}{c} \sim 1, \quad \tan \delta \to \tan \frac{\delta}{c} \sim \frac{\delta}{c} \qquad (43')$$
[see (43) and its analogues]. Thus, in the limit, the Euclidean law of sines
$$\frac{a}{\sin A} = \frac{b}{\sin B} = \frac{c}{\sin C} \qquad (33a)$$
yields
$$\frac{a}{A/m} = \frac{b}{B/m} = \frac{c}{C/m}$$
[here and up to the derivation of the relation (46) we use the letter m instead of c because we wish to reserve the letter c to denote the side AB of the triangle ABC], i.e., the law of sines of Galilean geometry:
$$\frac{a}{A} = \frac{b}{B} = \frac{c}{C} \qquad (33)$$
Similarly, starting with the Euclidean law of cosines
$$a^2 = b^2 + c^2 - 2bc \cos A, \qquad (45a)$$
we obtain
$$a^2 = b^2 + c^2 - 2bc \cdot 1 \quad [=(c-b)^2],$$
i.e., the familiar formula
$$c = a + b \qquad (34)$$

13. Galilean geometry as a limiting case of Euclidean and Minkowskian geometry

of Galilean geometry (here we assume that $c>a$, $c>b$); from the Euclidean formula
$$S = \tfrac{1}{2} ab \sin C$$
for the area of a triangle we obtain via the replacements $\sin C \to C/m$ and $S \to S/m$ (the latter amounts to a change of unit of area), the formula
$$S = \tfrac{1}{2} abC \tag{46}$$
for the area of a triangle in Galilean geometry [see (17′) and (17), Sec. 4]; and so on. Finally, by making the replacements
$$x \to x, \quad y \to \frac{1}{c} y, \quad \sin \alpha \to \frac{\alpha}{c}, \quad \cos \alpha \to 1$$
[see (38) and (43′)] in the formulas
$$\begin{aligned} x' &= x \cos \alpha + y \sin \alpha + a \\ y' &= -x \sin \alpha + y \cos \alpha + b \end{aligned} \tag{47a}$$
for a Euclidean motion [see (6), Sec. 1] we obtain
$$x' = x \cdot 1 + \frac{y}{c} \cdot \frac{\alpha}{c} + a,$$
$$\frac{y'}{c} = -x \cdot \frac{\alpha}{c} + \frac{y}{c} \cdot 1 + b,$$

or

$$x' = x + \frac{1}{c^2} \alpha y + a,$$
$$y' = -\alpha x + y + cb.$$

Replacing cb by b and $-\alpha$ by v and letting c tend to infinity, we arrive at
$$\begin{aligned} x' &= x + a, \\ y' &= vx + y + b, \end{aligned} \tag{47}$$
i.e., the formulas for a motion of the Galilean plane [see (1), Sec. 3]. In much the same way, it is possible to obtain the relations (33), (34), (46), and (47) from the corresponding relations of Minkowskian geometry [for example, (47) from (23), Sec. 12]; we need only note that
$$\lim_{\delta \to 0} \frac{\sinh \delta}{\delta} = 1, \quad \lim_{\delta \to 0} \cosh \delta = 1, \quad \lim_{\delta \to 0} \frac{\tanh \delta}{\delta} = 1. \tag{43′}$$

Next we investigate the connection between a Euclidean circle and a Galilean cycle. We defined a cycle as the set of points from which a given segment AB can be seen at a constant angle α (Sec. 6). We must, therefore, define a Euclidean circle as *the set S of points $M(x,y)$ of the Euclidean plane from which a segment AB can be seen at a constant (directed) angle α* (Fig. 183). In accordance with Section 6, we denote the coordinates of A and B by (a_1, a_2) and (b_1, b_2); then the slopes of the lines AM and BM are
$$k = \frac{a_2 - y}{a_1 - x} \quad \text{and} \quad k_1 = \frac{b_2 - y}{b_1 - x}$$

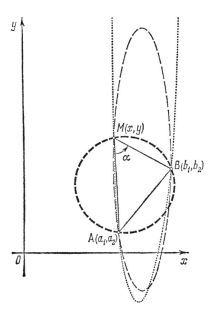

Figure 183

(see pp. 77–78). In view of the formula (42a) for the (Euclidean) angle between two lines, we see that the equation of the above set S of points $M(x,y)$ is

$$\frac{[(b_2-y)/(b_1-x)]-[(a_2-y)/(a_1-x)]}{[(b_2-y)/(b_1-x)]\cdot[(a_2-y)/(a_1-x)]+1}=\tan\alpha$$

or

$$(b_2-y)(a_1-x)-(b_1-x)(a_2-y)$$
$$-\tan\alpha[(b_2-y)(a_2-y)+(b_1-x)(a_1-x)]=0,$$

or finally,

$$(\tan\alpha)(x^2+y^2)+[(b_2-a_2)-\tan\alpha(b_1+a_1)]x$$
$$+[(a_1-b_1)-\tan\alpha(a_2+b_2)]y$$
$$+[(b_1a_2-b_2a_1)+\tan\alpha(a_1b_1+a_2b_2)]=0. \quad (48a)$$

This is clearly the equation of a circle (indicated by heavy dashes in Fig. 183). The replacements

$$x\to x, \quad y\to\frac{y}{c}, \quad a_1\to a_1, \quad a_2\to\frac{a_2}{c}, \quad b_1\to b_1, \quad b_2\to\frac{b_2}{c}, \quad \tan\alpha\to\frac{\alpha}{c}$$

13. Galilean geometry as a limiting case of Euclidean and Minkowskian geometry

transform (48a) into

$$\frac{\alpha}{c}\left(x^2 + \frac{y^2}{c^2}\right) + \left[\left(\frac{b_2}{c} - \frac{a_2}{c}\right) - \frac{\alpha}{c}(b_1 + a_1)\right]x$$

$$+ \left[(a_1 - b_1) - \frac{\alpha}{c}\left(\frac{a_2}{c} + \frac{b_2}{c}\right)\right]\frac{y}{c}$$

$$+ \left[\left(b_1\frac{a_2}{c} - \frac{b_2}{c}a_1\right) + \frac{\alpha}{c}\left(a_1 b_1 + \frac{a_2}{c}\frac{b_2}{c}\right)\right] = 0,$$

or

$$\alpha\left(x^2 + \frac{y^2}{c^2}\right) + \left[(b_2 - a_2) - \alpha(b_1 + a_1)\right]x$$

$$+ \left[(a_1 - b_1) - \frac{\alpha}{c^2}(a_2 + b_2)\right]y$$

$$+ \left[(b_1 a_2 - b_2 a_1) + \alpha\left(a_1 b_1 + \frac{a_2 b_2}{c^2}\right)\right] = 0, \quad (48b)$$

which is the equation of an ellipse (represented by light dashes in Fig. 183). Finally, transition to the limit as $c \to \infty$ transforms (48b) into

$$\alpha x^2 + \left[(b_2 - a_2) - \alpha(b_1 + a_1)\right]x + (a_1 - b_1)y + \left[(b_1 a_2 - b_2 a_1) + \alpha a_1 b_1\right] = 0, \quad (48)$$

which represents a (Euclidean) parabola, and thus a cycle in the Galilean plane (represented by light dots in Fig. 183; cf. pp. 77–79). All the results of Section 10 pertaining to Galilean geometry can also be obtained by a limiting process from the corresponding results of Euclidean geometry (proved in Sec. 10) or Minkowskian geometry (see Sec. 12).

It is now easy to discuss the relation between "classical" and "relativistic" mechanics or, equivalently, between the Galilean and Einsteinian principles of relativity. In Section 11 we always chose units of length and time in such a way that the speed of light was 1. With this choice of units, the version of Minkowskian geometry developed in Section 11 yielded a geometric picture of kinematics on the line o based on Einstein's principle of relativity. However, it is customary to choose a unit of length which is c times larger than the unit employed in Section 11. With this choice of unit of length the following replacements must be made in all the formulas in Section 11:

$$t \to t, \quad x \to \frac{x}{c}, \quad v \to \frac{v}{c} \quad (49)$$

[see (38), this section]. In particular, c is the speed of light in the system of units of length and time now under consideration.

After making the replacements (49) in the fundamental relations (7) in Section 11, we obtain the more familiar form of a Lorentz transformation:

$$\frac{x'}{c'} = \frac{(x/c)-(v/c)t}{\sqrt{1-(v/c)^2}}, \qquad t' = \frac{-(v/c)\cdot(x/c)+t}{\sqrt{1-(v/c)^2}},$$

or

$$\boxed{x' = \frac{x-vt}{\sqrt{1-(v^2/c^2)}}, \qquad t' = \frac{-(v/c^2)x+t}{\sqrt{1-(v^2/c^2)}},} \tag{50}$$

The latter is the form of a Lorentz transformation found in most books on relativity. It is clear that transition to the limit $c\to\infty$ transforms the formulas (50) into the formulas

$$x' = x - vt, \qquad t' = t \tag{50'}$$

of a Galilean motion. If $c\to\infty$, then the Fitzgerald contraction coefficient $\sqrt{1-(v^2/c^2)}$ and the (relativistic) time-delay coefficient $1/\sqrt{1-(v^2/c^2)}$ (cf. pp. 167–168) tend to 1, i.e., neither effect occurs in classical mechanics. Finally, the replacements (49) transform the formula

$$w = \frac{u+v}{uv+1} \tag{51}$$

of addition of velocities [cf. (9), Sec. 11] into the formula

$$\frac{w}{c} = \frac{(u/c)+(v/c)}{(u/c)(v/c)+1} \quad \text{or} \quad w = \frac{u+v}{1+(uv/c^2)}.$$

If $c\to\infty$, then the latter reduces to the classical formula of Galileo and Newton for addition of velocities:

$$w = u + v \tag{51'}$$

[cf. (8), Sec. 11].

PROBLEMS AND EXERCISES

15 Prove the relations (43').

16 Investigate the connection between inversive mappings in Euclidean, Galilean, and Minkowskian geometry.

17 Give examples of theorems of Galilean geometry that are limiting cases of theorems of Euclidean and Minkowskian geometry; describe the corresponding limiting processes.

VI Give the limiting process connecting the stereographic projections described in Sections 10 and 12 and the three inversive planes. Develop the analogies between the inversive geometries in the Euclidean, Galilean, and Minkowskian planes, including the theory of pencils of circles (cycles). Derive the fundamental theorem on cyclic transformations of the Galilean plane (cf.

13. Galilean geometry as a limiting case of Euclidean and Minkowskian geometry   213

p. 154) by applying a limiting process to the corresponding Euclidean theorem.

VII Compare the theory of quadric curves in the Euclidean, Galilean, and Minkowskian planes (see Problem **XVI** in Chap. II and Problem **IV** in Sec. 12).

VIII Outline the connections between three-dimensional semi-Galilean geometry (see Exercise **5** in the Introduction) and the three-dimensional Galilean and pseudo-Galilean geometries (see p. 231). [*Hint*: Three-dimensional semi-Galilean geometry occupies an intermediate position between three-dimensional Galilean geometry and three-dimensional pseudo-Galilean geometry; it can be obtained as a limiting case from each of them.]

Supplement A. Nine plane geometries

This book is essentially devoted to a comparative study of two geometric systems that can be introduced in the (ordinary or affine) plane, namely, the familiar *Euclidean geometry* and the simpler *Galilean geometry* which, in spite of its relative simplicity, confronts the uninitiated reader with many surprising results. In Section 12, we learned of a third geometric system, namely, Minkowskian geometry. Finally, most readers of this book are probably aware of the existence of, or are familiar with, a fourth geometric system, namely Bolyai–Lobachevskian (hyperbolic) geometry.

In this Supplement, we propose to describe a number of plane geometries including Euclidean, Galilean, Minkowskian, and Bolyai–Lobachevskian. Since these geometries were first introduced in 1871 by Klein,[1] who used the earlier work of Cayley,[2] it seems best to call them *Cayley–Klein geometries* (rather than non-Euclidean geometries; the latter term is obviously not applicable to Euclidean geometry and is applicable to a great many geometries not dealt with in this Supplement).

Following Cayley and Klein, we distinguish three fundamentally different geometries on a line o. They are *Euclidean geometry*, *elliptic geometry*, and *hyperbolic geometry*. Euclidean geometry is based on the familiar rule for measuring the length of segments. According to this rule we choose a unit of length OE on o and define the distance between points A and B on o by the formula[3]

$$d_{AB}^{(P)} = \frac{AB}{OE}. \tag{1}$$

This distance is called *parabolic* (hence the letter P in $d_{AB}^{(P)}$). If C is a point

---

[1] See Klein [73] (first published in 1871) or Klein [56].

[2] For an historical account of the discovery of these geometries, see Sec. 7, Chap. X of Klein [56] or the very interesting history of mathematics by Klein [79].

[3] The ratio of two segments on o can be defined in affine geometry without developing Euclidean geometry.

of the segment AB, then
$$d_{AC}^{(P)} + d_{CB}^{(P)} = d_{AB}^{(P)}$$
(Fig. 184a). The Euclidean motions of o, i.e., the transformations of o that preserve the distance $d_{AB}^{(P)}$ between points, are

(a) translations (along o);
(b) reflections (of o about its points).

To introduce elliptic geometry on o, we pick a point Q not on o and define the *elliptic distance* $d_{AB}^{(E)}$ between A and B on o as the usual Euclidean measure of $\angle AQB$:
$$d_{AB}^{(E)} = \angle AQB$$
(Fig. 184b). If A, B, and C are three consecutive points on o, then
$$d_{AC}^{(E)} + d_{CB}^{(E)} = d_{AB}^{(E)}$$
(Fig. 184b). The elliptic motions of o which preserve the elliptic distance $d_{AB}^{(E)}$ are:

(a) transformations induced by rotating the *pencil with center* Q (i.e., the lines through Q) through a fixed angle α; such a transformation takes A to the point A' such that $\angle AQA' = \alpha$;
(b) transformations induced by reflecting the pencil with center Q about a line l of the pencil; such a transformation takes A to the point A' such that l bisects $\angle AQA'$.

Hyperbolic geometry is introduced on o by fixing two of its points, I and J, and defining the *hyperbolic distance* $d_{AB}^{(H)}$ between A and B on o by the formula
$$d_{AB}^{(H)} = k\log\frac{AI/AJ}{BI/BJ} \tag{1b}$$
(Fig. 184c). The choice of k is arbitrary, and changing the value of k (with the base b of the logarithm fixed, say $b=10$) amounts to a change of the unit used in measuring distances (lengths of segments). It is convenient to assume that the segments on the right-hand side of the cross ratio
$$(A,B;I,J) = \frac{AI/AJ}{BI/BJ} \tag{2}$$
of A,B,I,J are directed. In view of the fact that a real-valued logarithm is defined only for positive reals, we must ensure that the cross ratio (2) is

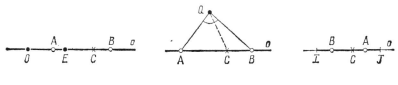

Figure 184a   Figure 184b   Figure 184c

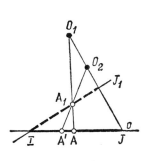
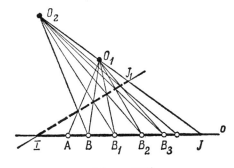

Figure 185a Figure 185b

positive. One way of doing this is to assume that A and B are points of the segment IJ. Then the segment IJ represents the whole *hyperbolic line*. It is easy to show that if A, C, B are three consecutive points on IJ (cf. Fig. 184c), then

$$d_{AC}^{(H)} + d_{CB}^{(H)} = d_{AB}^{(H)};$$

indeed,

$$(A,C; I,J) \cdot (C,B; I,J) = \frac{AI/AJ}{CI/CJ} \cdot \frac{CI/CJ}{BI/BJ}$$

$$= \frac{AI/AJ}{BI/BJ}$$

$$= (A,B; I,J),$$

and so

$$\log(A,C; I,J) + \log(C,B; I,J) = \log(A,B; I,J).$$

Another fact is that the cross ratio has the same value for two sets of quadruples on o related by a central projection.[4] It follows that the hyperbolic motions of o (more precisely, of the hyperbolic line, i.e., the segment IJ) which preserve the hyperbolic distance $d_{AB}^{(H)}$ are generated by

(a) reflections about the midpoint S of IJ [clearly, if such a reflection takes A, B to A', B', then $(A,B; I,J) = (A',B'; J,I)$];
(b) central projections of IJ to itself, where IJ is first projected to some segment IJ_1 from a point O_1, and IJ_1 is in turn projected to IJ from a point O_2 on the line J_1J (Fig. 185a).

Figure 185b shows a sequence of equal (in the sense of hyperbolic geometry) segments $AB, BB_1, B_1B_2, B_2B_3, \ldots$. This figure shows that if, beginning with some point A of the (hyperbolic) line IJ, we lay off a sequence of equal (in the sense of hyperbolic geometry) segments

$$AB = BB_1 = B_1B_2 = B_2B_3 = \cdots,$$

[4]See, for example, [12], [19], [32], or [67].

Supplement A. Nine plane geometries   217

we remain within the segment IJ, i.e., the hyperbolic line IJ is infinite. [This can also be seen from Fig. 184c. Indeed, if B tends to I, then $(A,B; I,J)=(AI/AJ)/(BI/BJ)\to\infty$ (since $BI\to 0$). Consequently, $d_{AB}^{(H)}\to\infty$.]

Just as there are three ways of measuring length that give rise to three geometries on a line o, so, too, are there three ways of measuring angles that give rise to three geometries in a pencil of lines with center O. The usual measure of angles in the pencil with center O given by the formula

$$\delta_{ab}^{(E)} = \angle aOb \tag{3a}$$

(Fig. 186a; the symbol $\angle$ is used here as well as below in the usual Euclidean sense) is called *elliptic*. The so-called *parabolic measure of angles* is very different from that used in Euclidean geometry; it can be defined by choosing a line q not passing through O and putting

$$\delta_{ab}^{(P)} = \overline{AB}\ (=d_{AB}^{(P)},\text{ the signed Euclidean length of } AB), \tag{3}$$

where $A=a\cap q$ and $B=b\cap q$ (Fig. 186b). It is clear that our pencil contains a unique line $\omega\|q$ which forms an "infinitely large angle" with every other line of the pencil. Finally, the *hyperbolic measure of angles* is introduced in our pencil as follows: we fix two lines i and j of the pencil, and for any two other lines a and b we put

$$\delta_{ab}^{(H)} = k\log\left(\frac{\sin\angle(a,i)/\sin\angle(a,j)}{\sin\angle(b,i)/\sin\angle(b,j)}\right) \tag{3b}$$

(Fig. 186c). The constant k depends on the choice of unit. It is convenient to regard the angles on the right-hand side of (3b) as directed. Define the cross ratio $(a,b;\ i,j)$ of four lines a,b,i,j as the quantity

$$(a,b;\ i,j) = \frac{\sin\angle(a,i)/\sin\angle(a,j)}{\sin\angle(b,i)/\sin\angle(b,j)}. \tag{2'}$$

For $\delta_{ab}^{(H)}$ to make sense we must ensure that $(a,b;\ i,j)>0$. One way of doing this is to assume that the lines $a,b,\ldots$ belong to a definite pair of vertical angles formed by i and j. Since the cross ratio (2') of four lines of the pencil is equal to the cross ratio of the four points in which these lines meet an arbitrary fixed line q (not passing through the center of the

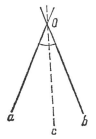

Figure 186a

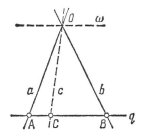

Figure 186b

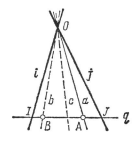

Figure 186c

Table I Nine Cayley–Klein geometries in the plane

Measure of angles	Measure of length		
	Elliptic	Parabolic	Hyperbolic
elliptic	elliptic geometry	Euclidean geometry	hyperbolic geometry
parabolic (Euclidean)	co-Euclidean geometry	Galilean geometry	co-Minkowskian geometry
hyperbolic	cohyperbolic geometry	Minkowskian geometry	doubly hyperbolic geometry

pencil),[5] we could also define $\delta_{ab}^{(H)}$ as

$$\delta_{ab}^{(H)} = d_{AB}^{(H)}, \tag{3c}$$

where A and B are the points in which the lines a and b meet an arbitrary fixed line q (see Fig. 186c). The equality (3c) implies that (in the hyperbolic measure of angles) each of the lines i,j forms an "infinitely large angle" with every other line.

The Cayley–Klein scheme yields nine plane geometries. Specifically, we can choose one of three ways (parabolic, elliptic, or hyperbolic) of measuring length on a line and one of three ways of measuring angles in a pencil with center O. This gives nine ways of measuring lengths and angles and thus the nine plane geometries listed in Table I.

We are familiar with the geometries of Euclid, Galileo, and Minkowski (middle column of Table I) in all of which *the metric on the line is parabolic*; the length of a segment AB is defined as a ratio AB/OE, where OE is a "unit segment" on the line.

In Euclidean geometry, the endpoints of all unit segments issuing from a point O lie on a (Euclidean) circle S (Fig. 187a). Angles in a pencil are measured in the elliptic, i.e., in the usual, way, which amounts to measuring the length of the arc of S cut off on S by the appropriate lines: in Figure 187a, $\delta_{ab} = \angle aOb = \text{arc } AB$.

In Galilean geometry, the endpoints of all unit segments issuing from a point O lie on a "Galilean circle" S (Fig. 187b) consisting of a pair of parallel ("special") lines (of which more will be said below). Angles in a pencil are measured in the parabolic, i.e., not in the usual, way: the angle δ_{ab} between the lines a and b is equal to the (Euclidean) length of the segment AB cut off by these lines on the Galilean circle, i.e., on the line q, in Figure 187b (compare this with Fig. 186b). We note that the circle S does not determine a unit segment on any line $\omega \| q$. In fact, comparing a segment on ω with other segments we are compelled to assign to it the length zero. But then, contrary to our insistence that length in a Cayley–Klein plane geometry must be measured in the same way on every line, we end up with a metric in which the distance between any points P and Q on

---

[5]See footnote 4.

Supplement A. Nine plane geometries

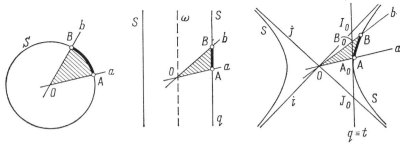

Figure 187a   Figure 187b   Figure 187c

ω is zero. Thus, strictly speaking, the "special lines" (i.e., the lines parallel to q) must be excluded from the class of lines (which is what we often did!). On the other hand, if we restrict Galilean geometry to a study of the affine plane and its ordinary lines (briefly, the Galilean plane), then we end up with a "plane" which fully satisfies the restrictions imposed by our table.

In Minkowskian geometry, the "unit circle" S—the locus of the end points of all unit segments issuing from a point O—is a Euclidean hyperbola (Fig. 187c). Earlier (see p. 182) we defined the angle δ_{ab} between two lines a and b as

$$\delta_{ab} = 2s_{AOB},$$

where s_{AOB} is the area of the shaded sector of the Minkowskian circle S (Fig. 187c), an idea suggested by Euclidean and Galilean geometry. However this quantity can also be defined as

$$\delta_{ab} = k \log\left(\frac{\sin \angle (a,i)/\sin \angle (a,j)}{\sin \angle (b,i)/\sin \angle (b,j)} \right), \tag{3b}$$

where i and j are the asymptotes of the hyperbola S, or

$$\delta_{ab} = k \log\left(\frac{A_0 I_0 / A_0 J_0}{B_0 I_0 / B_0 J_0} \right), \tag{3c}$$

where A_0, B_0, I_0, and J_0 are the points in which the lines a, b, i, and j meet a fixed line q, say, the line t tangent to S at its vertex (try to prove this result)[6]; the number k in (3b) and (3c) is determined by the choice of base

---

[6]In translator's footnote 12 of Section 12 we showed that $\varphi = \delta_{OA,OM} = \log(x + \sqrt{x^2 - 1}) =$ Minkowskian length of arc AM (φ and arc AM appear in Fig. 166a). Now

$$x + \sqrt{x^2 - 1} = x + y = \frac{1 + (y/x)}{1/x} = \frac{1 + (y/x)}{\sqrt{1 - (y^2/x^2)}} = \sqrt{\frac{1 + (y/x)}{1 - (y/x)}} = \sqrt{\frac{1 + m}{1 - m}},$$

where m is the slope of OM. But then $\delta_{OA,OM}$ = Minkowskian length of arc $AM = \log(x + \sqrt{x^2 - 1}) = \frac{1}{2}\log[(1 + m)/(1 - m)] = \frac{1}{2}\log(A, M; I, J)$, where I, J are the points of intersection of the line AN in Figure 166a with the lines $y = x$ and $y = -x$, respectively. The proof of the general case is left to the reader. (Translator's note.)

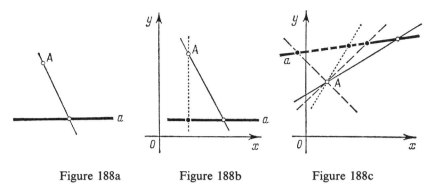

Figure 188a   Figure 188b   Figure 188c

of the logarithms. It follows that in Minkowskian geometry the measure of angles in a plane pencil with center O is hyperbolic.

In accordance with the requirement of congruence of any two lines in the plane, it is natural to define the "Minkowskian plane" as the points of the (ordinary) plane and the lines of one kind—say, the first. In this connection we note that in Euclidean geometry every point on a line a can be joined by a line to a point A not on a (Fig. 188a); in Galilean geometry, each line a contains a unique point, "parallel" to A, which cannot be joined by an (ordinary) line to A (Fig. 188b); in Minkowskian geometry, there are infinitely many points on a line a of the first kind which cannot be joined to A by means of lines of the first kind (Fig. 188c). On the other hand, all three geometries share the property that through any point A not on a there passes a unique line not intersecting a (a unique line parallel to a).

We must still describe the remaining six Cayley–Klein geometries in the plane (see Table I). Of these, *elliptic geometry* closely resembles spherical geometry, the oldest "non-Euclidean" geometry known to mankind. Nevertheless, Riemann was the first person to juxtapose this geometry, classical Euclidean geometry, and the (then just discovered) hyperbolic geometry in his famous memoir [74] "*On the hypotheses underlying the foundations of geometry*" (1854).[7]

Let Σ be a unit sphere in three-dimensional Euclidean space. We call a great circle of Σ (i.e., a plane section of Σ passing through its center O; Fig. 189a) a *line*. If we defined a "point" to be simply a point of Σ, then the intersection of two lines would consist of two points rather than one. To avoid this awkward situation we define a *point* of the elliptic plane to be a pair of antipodal points of Σ. Thus the elliptic plane is not the sphere, but rather a hemisphere Σ' with identified ("glued together") antipodal points of its boundary circle (Fig. 189b). We define the distance d_{AB} between two points of the elliptic plane (i.e., two points on a great circle, or rather semicircle) as the length of the arc AB. This measure of length is clearly elliptic. By the angle δ_{ab} between the lines a and b of the elliptic

---

[7]See footnote 4 of the Preface; cf. [79].

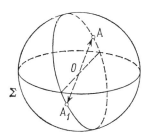

Figure 189a

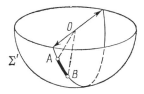   

Figure 189b   Figure 189c

plane we shall mean the dihedral angle between the planes containing the great semicircles a and b (Fig. 189c); this is the elliptic, i.e., the usual measure of angle in a pencil of lines. Finally, by the motions of the elliptic plane we shall mean the rotations of Σ (viewed as a set of pairs of antipodal points) about its center O.[8]

Hyperbolic geometry and doubly hyperbolic geometry can be described in a similar manner. In three-dimensional Euclidean space we replace the usual distance between $A(x,y,z)$ and $A_1(x_1,y_1,z_1)$ given by[9]

$$d^2 = (x_1-x)^2 + (y_1-y)^2 + (z_1-z)^2 \qquad (4')$$

by the distance

$$d^2 = (x_1-x)^2 + (y_1-y)^2 - (z_1-z)^2, \qquad (4)$$

and call the resulting space *Minkowskian*. Formula (4) shows that in Minkowskian three-dimensional space there are two types of "unit spheres," namely, a "sphere" Σ_1 with center $O(0,0,0)$ and imaginary radius i given by

$$x^2 + y^2 - z^2 = -1 \qquad (5)$$

(a Euclidean hyperboloid of two sheets shown in Fig. 190a), and a "sphere" Σ_2 of radius 1 given by

$$x^2 + y^2 - z^2 = 1 \qquad (5')$$

(a Euclidean hyperboloid of one sheet shown in Fig. 191a). By *points* of the

---

[8] Elliptic geometry is obviously very close to spherical geometry, which investigates figures on the surface of a sphere (see, for example, [76] and [25]). An introduction, in English, to these and other geometries is found in [72b].

[9] See footnote 3 of Section 1.

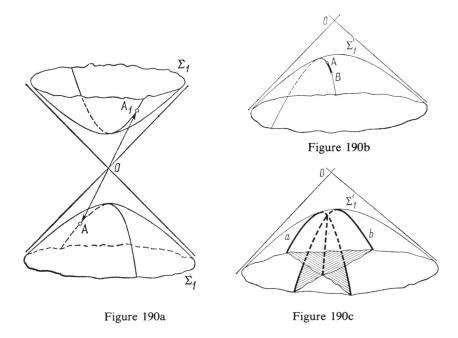

Figure 190b

Figure 190a

Figure 190c

hyperbolic plane we mean pairs of antipodal points of Σ_1 (which means that the hyperbolic plane is represented by a single sheet Σ'_1 of Σ_1; Fig. 190b), and by *lines* we mean "great circles of the sphere" Σ_1, i.e., plane sections of Σ_1 passing through its center O. It is clear that a line is a Euclidean hyperbola (or rather a branch of such a hyperbola; see Figs. 190a and 190b) on which distance can be defined in the manner in which we defined angular measure in a pencil of lines in the Minkowskian plane; this metric is, of course, hyperbolic.[10] The nature of a "pencil of lines" (the set of sections of Σ'_1 by a pencil of planes; see Fig. 190c) is very much like that of a pencil of lines in the Euclidean plane, and it is therefore natural to introduce in such a pencil the elliptic (i.e., the usual) measure of angles. This agrees fully with the requirements in Table I pertaining to hyperbolic geometry.[11] The motions of the hyperbolic plane are defined as rotations of the "sphere" Σ_1 (or "hemisphere" Σ'_1), i.e., as transformations of

---

[10] The distance d_{AB} between points A and B can also be defined as the Minkowskian length of the segment AB of the corresponding line, i.e., as the limit of the lengths of the polygonal lines inscribed in AB, where the length of a polygonal line is the sum of the lengths of its segments, and the length of a segment with endpoints (x,y,z) and (x_1,y_1,z_1) is given by the formula

$$d^2 = |(x_1-x)^2 + (y_1-y)^2 - (z_1-z)^2|.$$

[11] In 1829, the eminent Russian mathematician N. I. Lobachevsky discovered this first "non-Euclidean" geometry. The new geometric system was also discovered, independently and almost simultaneously, by the great German mathematician C. F. Gauss and the Hungarian J. Bolyai; cf. footnote 3 of the Preface.

Supplement A. Nine plane geometries

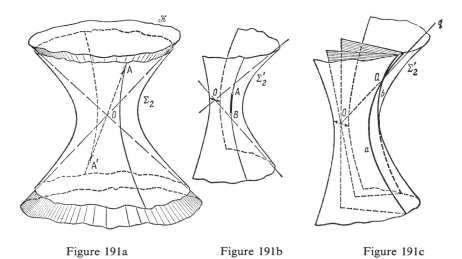

Figure 191a   Figure 191b   Figure 191c

three-dimensional space $\{x,y,z\}$ which keep the origin O fixed and preserve the distance (4) (it is clear that these transformations map Σ_1 to itself).[12]

Next we turn to the "sphere" Σ_2 (Fig. 191a). Here we also identify antipodal points, i.e., we consider the "hemisphere" Σ_2' with identified antipodal points on its boundary "circle" (Fig. 191b). By the *points* of the "plane" we are about to construct we mean the points of Σ_2' (pairs of antipodal points of Σ_2), and by its *lines* we mean those sections of Σ_2 (Σ_2') by planes through O which are hyperbolas (rather than ellipses!). On such a "line" (i.e., hyperbola) we can introduce a hyperbolic measure of length.[13] A pencil of lines with center Q of our geometry is determined by the planes of the pencil with axis $OQ \equiv q$ which intersect the cone $\mathcal{K}$ given by

$$x^2 + y^2 - z^2 = 0 \qquad (5a)$$

(see Figs. 191a and 191b). The structure of such a pencil is analogous to the structure of a pencil of lines in the Minkowskian plane—for example, its determining pencil of planes contains two "limiting" planes i,j tangent to $\mathcal{K}$, and consists of the planes in the interior of the dihedral angle formed by i and j. In such a pencil we can introduce a hyperbolic measure of angles by defining the angle δ_{ab} between lines a and b by means of the by now familiar formula

$$\delta_{ab} = k \log \left(\frac{\sin \angle (a,i)/\sin \angle (a,j)}{\sin \angle (b,i)/\sin \angle (b,j)} \right) \qquad (3b)$$

---

[12] This approach to hyperbolic geometry enables us to develop it extensively by relying on the analogy between Euclidean and Minkowskian space (see, for example, [25]).

[13] Cf. footnote 10.

[where, for example, $\angle(a,i)$ denotes the usual, i.e., Euclidean, dihedral angle between the plane i and the plane determined by the "line" a]. The resulting "plane" with its measures of length and angle is the doubly hyperbolic plane, and its motions are again rotations of the "sphere" Σ_2 (regarded as a set of pairs of antipodal points) of Minkowskian space about its center O. We note that in hyperbolic and in doubly hyperbolic geometry there are infinitely many lines through a given point not on a line a which do not intersect a (Figs. 192a and 192b). On the other hand, whereas in hyperbolic geometry any two points A and B determine a unique line (since any three points A, B, O in space determine a unique plane), this is not the case in doubly hyperbolic geometry. Specifically, each line a contains infinitely many points which cannot be joined to a point A not on a (for not all planes AOB, where B varies on the hyperbola a, intersect Σ_2 along a hyperbola; cf. Fig. 192c).

There is a simple way of obtaining from each of our geometries a new (but closely related) geometry. All we need do is call a point a "line" and a line a "point" (and, of course, call the distance between points the "angle between lines," and call the angle between lines the "distance between points"). We know (see Sec. 5) that such a verbal transformation takes Galilean geometry to itself. The same is true of elliptic geometry and of doubly hyperbolic geometry (in both of which the principle of duality is also valid). Our verbal transformation takes each of the geometries of Euclid, Minkowski, and Bolyai–Lobachevsky to its "dual" or its "cogeometry"; Euclidean to co-Euclidean, Minkowskian to co-Minkowskian, and hyperbolic to cohyperbolic.

In conclusion, we present Tables II and III, which characterize the nine Cayley–Klein geometries from the point of view of parallelism of points and lines, and Table IV which summarizes the fundamental relations connecting the elements of a triangle ABC in each of these geometries. These relations are stated without proof (see, however, pp. 280–282). The letters "E", "P", and "H" in Tables II–IV refer to the elliptic, parabolic, or hyperbolic nature of the metric. The letters a and A in Tables II and III

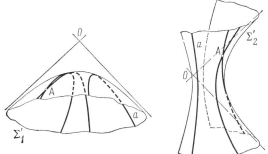

Figure 192a   Figure 192b   Figure 192c

Supplement A. Nine plane geometries

Table II Number of lines passing through the point A which do not intersect the line a

Measure of angles	Measure of lengths		
	E	P	H
E	0	1	∞
P	0	1	∞
H	0	1	∞

Table III Number of points on the line a which cannot be joined by a line to the point A

Measure of angles	Measure of lengths		
	E	P	H
E	0	0	0
P	1	1	1
H	∞	∞	∞

denote an arbitrary line and an arbitrary point not on that line. The letters A, B, C in Table IV denote the magnitudes of the angles of the triangle ABC, and $a \equiv BC$, $b \equiv CA$, $c \equiv AB$ represent the lengths of its sides.[14]

The interpretation of plane hyperbolic geometry by means of Minkowskian geometry in three-dimensional space yields a curious and sometimes overlooked connection between hyperbolic geometry and Einstein's theory of relativity. Thus, consider the relativistic kinematics of plane (more accurately, plane-parallel) motions (cf. p. 16). It is not difficult to see that its geometric equivalent is the study of three-dimensional space $\{x, y, t\}$ whose motions are defined by formulas related to formulas (7) and (10) in Section 11; for example, if the reference frame $\{x', y', t'\}$ moves with respect to the reference frame $\{x, y, t\}$ with uniform velocity v in the direction of the x-axis, then the connection between the two coordinate systems is given by

$$
\begin{aligned}
x' &= \frac{1}{\sqrt{1-v^2}} x - \frac{v}{\sqrt{1-v^2}} t + a, \\
y' &= \phantom{\frac{1}{\sqrt{1-v^2}}} y \phantom{- \frac{v}{\sqrt{1-v^2}} t} + b, \\
t' &= -\frac{v}{\sqrt{1-v^2}} x + \frac{1}{\sqrt{1-v^2}} t + a.
\end{aligned}
\quad (6)
$$

It is easy to see that if (x, y, t) and (x_1, y_1, t_1) are the coordinates of two events in the first reference frame and (x', y', t') and (x'_1, y'_1, t'_1) are their coordinates in the

---

[14]In many cases it is convenient to think of a, b, c; A, B, C as signed quantities, i.e., to regard the sides and angles of the triangle as directed entities. However, we shall not go into this matter here. Also, we shall not discuss the extended interpretation given to the terms "side" and "angle" in Table IV which allows us to assert, for example, that in Euclidean geometry $A = B + C$ (the theorem of the exterior angle in a triangle).

Table IV Metric relations between the elements of a triangle

	Measure of lengths		
Measure of angles	E	P	H
E	$\cos a = \cos b \cos c + \sin b \sin c \cos A$ $\dfrac{\sin A}{\sin a} = \dfrac{\sin B}{\sin b} = \dfrac{\sin C}{\sin c}$ $\cos A = \cos B \cos C + \sin B \sin C \cos a$	$a^2 = b^2 + c^2 - 2bc \cos A$ $\dfrac{\sin A}{a} = \dfrac{\sin B}{b} = \dfrac{\sin C}{c}$ $A = B + C$	$\cosh a = \cosh b \cosh c + \sinh b \sinh c \cos A$ $\dfrac{\sin A}{\sinh a} = \dfrac{\sin B}{\sinh b} = \dfrac{\sin C}{\sinh c}$ $\cos A = \cos B \cos C + \sin B \sin C \cosh a$
P	$a = b + c$ $\dfrac{A}{\sin a} = \dfrac{B}{\sin b} = \dfrac{C}{\sin c}$ $A^2 = B^2 + C^2 + 2BC \cos a$	$a = b + c$ $\dfrac{A}{a} = \dfrac{B}{b} = \dfrac{C}{c}$ $A = B + C$	$a = b + c$ $\dfrac{A}{\sinh a} = \dfrac{B}{\sinh b} = \dfrac{C}{\sinh c}$ $A^2 = B^2 + C^2 + 2BC \cosh a$
H	$\cos a = \cos b \cos c + \sin b \sin c \cosh A$ $\dfrac{\sinh A}{\sin a} = \dfrac{\sinh B}{\sin b} = \dfrac{\sinh C}{\sin c}$ $\cosh A = \cosh B \cosh C + \sinh B \sinh C \cos a$	$a^2 = b^2 + c^2 + 2bc \cosh A$ $\dfrac{\sinh A}{a} = \dfrac{\sinh B}{b} = \dfrac{\sinh C}{c}$ $A = B + C$	$\cosh a = \cosh b \cosh c + \sinh b \sinh c \cosh A$ $\dfrac{\sinh A}{\sinh a} = \dfrac{\sinh B}{\sinh b} = \dfrac{\sinh C}{\sinh c}$ $\cosh A = \cosh B \cosh C + \sinh B \sinh C \cosh a$

Supplement A. Nine plane geometries

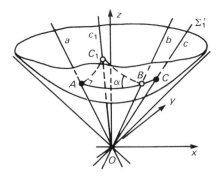

Figure 193

second, then

$$(x'_1 - x')^2 + (y'_1 - y')^2 - (t'_1 - t')^2$$
$$= \frac{1}{1-v^2}[(x_1 - x) - v(t_1 - t)]^2 + (y_1 - y)^2 - \frac{1}{1-v^2}[-v(x_1 - x) + (t_1 - t)]^2$$
$$= \frac{1}{1-v^2}\left[(1-v^2)(x_1 - x)^2 + (v^2 - 1)(t_1 - t)^2\right] + (y_1 - y)^2$$
$$= (x_1 - x)^2 + (y_1 - y)^2 - (t_1 - t)^2.$$

Quite generally, one can show that two-dimensional relativistic kinematics reduces to the geometry of three-dimensional Minkowskian space $\{x,y,z\}$ with the metric

$$d^2 = (x_1 - x)^2 + (y_1 - y)^2 - (z_1 - z)^2, \qquad (4)$$

where time is denoted by z instead of t.

Next we consider the totality of plane uniform motions. As above, we can show that to every such motion there corresponds a line in three-dimensional Minkowskian space representing the totality of events each characterized by two space coordinates and one time coordinate. Thus, from the point of view of the theory of relativity, the set of plane uniform motions is equivalent to the set of lines of three-dimensional Minkowskian space. With every (ordered) pair of uniform motions there is associated in a natural way a "deflection" given by the velocity of the first motion with respect to the second—the latter being regarded as defining the state of rest (we are obviously describing the concept of "relative velocity" treated in Secs. 3 and 11). This distance between uniform motions requires us to identify all uniform motions with the same velocity, since the distance between two such motions is zero. Thus the totality of uniform motions can be reduced to the uniform motions which include the event $O(0,0,0)$ (since each uniform motion with velocity v can be represented by a motion with velocity v passing through the point $O(0,0,0)$. These uniform motions are represented by the lines of three-dimensional space passing through the origin and contained in the interior of the cone $\mathcal{K}$ given by Eq. (5a) (Fig. 193). The latter condition is equivalent to the condition $v < 1$, where v is the speed of the given motion in some inertial frame of reference (here, as in Sec. 11, we take as our unit the speed of light[15]).

---

[15]We recall that according to the theory of relativity there are no motions with speed $v \geqslant 1$.

We now replace each of the lines under consideration by the point where it meets the "hemisphere" Σ'_1—one sheet of the (Euclidean) hyperboloid

$$x^2 + y^2 - z^2 = -1. \tag{5}$$

We recall that if a and b are lines representing uniform motions, then v_{ba}, the relative velocity of b with respect to a, is given by $\tanh\varphi$, where $\varphi = \delta_{ab}$ is the Minkowskian measure of $\angle(a,b)$.[16] In turn, φ is equal to the distance, in the sense of the hyperbolic geometry on Σ'_1, between the points where a and b meet Σ'_1.[17] These facts enable us to conclude that, *from the point of view of Einstein's theory of relativity, the set of plane uniform motions can be identified with the hyperbolic plane provided that the distance between two motions a and b is taken to be the relative velocity v_{ba} of b with respect to a.*

We see that the facts and concepts of relativistic mechanics can be thought of not only in terms of Minkowskian geometry (which was originally devised to provide a geometric interpretation of Einstein's special theory of relativity), but also in terms of hyperbolic geometry. Here we can also construct a dictionary, analogous to the dictionary in Sec. 3 connecting classical mechanics and Galilean geometry, which enables us to translate every theorem of hyperbolic geometry into a theorem of relativistic kinematics, and conversely. In much the same way, we can show that *from the point of view of Einstein's theory of relativity the set of uniform motions in space can be identified with hyperbolic space*,[18] *provided the distance between two motions a and b is taken to be the relative velocity of b with respect to a.*

This connection between the theory of relativity and hyperbolic geometry enables us to derive by purely geometric means many results connected with the Einsteinian principle of relativity. Here we need only bear in mind that the relative velocity v_{ba} of the motion b with respect to the motion a (the latter representing the state of rest!) is $\tanh d_{AB}$, where d_{AB} is the hyperbolic distance between the points A and B of Figure 193.[19] Two such derivations follow.

Keeping the setting of Figure 193, put $v_{ba} = u$, $v_{cb} = v$, $v_{ca} = w$, $d_{AB} = d$, $d_{BC} = \delta$, and $\delta_{AC} = \Delta$. Then [cf. Eq. (21c), Sec. 12] we obtain the relation

$$w = \tanh\Delta = \tanh(d+\delta) = \frac{\tanh d + \tanh\delta}{1 + \tanh d \tanh\delta} = \frac{u+v}{1+uv},$$

which is identical with Eq. (9) of Section 11. Again, in the setting of Figure 193, let A, B, and C_1 be three points which do not lie on the same (hyperbolic) line but form, say, a right triangle. In physical terms, let an observer H move horizontally with velocity $u = \tanh d_{AB}$, and let a particle U move vertically with velocity $v_1 = \tanh d_{AC_1}$. Select an inertial coordinate system in which the line a through A represents the state of rest. Then the velocity $w_1 = \tanh d_{BC_1}$ of the particle U relative to the observer H, and the angle α—as observed by H—between the

---

[16]See translator's footnote 16 in Section 12.

[17]See translator's footnote 6 in this Supplement.

[18]Hyperbolic space is discussed in a number of the bibliographical items dealing with non-Euclidean geometry (for example, [64], [65], [66] in [67] or [72]).

[19]The relation $v_{ba} = \tanh d_{AB}$ implies the fundamental inequality $v \leq 1$ of the theory of relativity as well as the constancy of the speed of light in all inertial reference frames. The corresponding relation associated with the classical Galilean principle of relativity is $v_{ba} = \tan g\delta_{ab} = \text{sing}\delta_{ab}/\cos g\delta_{ab}$, where δ_{ab} is the Galilean angle between a and b (which may be identified with the "non-Euclidean distance" between the points A and B where the lines a and b intersect the "unit sphere" of Galilean space; see below), and $\text{sing}\delta$, $\cos g\delta$ and $\tan g\delta$ are, respectively, the Galilean sine, cosine, and tangent of the angle δ, so that $\text{sing}\delta = \tan g\delta = \delta$ and $\cos g\delta \equiv 1$ (cf. Exercise 3, Sec. 3).

horizontal and the direction of the motion of U, can be found by "solving" the triangle ABC_1 in the hyperbolic plane, with $AB=d$ (and $\tanh d = u$), $AC_1 = \delta_1$ (and $\tanh \delta_1 = v_1$), $\angle BAC_1 = 90°$, and $\angle ABC_1 = \alpha$ (make a drawing). Putting $BC_1 = \Delta_1$ and $\tanh \Delta_1 = w_1$ and using the appropriate formulas of hyperbolic geometry (cf. Table IV, when $\triangle ABC$ is a right triangle), we get $\cosh \Delta_1 = \cosh d \cosh \delta_1$ (hyperbolic version of Pythagoras' Theorem) and $\tan \alpha = \tanh \delta_1 / \sinh d$. (For small d and δ_1 these formulas are very close to the Euclidean formulas $\Delta_1^2 = d^2 + \delta_1^2$ and $\tan \alpha = \delta_1 / d$.) Using, say, the relations $\sinh d = \tanh d / \sqrt{1 - \tanh^2 d}$ and $\cosh d = 1/\sqrt{1 - \tanh^2 d}$, we obtain $w_1^2 = u^2 + v_1^2 - u^2 v_1^2$ and $\tan \alpha = (v_1/u)\sqrt{1 - u^2}$. (For small u and v_1 these formulas are close to the classical formulas $w_1^2 = u^2 + v_1^2$ and $\tan \alpha = v_1 / u$.)

Of special interest is the case of a photon U, when $v_1 = 1$ and $AC_1 \| BC_1$ (in the sense of hyperbolic geometry). Then $w_1 = 1$ and $\cos \alpha = u$.

Euclidean, Galilean, and Minkowskian geometry are defined in the usual affine[20] plane. Elliptic geometry is defined on a unit sphere of three-dimensional Euclidean space with the metric

$$d^2 = (x_1 - x)^2 + (y_1 - y)^2 + (z_1 - z)^2. \tag{4'}$$

Hyperbolic and doubly hyperbolic geometry are defined on unit spheres in three-dimensional Minkowskian space with the metric

$$d^2 = (x_1 - x)^2 + (y_1 - y)^2 - (z_1 - z)^2. \tag{4}$$

In each case, a "point" is a pair of antipodal points. This method of constructing plane Cayley–Klein geometries in three-dimensional space can be used in other cases. We note first of all that a domain of definition for cohyperbolic geometry can be obtained by identifying antipodal points of the sphere Σ_2 of Minkowskian space given by the equation

$$x^2 + y^2 - z^2 = 1. \tag{5'}$$

To do this we designate the points of the hemisphere Σ_2' (Fig. 191b) as the *points* of cohyperbolic geometry, and the sections of Σ_2 by planes through O which yield ellipses[21] on Σ_2 as the *lines* of cohyperbolic geometry. As before, the motions of the cohyperbolic plane are induced by "rotations" of Minkowskian space about O.

In addition to Euclidean and Minkowskian space we consider semi-Euclidean space, i.e., three-dimensional space $\{x,y,z\}$ in which the distance d between points (x,y,z) and (x_1,y_1,z_1) is given by

$$d^2 = (x_1 - x)^2 + (y_1 - y)^2, \tag{7}$$

and semi-Minkowskian space, i.e., three-dimensional space with metric

$$d^2 = (x_1 - x)^2 - (y_1 - y)^2. \tag{7'}$$

The motions of semi-Euclidean and semi-Minkowskian space are given,

---
[20]See, for example, [19] or [57].
[21]In other words, the planes involved must not cut the cone $\mathcal{K}$ given by (5a).

respectively, by

$$x' = \cos\alpha \cdot x + \sin\alpha \cdot y \quad + a,$$
$$y' = -\sin\alpha \cdot x + \cos\alpha \cdot y \quad + b, \quad (8)$$
$$z' = \quad ux + \quad vy + wz + c,$$

and

$$x' = \cosh\alpha \cdot x + \sinh\alpha \cdot y \quad + a,$$
$$y' = \sinh\alpha \cdot x + \cosh\alpha \cdot y \quad + b, \quad (8')$$
$$z' = \quad ux + \quad vy + wz + c.$$

In fact, it is easy to see that if the coordinates (x',y',z') and (x,y,z) are connected by the formulas (8), say, then

$$(x_1' - x')^2 + (y_1' - y')^2 = [(x_1 - x)\cos\alpha + (y_1 - y)\sin\alpha]^2$$
$$+ [-(x_1 - x)\sin\alpha + (y_1 - y)\cos\alpha]^2$$
$$= (x_1 - x)^2 + (y_1 - y)^2$$

[compare (8) and (8') with (6) of Sec. 1 and (23) of Sec. 12; the latter describe plane Euclidean and Minkowskian motions, respectively].

The "unit spheres" of semi-Euclidean and semi-Minkowskian space are given by the equations

$$x^2 + y^2 = 1 \quad (9)$$

and

$$x^2 - y^2 = 1, \quad (9')$$

and represent, from the Euclidean viewpoint, a circular and hyperbolic cylinder, respectively (Figs. 194a and 195a). We call the points $O(0,0,0)$ on the axis of each of these cylinders its center, and identify antipodal points of the cylinders. Then the resulting hemicylinders σ' and σ_1' (see Figs. 194b and 195b) become models of the co-Euclidean and the co-Minkowskian plane, respectively. The *points* of these planes are the points of the hemispheres σ' and σ_1' and their *lines* are sections of σ' and σ_1' by planes passing through $O(0,0,0)$; the distance between points A and B is the

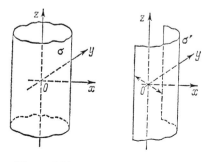

Figure 194a   Figure 194b

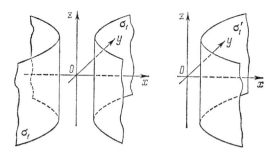

Figure 195a   Figure 195b

length of the segment AB of the "line" joining A and B, measured in terms of semi-Euclidean or semi-Minkowskian geometry, and the angle between lines a and b is the angle between the corresponding planes measured in terms of the semi-Euclidean or semi-Minkowskian metric in three-dimensional space (we forego a precise definition of the latter concept here).[22] The motions of the semi-Euclidean and semi-Minkowskian plane are induced by the motions (8) or (8') of three-dimensional space, which leave fixed the center $O(0,0,0)$ of the sphere σ or σ_1.

It is possible to describe the remaining three Cayley–Klein geometries (Euclidean, Galilean, and Minkowskian) in similar terms. In addition to the Euclidean metric (4'), the Minkowskian metric (4), the semi-Euclidean metric (7), and the semi-Minkowskian metric (7'), we introduce the metric in which the distance d between the points $A(x,y,z)$ and $A_1(x_1,y_1,z_1)$ of three-dimensional space is given by the formula

$$d^2 = (x_1 - x)^2, \tag{10}$$

analogous to the formula (37) of Sec. 13, which expresses the distance between points $A(x,y)$ and $A_1(x_1,y_1)$ in the Galilean plane. We will find it convenient to write this metric in the form

$$d^2 = (z_1 - z)^2. \tag{10a}$$

When combined with suitable groups of motions, this metric gives rise to three geometries in space. The three groups of motions are defined by

$$\begin{aligned} x' &= \cos\alpha \cdot x + \sin\alpha \cdot y + v_1 z + a, \\ y' &= -\sin\alpha \cdot x + \cos\alpha \cdot y + v_2 z + b, \\ z' &= \qquad\qquad\qquad\qquad\quad z + c \end{aligned} \tag{11}$$

[three-dimensional Galilean space; cf. (11) and (12) of Sec. 2],

$$\begin{aligned} x' &= \cosh\alpha \cdot x + \sinh\alpha \cdot y + v_1 z + a, \\ y' &= \sinh\alpha \cdot x + \cosh\alpha \cdot y + v_2 z + b, \\ z' &= \qquad\qquad\qquad\qquad\quad z + c \end{aligned} \tag{11'}$$

[22]But see p. 240 for more details. (Translator's note.)

[pseudo-Galilean space; cf. (11′) and (23) of Sec. 12], and

$$\begin{aligned} x' &= x + uy + vz + a, \\ y' &= y + wz + b, \\ z' &= z + c \end{aligned} \tag{11a}$$

(semi-Galilean space; cf. Exercise 5, Sec. 2). In all three geometries, the equation of the "unit sphere" σ_0 is

$$z^2 = 1 \quad \text{or} \quad z = \pm 1, \tag{12}$$

i.e., the unit sphere is a pair of parallel planes (Fig. 196a).

As usual, we call $O(0,0,0)$ the center of the sphere σ_0, and identify antipodal points of σ_0; this leads to the consideration of the "hemisphere" σ_0' (a plane; Fig. 196a). If we define *lines* as sections of σ_0' by planes through $O(0,0,0)$, and *motions* as the rotations (11), (11′), and (11a) which leave $O(0,0,0)$ fixed, then we obtain, respectively, Euclidean, Minkowskian, and Galilean geometry in σ_0'.

The representation of six plane Cayley–Klein geometries as spheres in Euclidean, Minkowskian, and semi-Minkowskian space makes it simple to identify the curves which in these geometries play the role of (Euclidean) circles. By a *cycle* of a Cayley–Klein plane we shall mean a curve which "has the same structure at each of its points," i.e., a curve such that for each pair A, A' of its points it admits a glide along itself which takes A to A' (cf. Sec. 8). With this definition of cycle we see that the Euclidean cycles are circles and lines [given by Eq. (35a), Sec. 13]; the Galilean cycles are the curves which we earlier called cycles (and shall now call "proper cycles"), as well as circles and both ordinary and special lines [given by Eq. (35), Sec. 13]; and the Minkowskian cycles are "circles': and lines [given by Eq. (35b), Sec. 13]. It can be shown that *the cycles of the six Cayley–Klein geometries— elliptic, hyperbolic, doubly hyperbolic, cohyperbolic, co-Euclidean and co-Minkowskian—are plane sections of appropriate "spheres"* (Σ, Σ_1, Σ_2, σ, or σ_1). Thus, for example, the cycles of elliptic geometry are *lines* and *circles*, where the term circle is defined, just as in Euclidean geometry, as the locus of points equidistant from a point called the center of the circle. If the elliptic plane is represented by a Euclidean sphere Σ with antipodal points identified, then its cycles are the plane

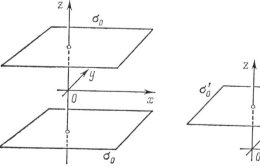

Figure 196a Figure 196b

Supplement A. Nine plane geometries

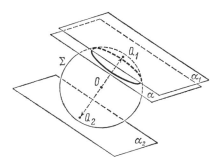

Figure 197

sections of Σ passing through the center $O(0, 0, 0)$ of Σ (the elliptic lines) as well as the plane sections not passing through O (elliptic circles; Fig. 197). [We note that in *spherical geometry* each circle has two centers. They are the points Q_1 and Q_2, where planes α_1 and α_2, parallel to the plane α of the circle, touch Σ. On the other hand, in elliptic geometry the antipodal points Q_1 and Q_2 represent the same point, so that in *elliptic geometry* a circle has a single center.]

We turn now to hyperbolic geometry. Here the cycles are: *lines*, *circles* (a circle of radius r is the locus of points at a distance r from a point Q called its center), *equidistant curves* (an equidistant curve of width h is the locus of points at a distance h from a line q called its base), and *horocycles* (a horocycle is the curve obtained by letting the radius r of a circle tangent to a line a at a point A tend to infinity, or the curve obtained by letting the width h of an equidistant curve tangent to a line a at a point A tend to infinity[23]). We know that in the model of the hyperbolic plane on the "sphere" Σ_1, lines are represented by plane sections passing through the center $O(0, 0, 0)$ of Σ_1 and intersecting the cone $\mathcal{K}$ given by Eq. (5a). (Planes through O which do not intersect $\mathcal{K}$ do not intersect Σ_1.) It can be shown that the circles of the hyperbolic plane are sections of Σ_1 by planes α not passing through O and intersecting $\mathcal{K}$ (and Σ_1) in ellipses (Fig. 198a); that the equidistant curves are sections of Σ_1 by planes β not passing through O and intersecting $\mathcal{K}$ along hyperbolas (Fig. 198b); and that the horocycles are sections of Σ_1 by planes γ not passing through O and intersecting $\mathcal{K}$ in parabolas (Fig. 198c). We note that in the hyperbolic plane the center of the circle represented by the section of Σ_1 by α is the point where the plane $\alpha_0 \| \alpha$ touches Σ_1 (Fig. 198a), and the base of the equidistant curve represented by the section of Σ_1 by β is the curve q (a hyperbolic line) in which the plane $\beta_0 \| \beta$ passing through O intersects Σ_1 (Fig. 198b).

The cycles of doubly hyperbolic, cohyperbolic, co-Euclidean and co-Minkowskian geometry can be described in a similar manner. Our discussion implies that the circular transformations of each of our six geometries, i.e., transformations of each of the six planes which take cycles to cycles, are represented by transformations of three-space that carry planes to planes[24] and the appropriate "sphere" (Σ, Σ_1, Σ_2, σ, or σ_1) to itself.

We illustrate the above discussion by considering co-Euclidean geometry on the sphere σ (or rather on the hemisphere σ'). We know that we can think of Euclidean lines as the "points" of co-Euclidean geometry. The equation of a Euclidean line l

---

[23] See, for example, [19] and [64]; or [56], [12] and [13]; or [25], [65]–[72a], [80], [81], and [78].
[24] These transformations are called *projective transformations* of three-dimensional space (see, for example, [19], [31], [32] or [67]).

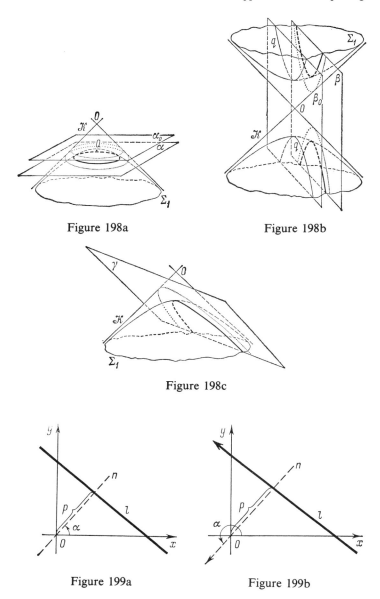

Figure 198a

Figure 198b

Figure 198c

Figure 199a

Figure 199b

can be written in the so-called normal form

$$x\cos\alpha + y\sin\alpha - p = 0, \qquad (13)$$

where α is the angle between the perpendicular to l and the positive direction of the x-axis, and p is the distance from the origin O to l (Fig. 199a). One advantage of the normal form of a line is that it permits us to evaluate easily the distance $d = d_{Ml}$ from a point $M(x_0, y_0)$ to l by simply substituting x_0, y_0 for x, y in (13) and taking the absolute value:

$$d = |x_0 \cos\alpha + y_0 \sin\alpha - p|. \qquad (13a)$$

Supplement A. Nine plane geometries 235

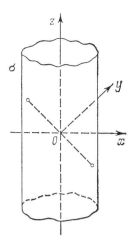

Figure 200

If we regard l as a directed line, then we can make n, the perpendicular from O to l, into a directed line by rotating l counterclockwise, say, through 90° about the point $l \cap n$, and transferring the positive direction on l to n (Fig. 199b). Then the angle α can take on any of the values $0° \leq \alpha < 360°$. If, in addition, we agree to consider the distance from a point to a directed line as positive or negative according as the point is on the right- or left-hand side of the line, then the distance p from O to l can take on any real value, $-\infty < p < \infty$. We can thus establish a one-to-one correspondence between *the set of directed lines l in the (Euclidean) plane and the set of triples*

$$\cos\alpha, \sin\alpha, -p \qquad (14)$$

of coefficients of l in (13), i.e., *the set of points $(\cos\alpha, \sin\alpha, -p)$ of the cylinder σ in the three-dimensional space $\{x,y,z\}$* (Fig. 200). Since the equation of σ is

$$x^2 + y^2 = 1, \qquad (9)$$

we may regard σ as a sphere in the semi-Euclidean space $\{x,y,z\}$. Note that two lines l and l_1 which differ only in direction are characterized by the number triple (14) and the number triple

$$(\cos(\alpha+180°), \sin(\alpha+180°), -(-p)) = (-\cos\alpha, -\sin\alpha, p)$$

(see Fig. 199b); i.e., l and l_1 correspond to antipodal points of the cylinder σ (Fig. 200). It follows that if we think of a "point" of the co-Euclidean plane as a nondirected line in the Euclidean plane, then such a point is represented by a pair of antipodal points of σ.

It turns out, however, that in studying cycles of co-Euclidean geometry it is more convenient to take "points" to be directed Euclidean lines.[25] If we retain the

---

[25]Moreover, in studying circular transformation of co-Euclidean geometry, it is natural to regard the points of the co-Euclidean plane as being "directed," i.e., to consider a point as a pair of coincident points of the (co-Euclidean) plane having different "directions." One way of assigning a direction to a point is to draw about it a circular arc with an arrow which indicates which of the two possible rotations about the point is to be viewed as positive; cf. Figures 225–227. Similarly, in many cases it is convenient to assign directions to points in the elliptic, hyperbolic, doubly hyperbolic, cohyperbolic, and co-Minkowskian plane: see pp. 284–286 or [13] and [80].

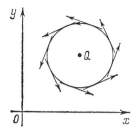

Figure 201a

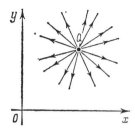

Figure 201b

above convention for the sign of the distance from a point to a line, then the (signed) distance $d = d_{Ml}$ from $M(x_0, y_0)$ to l [given, as before, by (13)] is equal to

$$d = x_0 \cos \alpha + y_0 \sin \alpha - p. \tag{13b}$$

Then the section of the "sphere" σ [i.e., the cylinder (9)] by the plane[26]

$$Ax + By + Cz = D, \quad C \neq 0, \tag{15}$$

or, equivalently, the plane

$$ax + by + z = d \tag{15'}$$

with $a = A/C$, $b = B/C$, $d = D/C$, is the set of points of the cylinder $x^2 + y^2 = 1$ which represent the (directed) lines of the Euclidean plane at (positive or negative) distance d from the point $Q(a,b)$. This means that the cycles of co-Euclidean geometry represented by the plane sections of the "sphere" σ are just directed *circles*, each viewed as the totality of its (directed) tangents (Fig. 201a); these circles include points (circles of radius zero) or, more accurately, pencils of lines through these points (Fig. 201b). In our model of the co-Euclidean plane, points are represented by sections (15') of the cylinder σ for which $d = 0$, i.e., by sections of σ by planes passing through the origin $O(0, 0, 0)$.

Circular transformations of the co-Euclidean plane are transformations of the set of (directed) lines of the Euclidean plane (points of the co-Euclidean plane!) which carry *circles to circles*. Such transformations of the Euclidean plane were first considered by the eminent French mathematician E. N. LAGUERRE (1834–1886); hence the name *Laguerre transformations*.[27] Laguerre transformations are represented by mappings of the cylilnder σ onto itself which take plane sections to plane sections. An example of such a transformation is a central projection of σ to itself from some center M which takes a point A of σ to the second point A' of

---

[26]For $C = 0$, the plane (15) is parallel to the axis of the cylinder.
[27]See [13], [80] or [29].

Supplement A. Nine plane geometries

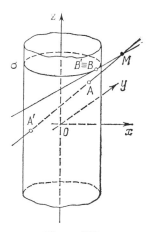

Figure 202

intersection of the line MA with σ (Fig. 202); if MB is tangent to σ (at B), then $B' = B$. (This transformation is called an *axial inversion*, and is similar to an ordinary inversion.)

Starting with three-dimensional models of plane Cayley–Klein geometries, it is easy to obtain representations of these geometries in the plane. To this end, we project each of the "spheres" Σ, Σ_1, Σ_2, σ, and σ_1 from its center $O(0,0,0)$ to the plane $z=1$ (Fig. 203). Then each pair of antipodal points of the "sphere" is mapped to the same point of $z=1$, and we can say that the points of our "Cayley–Klein plane" are represented by points of either an ordinary or, more precisely, a projective plane. Specifically, the elliptic plane is represented by the *whole plane* $z=1$ (Fig. 203a), the hyperbolic plane by the *interior of the circle* $x^2+y^2=1$ (Fig. 203b), the doubly hyperbolic plane and the cohyperbolic plane by the *exterior of that circle* (Fig. 203c), the co-Euclidean plane by the plane $z=1$ *punctured* at the point $x=y=0$ (Fig. 203d), and the co-Minkowskian plane by the *region* $|x/y|<1$ (Fig. 203e). It is clear that our central projection of each of the spheres Σ, Σ_1, Σ_2, σ, and σ_1 from $O(0,0,0)$ to $z=1$ takes a plane section through O to a line. Hence in our models of the six plane Cayley–Klein geometries, *the lines of these planes are represented by lines of the plane $z=1$*. It is these models which were the starting point of the algebraic constructions of Cayley and the geometric constructions of Klein. They are called *Klein models* of the geometries in question.[28]

By using stereographic projections (see pp. 142, 149 and 198) of the spheres Σ, Σ_1, Σ_2, σ, and σ_1 to planes we obtain other models of the same six plane Cayley–Klein geometries. The spheres Σ and Σ_1 contain the point $Q(0,0,-1)$. We project each sphere to the plane $z=1$ tangent to it at the point $Q_1(0,0,1)$ antipodal to Q (Figs. 204a and 204b). The image of Σ is the whole plane $z=1$. On the other hand, if we restrict ourselves to, say, the upper hemisphere Σ', then the plane model obtained for elliptic geometry is the (rather odd) region of $z=1$ consisting of the interior and half the points of the circle $x^2+y^2=4$ (for the boundary of Σ' is a semicircle; cf. p. 220, in particular, Fig. 189b). It is often convenient to assume that elliptic geometry "acts" in all of $z=1$. But then the points of that plane are images of the "directed" points of the elliptic plane (cf. footnote 25) represented by all the

---

[28]See [56] and [73]. In connection with Klein's model of hyperbolic geometry, see [64], [37]; [69], and [12].

238   Supplement A. Nine plane geometries

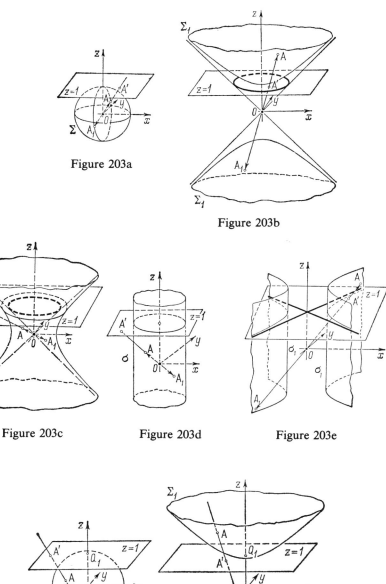

Figure 203a

Figure 203b

Figure 203c   Figure 203d   Figure 203e

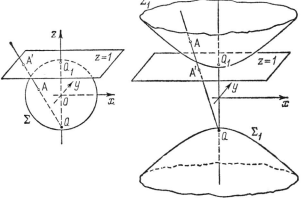

Figure 204a   Figure 204b

Supplement A. Nine plane geometries

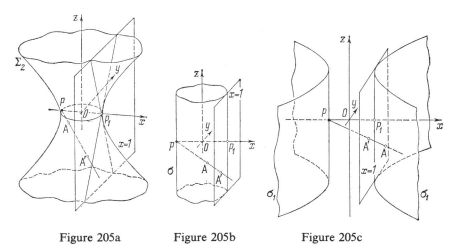

Figure 205a   Figure 205b   Figure 205c

points of the sphere Σ. We note that as a result of the properties of stereographic projection (cf. pp. 142–146) *elliptic lines and circles* (plane sections of Σ) *are represented in our model by* (*Euclidean*) *line and circles*; in particular, the elliptic lines—the great circles of Σ—are represented by the circles of the plane $z=1$ which intersect the circles $x^2+y^2=4$ in antipodal points. Similarly, the image under stereographic projection of the "upper hemisphere" Σ_1' is the disk $x^2+y^2<4$ of the plane $z=1$, which is thus the natural domain of hyperbolic geometry. Nevertheless, in some cases (e.g., in developing the theory of circular transformations of the hyperbolic plane) it is convenient to regard the points of the hyperbolic plane as directed and to assume that the "field of action" of hyperbolic geometry is the whole plane $z=1$, which is the image under stereographic projection of the whole sphere Σ_1. Since stereographic projection of the sphere Σ_1 to the plane $z=1$ takes plane sections of Σ_1 to circles (cf. pp. 142–144), it follows that in our model of hyperbolic geometry *all lines and cycles* (circles, horocycles, and equidistant curves) *are represented by* (*Euclidean*) *lines and circles*.

The spheres Σ_2, σ, and σ_1 contain the point $P(-1,0,0)$. We project them to the plane $x=1$ which touches each sphere at the point $P_1(1,0,0)$ antipodal to P (Figs. 205a–c). The resulting plane maps enable us to construct plane models of the geometries on Σ_2, σ, and σ_1, namely, doubly hyperbolic, cohyperbolic, co-Euclidean, and co-Minkowskian geometry. Again, it is convenient to regard the points of the various Cayley–Klein geometries as directed, for only then are all of the points of the spheres Σ_2, σ, and σ_1 images of the points of the Cayley–Klein planes.[29] We note that stereographic projection takes plane sections of the sphere Σ_2 to hyperbolas in the plane $x=1$, and plane sections of the spheres σ and σ_1 into parabolas. It follows that in our models *the lines and cycles of the doubly hyperbolic and cohyperbolic plane are represented by hyperbolas, and the lines and cycles* of the co-Euclidean and co-Minkowskian plane by parabolas.

---

[29]We note that the fields of action of the models of the six Cayley–Klein geometries obtained in the above manner are not copies of the familiar Euclidean (or affine) plane but more complex constructs. Specifically, stereographic projection of the sphere Σ and of the two-sheeted hyperboloid Σ_1 from P to the plane $z=1$ maps Σ and Σ_1 to the inversive Euclidean plane (see p. 142); stereographic projection of the cylinders σ and σ_1 to the plane $x=1$ maps σ and σ_1 to the inversive Galilean plane (p. 149); and stereographic projection of the one-sheeted hyperboloid Σ_2 to the plane $x=1$ maps it to the inversive Minkowskian plane (see p. 198).

The models of plane hyperbolic and elliptic geometry obtained by means of stereographic projection (models closely related to those studied in Supplement C below) were first investigated by Poincaré. That is why all of these models are called *Poincaré models* of Cayley–Klein geometries.[30]

We conclude with a few remarks on three-dimensional Cayley–Klein geometries. Just as there are $3^2 = 9$ plane Cayley–Klein geometries, so there are $3^3 = 27$ such three-dimensional systems, corresponding to all possible combinations of measures of *distance between points* on a line, *plane angles* in a pencil of lines, and *dihedral angles* in a pencil of planes. Each of these measures can be elliptic, parabolic (Euclidean), or hyperbolic. We have already defined the three measures of distance and the three measures of plane angles. It remains to define the three measures of dihedral angles. The usual measure of dihedral angles (i.e., the one used in Euclidean solid geometry) is called *elliptic*. The measure defined by the requirement that the magnitude of the angle between two planes α and β is equal to the (Euclidean) length of the segment AB cut off by these planes on a fixed line q (not parallel to the edge of the angle) is called *parabolic*. Finally, the measure defined by the formula

$$\delta_{\alpha\beta} = k \log\left(\frac{\angle(\alpha,\iota)/\angle(\alpha,\varepsilon)}{\angle(\beta,\iota)/\angle(\beta,\varepsilon)} \right),$$

where $\angle$ denotes the usual dihedral angle between two planes, and ι and ε denote two fixed planes of the pencil, is called *hyperbolic*.

Examples of three-dimensional Cayley–Klein geometries are the usual Euclidean geometry, Minkowskian geometry, semi-Euclidean and semi-Minkowskian geometry, Galilean, pseudo-Galilean, and semi-Galilean geometry (see pp. 229–230 and pp. 231–232). We shall denote the metric on a line, in a pencil of lines and in a pencil of planes by means of one of the letters E, P, and H, depending on whether the metric in question is elliptic, parabolic, or hyperbolic. We shall use ordered triples of these letters to denote various geometries. The first letter of the triple will refer to the metric of distances, the second to the metric of plane angles, and the third to the metric of dihedral angles. Thus Euclidean geometry is of type PEE. A *pseudo-Euclidean* geometry (i.e., a non-Euclidean geometry in which the measure of length is parabolic but the two measures of angles are not parabolic) may be of type PEH, PHE, or PHH. In classifying such a geometry, we must decide whether its "planes" are planes which intersect the cone $\mathcal{K}$ given by

$$x^2 + y^2 - z^2 = 0 \tag{5a}$$

in ellipses or in hyperbolas, and consider separately the space whose "lines" are lines parallel to those included in the interior of $\mathcal{K}$ and the space whose "lines" are parallel to the lines through $O(0,0,0)$ in the exterior of $\mathcal{K}$.[31] Semi-Euclidean geometry is of type PEP. Semi-Minkowskian geometry is of type PHP; as "lines" of this geometry we must consider (Euclidean) lines parallel to lines passing through $O(0,0,0)$ and belonging to any (rather than just one) pair of vertical angles formed by the planes

$$x^2 - y^2 = 0 \quad \text{or} \quad (x+y)(x-y) = 0.$$

---

[30]See [80]. In connection with the Poincaré models of hyperbolic and elliptic geometry, see also [56] or [13].

[31]If all the "planes" of our Cayley–Klein geometry intersect the cone $\mathcal{K}$ in (Euclidean) ellipses, then its "lines" must be (Euclidean) lines parallel to lines through O in the exterior of $\mathcal{K}$ (for our "planes" contain only such "lines").

Supplement A. Nine plane geometries

Finally, Galilean, pseudo-Galilean, and semi-Galilean geometry are of type PPE, PPH, and PPP, respectively.

By now we have accounted for nine of the 27 three-dimensional Cayley–Klein geometries, namely those in which the metric of distance is parabolic. It is natural to compare these geometries with the three plane Cayley–Klein geometries which are the main concern of this book, i.e., Euclidean, Galilean, and Minkowskian geometry, in which the distance metric is also parabolic. The remaining $27-9=18$ three-dimensional Cayley–Klein geometries can be realized on spheres in four-dimensional[32] Euclidean, semi-Euclidean, pseudo-Euclidean, and semipseudo-Euclidean spaces. However, these constructions[33] are beyond the level of this book.

---

[32]See [19]–[21].
[33]See, for example, [77] or [78].

Supplement B. Axiomatic characterization of the nine plane geometries

In Supplement A, we gave a rather detailed exposition of the nine plane Cayley–Klein geometries. One reason for including the present supplement is to comply with a tradition going back to Euclid [1–3] and Hilbert [4] which requires that each geometric system be described by a complete list of axioms. Another reason is the need to prepare a basis for proving the duality principle of Galilean geometry (cf. Sec. 6).

The axioms of a geometry are the assumed properties of its undefined terms (such as point and line); these terms are used to define more complex geometric entities (such as triangle and circle). In turn, their properties are formulated as theorems deduced from the (unproved) axioms by the use of logical rules of deduction.

Every geometric system can be characterized by many (equivalent) systems of axioms. Two equivalent systems of axioms may be based on different sets of undefined terms, and what is an undefined term in one axiomatization may be defined in terms of other concepts in an equivalent axiomatization. Thus, at the end of the 19th and the beginning of the 20th century there appeared a number of equivalent axiomatizations of Euclidean geometry. The axiomatization of plane geometry by D. HILBERT (1862–1943) used as undefined terms *point* and *line*, as well as the undefined relations of *incidence* of points and lines, of a point lying *between* two other points, and of *congruence* of segments and angles (which were defined in terms of points, lines, and betweenness within the geometric system). The axiomatization of Euclidean geometry published by the Italian mathematician M. PIERI (1860–1913) at about the same time as Hilbert's book [4] was based on the undefined terms *point* and *motion*, and the system of axioms also formulated at about that time by the Russian geometer B. F. KAGAN (1860–1953) was based on concepts of *point* and *distance*. (The "metric" approach to geometry based on the concept of distance was subsequently perfected by the American mathematicians O. VEBLEN (1880–1960) and G. D. BIRKHOFF (1884–1947); in connection with Birkhoff's axiomatization, see, for example, [5], [6], or [20].)

Supplement B. Axiomatic characterization of the nine plane geometries 243

Axiomatizations of (plane and solid) Euclidean geometry developed in the last few decades are usually based on the undefined terms *point* and *vector*. This approach is usually, but perhaps without sufficient justification, attributed to the great German mathematician H. WEYL (1885–1955) whose famous book [39] on the theory of relativity entitled *Space, Time, Matter* (first published in 1918), opens with this axiomatization. The simplicity and universality of the vector approach which, when suitably modified, encompasses not only classical Euclidean geometry but also many non-Euclidean geometries (including all the Cayley–Klein geometries) gave rise to the view that all other axiomatizations of geometry are hopelessly out of date and deserve no attention (cf., for example, [8]). Our own axiomatization of **plane Euclidean geometry** will also be vectorial and will involve five groups of axioms, I, II, III$^{(2)}$, IV$^{(E)}$, and V.

As indicated above, our undefined terms will be *point* and *vector*. We take for granted the *real numbers* (which, in turn, can be based on a suitable system of axioms; see, for example, Chap. I of [8]). The vectors are required to form a *vector space*, i.e., a system closed under the operations of addition (to each pair of vectors **a** and **b** there is associated a third vector **c** called their sum and denoted by **a**+**b**) and *multiplication of a vector by a number* (to each vector **a** and real number α there is associated a vector **d** called the product of **a** by α and denoted by α**a**) subject to the following axioms.

I. Axioms of addition of vectors

I_1. Vector addition is commutative[1]:

a+**b**=**b**+**a** *for all pairs of vectors* **a** *and* **b**.

I_2. Vector addition is associative:

(**a**+**b**)+**c**=**a**+(**b**+**c**) *for any three vectors* **a**, **b**, *and* **c**.

I_3. Existence of a zero vector: *There exists a vector* **0** *such that*

a+**0**=**a** *for each vector* **a**.

I_4 Existence of an additive inverse: *For every vector* **a** *there exists a vector* **a**′ *such that*

a+**a**′=**0**.

II. Axioms of multiplication of a vector by a number

II_1. For every vector **a**

1**a**=**a**.

II_2 Multiplication of a vector by a number is associative:

$\alpha(\beta$**a**$)=(\alpha\beta)$**a** *for every vector* **a** *and numbers* α, β.

II_3. Multiplication of a vector by a number is distributive over addition of numbers:

$(\alpha+\beta)$**a**$=\alpha$**a**$+\beta$**a** *for all numbers* α, β *and for each vector* **a**.

---
[1] Axiom I_1 can be deduced form axioms I_{2-4} and II_{1-4} (see, for example, W. Nef, *Linear Algebra*, McGraw Hill, London, 1967, p. 18).

II$_4$. Multiplication of a vector by a number is distributive over vector addition:

$\alpha(\mathbf{a}+\mathbf{b}) = \alpha\mathbf{a} + \alpha\mathbf{b}$ *for each number α and all vectors* **a** *and* **b**.

The axioms in groups I and II imply the uniqueness of the zero vector **0** as well as the uniqueness of the additive inverse $\mathbf{a}' = -\mathbf{a}$ of a given vector **a**. Also, it is easy to establish the possibility of subtraction of vectors, i.e., the fact that given two vectors **a** and **b** there exists a unique vector x such that

$$\mathbf{x} + \mathbf{a} = \mathbf{b};$$

the vector x [equal to $\mathbf{b} + (-\mathbf{a})$] is called the *difference* of **b** and **a** and is denoted by $\mathbf{b} - \mathbf{a}$. Other simple results are

$$0\mathbf{a} = \mathbf{0} \quad \text{and} \quad \alpha\mathbf{0} = \mathbf{0}$$

for every vector **a** and number α, and

$$(-\alpha)\mathbf{a} = -(\alpha\mathbf{a}),$$

for every number α and vector **a**. Indeed,

$$(-\alpha)\mathbf{a} + \alpha\mathbf{a} = (-\alpha + \alpha)\mathbf{a} = 0 \cdot \mathbf{a} = \mathbf{0},$$
$$(-\alpha)\mathbf{a} = (-1 \cdot \alpha)\mathbf{a} = (-1)(\alpha\mathbf{a}) = -(\alpha\mathbf{a}).$$

The fact that we are dealing with *plane geometry* (or two-dimensional geometry) is determined by the following axioms.

III$^{(2)}$. Dimension axioms

III$_1^{(2)}$. *For every three vectors* **a,b,c** *there exist three numbers, not all zero, such that*

$$\alpha\mathbf{a} + \beta\mathbf{b} + \gamma\mathbf{c} = \mathbf{0}. \tag{16}$$

III$_2^{(2)}$. *There exist two vectors* **a** *and* **b** *such that*

$$\alpha\mathbf{a} + \beta\mathbf{b} = \mathbf{0} \quad \text{only if} \quad \alpha = 0 \quad \text{and} \quad \beta = 0.$$

The usual formulation of the dimension axioms involves the concept of linear dependence of vectors. We say that vectors $\mathbf{a}_1, \mathbf{a}_2, \ldots, \mathbf{a}_k$ are *linearly dependent* if there exist numbers $\alpha_1, \alpha_2, \ldots, \alpha_k$, not all zero, such that

$$\alpha_1\mathbf{a}_1 + \alpha_2\mathbf{a}_2 + \cdots + \alpha_k\mathbf{a}_k = \mathbf{0}. \tag{16'}$$

If this is not the case, then we say that the vectors $\mathbf{a}_1, \mathbf{a}_2, \ldots, \mathbf{a}_k$ are linearly independent. We can now restate the axioms III$_1^{(2)}$ and III$_2^{(2)}$ as follows.

III$_1^{(2)}$. *Any three vectors are linearly dependent.*
III$_2^{(2)}$. *There exist two linearly independent vectors.*

The dimension axioms enable us to introduce the important concept of coordinates of a vector. Call a pair of linearly independent vectors (whose

Supplement B. Axiomatic characterization of the nine plane geometries

existence is guaranteed by III$_2^{(2)}$) a *basis*. Let **e,f** be a basis. Then it is easy to show that for every vector **a** (which may coincide with **e** or **f**) there exists a unique pair of (real) numbers x,y such that

$$\mathbf{a} = x\mathbf{e} + y\mathbf{f}. \tag{17}$$

The numbers x,y are called the *coordinates of* **a** relative to the basis $\{\mathbf{e},\mathbf{f}\}$.

To prove the existence of the decomposition (17) we note that axiom III$_1^{(2)}$ guarantees the existence of numbers α, ε, and ζ, not all zero, such that

$$\alpha\mathbf{a} + \varepsilon\mathbf{e} + \zeta\mathbf{f} = \mathbf{0}. \tag{18}$$

Also, the number α in (18) is not zero; otherwise, we would have the relation

$$\varepsilon\mathbf{e} + \zeta\mathbf{f} = \mathbf{0}$$

with ε and ζ not both zero, and this would contradict the fact that the pair **e,f** is a basis. Now $-(\varepsilon\mathbf{e}) = (-\varepsilon)\mathbf{e}$ and $-(\zeta\mathbf{f}) = (-\zeta)\mathbf{f}$. By adding to both sides of (18) the sum $(-\varepsilon\mathbf{e}) + (-\zeta\mathbf{f})$ we can rewrite (18) as

$$\alpha\mathbf{a} = (-\varepsilon)\mathbf{e} + (-\zeta)\mathbf{f}. \tag{18'}$$

Multiplying both sides of (18') by $1/\alpha$ and denoting the numbers $-\varepsilon/\alpha$ and $-\zeta/\alpha$ by x and y, we arrive at (17). The uniqueness of the coordinates (x,y) of **a** follows from the fact that if, in addition to (17), we had an equality

$$\mathbf{a} = x_1\mathbf{e} + y_1\mathbf{f}, \tag{17'}$$

then subtraction of (17') from (17) would yield

$$(x - x_1)\mathbf{e} + (y - y_1)\mathbf{f} = \mathbf{0}.$$

Bearing in mind that **e,f** is a basis we conclude that

$$x - x_1 = 0, \quad y - y_1 = 0, \quad \text{i.e.,} \quad x = x_1, \quad y = y_1.$$

Formula (17) and the properties of addition of vectors and multiplication of a vector by a number (given by axioms I and II) imply that *if the vectors* **a,b** *have coordinates* (x,y) *and* (x_1,y_1), *then the vectors* **a** + **b** *and* $\alpha\mathbf{a}$ (where α is any number) *have coordinates* $(x + x_1, y + y_1)$ *and* $(\alpha x, \alpha y)$ [and **a** − **b** has coordinates $(x - x_1, y - y_1)$].

In order to make the two-dimensional vector space governed by the groups of axioms I, II, and III$^{(2)}$ (the vector plane) into a *Euclidean vector space*, we must add to the undefined relations connecting vectors and (real) numbers a binary operation called the *scalar product* of vectors, which associates to a pair of vectors **a,b** a number σ called the scalar product of **a** and **b** and denoted by **ab**. The scalar product is governed by the following axioms.

IV$^{(E)}$. Axioms of the scalar product of vectors

IV$_1$. *The scalar product of vectors is commutative:*

$$\mathbf{ab} = \mathbf{ba} \text{ for any two vectors } \mathbf{a} \text{ and } \mathbf{b}.$$

IV$_2$. *The scalar product of vectors is associative relative to the operation of multiplication of a vector by a number:*

$$(\alpha \mathbf{a})\mathbf{b} = \alpha(\mathbf{ab}) \text{ for arbitrary } \mathbf{a}, \mathbf{b} \text{ and } \alpha.$$

IV$_3$. *The scalar product of vectors is distributive over vector addition:*

$$(\mathbf{a} + \mathbf{b})\mathbf{c} = \mathbf{ac} + \mathbf{bc} \text{ for arbitrary } \mathbf{a}, \mathbf{b}, \mathbf{c}.$$

IV$_4^{(E)}$. *The scalar product is positive semi-definite:*

$$\mathbf{aa} \geqslant 0 \text{ for any } \mathbf{a}.$$

The number $\mathbf{aa}$ is called the (scalar) square of $\mathbf{a}$ and is denoted by $\mathbf{a}^2$. Thus IV$_4^{(E)}$ asserts that the square of any vector $\mathbf{a}$ is nonnegative. Finally, we require

IV$_5^{(E)}$. $\mathbf{a}^2 = 0$ *only if* $\mathbf{a} = \mathbf{0}$.

Together, axioms IV$_4^{(E)}$ and IV$_5^{(E)}$ assert that *the scalar product is positive-definite*. (Of course, the last statement is merely a definition of the term "positive-definite.")

[We note that $\mathbf{0a} = 0$ for every $\mathbf{a}$; indeed, by IV$_3$, $(\mathbf{c}+\mathbf{0})\mathbf{a} = \mathbf{ca} + \mathbf{0a}$, i.e., $\mathbf{ca} = \mathbf{ca} + \mathbf{0a}$, so that $\mathbf{0a} = 0$, as asserted.]

The axioms IV$^{(E)}$ imply the existence of *orthonormal bases*, i.e., bases $\mathbf{i}, \mathbf{j}$ such that

$$\mathbf{i}^2 = \mathbf{j}^2 = 1 \quad \text{and} \quad \mathbf{ij} = 0. \tag{19e}$$

Indeed, let $\mathbf{e}, \mathbf{f}$ be a basis. Since $\mathbf{e} \neq \mathbf{0}$ and thus $\mathbf{e}^2 = \alpha > 0$, we can form the vector $\mathbf{i} = (1/\sqrt{\alpha})\mathbf{e}$. For this vector,

$$\mathbf{i}^2 = \left[(1/\sqrt{\alpha})\mathbf{e}\right]^2 = (1/\alpha)\mathbf{e}^2 = (1/\alpha)\cdot\alpha = 1.$$

Now let $\mathbf{if} = \beta$. Put $\mathbf{f}_1 = \mathbf{f} - \beta\mathbf{i}$. Then

$$\mathbf{if}_1 = \mathbf{i}(\mathbf{f} - \beta\mathbf{i}) = \mathbf{if} - \beta\mathbf{i}^2 = \beta - \beta = 0.$$

If we put $\mathbf{f}_1^2 = \gamma$ (where, clearly, $\gamma > 0$), then $\mathbf{i}$ and $\mathbf{j} = 1/\sqrt{\gamma}\,\mathbf{f}_1$ satisfy the conditions (19e). Indeed,

$$\mathbf{i}^2 = 1, \quad \mathbf{ij} = (1/\sqrt{\gamma})(\mathbf{if}_1) = 0$$

and

$$\mathbf{j}^2 = \left[((1/\sqrt{\gamma})\mathbf{f}_1)\right]^2 = (1/\gamma)\mathbf{f}_1^2 = (1/\gamma)\cdot\gamma = 1.$$

Now let $\mathbf{a}, \mathbf{b}$ be two vectors, $\{\mathbf{i}, \mathbf{j}\}$ an orthonormal basis, and (x, y) and (x_1, y_1) the coordinates of $\mathbf{a}$ and $\mathbf{b}$ relative to $\{\mathbf{i}, \mathbf{j}\}$. Then

$$\mathbf{ab} = (x\mathbf{i} + y\mathbf{j})(x_1\mathbf{i} + h_1\mathbf{j}) = (xx_1)(\mathbf{i}^2) + (xy_1)(\mathbf{ij}) + (y_1 x)(\mathbf{ji}) + (yy_1)(\mathbf{j}^2)$$
$$= xx_1 \cdot 1 + xy_1 \cdot 0 + yx_1 \cdot 0 + yy_1 \cdot 1,$$

or

$$\mathbf{ab} = xx_1 + yy_1 \tag{20e}$$

—a formula for the scalar product of $\mathbf{a}$ and $\mathbf{b}$ expressed in terms of their coordinates relative to an orthonormal basis. In particular, (20e) implies

that
$$\mathbf{a}^2 = x^2 + y^2. \tag{20'e}$$
The *norm* $\|\mathbf{a}\|$ and *length* $|\mathbf{a}|$ of the vector $\mathbf{a}$ are defined by
$$\|\mathbf{a}\| = x^2 + y^2, \quad |\mathbf{a}| = \sqrt{x^2 + y^2}. \tag{20''e}$$

Finally, the (undefined) concept of a *point* is linked to the concept of a *vector* by a relation which associates to each (ordered) pair of points A, B a vector $\mathbf{a}$ denoted by $\overline{AB}$, where A is called the *beginning* and B the *end* of $\overline{AB}$. This (undefined) "point-vector" relation is subject to the following two axioms.

V. Axioms of the point-vector relation

$\mathbf{V_1}$. *For every point A and vector $\mathbf{a}$ there is a unique point B such that*
$$\overline{AB} = \mathbf{a}.$$
$\mathbf{V_2}$. *For any three points A, B, C we have $\overline{AB} + \overline{BC} = \overline{AC}$.*

By the *line AB* $(A \neq B)$ we mean the set of points M such that $\overline{AM}$ and $\overline{AB}$ are linearly dependent (i.e., $\overline{AM} = \lambda \overline{AB}$ for some number λ). The vector $\overline{AB}$ is called the *direction vector* of the line and determines it uniquely. If we put $\overline{AB} = \mathbf{t}$, then the line AB can be described as the set of points M such that
$$\overline{AM} = \lambda \mathbf{t}, \tag{21}$$
or, if O is any point of the plane, as the set of points M such that
$$\overline{OM} = \overline{OA} + \lambda \mathbf{t} \tag{21a}$$
(Fig. 206).

Let $\{\mathbf{e}, \mathbf{f}\}$ be a basis and let O be a point of the plane. By the *coordinates of a point M* of the plane we shall mean the coordinates of (the vector) $\overline{OM}$ relative to the basis $\{\mathbf{e}, \mathbf{f}\}$. The point O is called the origin. If we denote the coordinates of A by (p, q) and the coordinates of $\mathbf{t}$ by $(-b, a)$, then the coordinates (x, y) of a point M on the line l passing through A and having direction vector $\mathbf{t}$ satisfy the relations
$$x = p - \lambda b, \quad y = q + \lambda a,$$
or
$$ax + by = ap + bq,$$
or briefly
$$ax + by + c = 0, \tag{22}$$
the equation of a line with direction vector $\mathbf{t} = (-b, a)$ passing through the point $A(p, q)$. The constant c in (22) has the value $c = -ap - bq$.

The *distance* between points A and B is defined as the length of the vector $\overline{AB}$. By V_2, we have
$$\overline{AB} = \overline{OB} - \overline{OA},$$

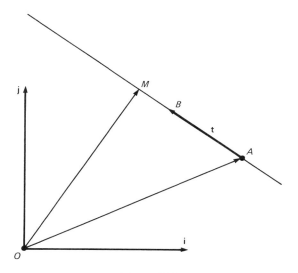

Figure 206

so that the coordinates of $\overline{AB}$ are equal to the differences of the coordinates of the points B and A. It follows that if $\{\mathbf{i},\mathbf{j}\}$ is an orthonormal (rather than an arbitrary) basis, then by (20″e) the distance d_{AA_1} between the points $A(x,y)$ and $A_1(x_1,y_1)$ is given by

$$d_{AA_1} = \sqrt{(x_1-x)^2 + (y_1-y)^2}. \tag{23e}$$

Lines l and l_1 with direction vectors $\mathbf{t}$ and $\mathbf{t}_1$ are said to be *parallel* if the vectors $\mathbf{t}$ and $\mathbf{t}_1$ are linearly dependent (i.e., proportional, $\mathbf{t}_1 = \lambda \mathbf{t}$ for some λ), and *perpendicular* if $\mathbf{t}$ and $\mathbf{t}_1$ are *orthogonal*, i.e.,

$$\mathbf{tt}_1 = 0. \tag{24}$$

Finally, the *angle* δ_{ll_1} between l and l_1 is defined by the formula

$$\cos \delta_{ll_1} = \frac{\mathbf{tt}_1}{|\mathbf{t}||\mathbf{t}_1|}, \tag{25e}$$

[It is easy to see that the absolute value of the number on the right-hand side of (25e) does not exceed 1, so that the formula (25e) actually determines an angle δ.]

The development of Euclidean geometry based on two undefined terms—*vector* and *point*—and on the groups of axioms I, II, III, IV, and V is very convenient. (It is usually associated with the name of H. Weyl.) Nevertheless, most of the expositions of the subject in the scientific literature (and often in textbooks) rely on the single undefined term *vector*. Such presentations dispense with the axioms in group V and identify *point* with *vector*. In other words, it is assumed that to the vector **0** (which is "special" by virtue of the fact that its existence is given by axiom I_3) there corresponds a point O (more precisely, the vector **0** is also called a point; as such it is denoted by the letter O and is referred to as the *center* of the plane or

Supplement B. Axiomatic characterization of the nine plane geometries

of space).² Further, to every vector **a** there is associated a point A (more precisely, each vector **a** is called a point, and as such it is denoted by A). The vector **a** can also be given by the pair of points O, A and can be denoted by $\overline{OA}$. Finally, to each pair of points B, C we associate a new vector $\mathbf{c} - \mathbf{b}$ which can be denoted by $\overline{BC}$. It is then easy to verify axioms V_1 and V_2. This method of presenting geometry, which ignores the concept of "point" is adopted in many university texts (as well as in, say, the book [8]; cf. also pp. 254–256 below).

A kind of "dual" development of geometry uses the single undefined term *point* and a quaternary relation $(ABCD)$ on points governed by the following axioms (which replace the axioms of group I and are therefore denoted by $\tilde{I}$ with appropriate subscripts).

$\tilde{I}_1$. *If* $(ABCD)$ *then* $(ACBD)$.
$\tilde{I}_2$. *If* $(ABCD)$ *then* $(CDAB)$.
$\tilde{I}_3$. *If* $(ABCD)$ *and* $(CDEF)$ *then* $(ABEF)$.
$\tilde{I}_4$. *For every triple of points* A, B, C, *there is a unique point* D *such that* $(ABCD)$.

The relation $(ABCD)$ is often written in the form $A, B \sim C, D$ and is called a relation of *equipollence* of pairs of points A, B and C, D. A class of equipollent pairs of points³ is called a *vector*. The undefined relations connecting vectors and numbers are the relations of multiplication of a vector by a number and scalar multiplication of vectors satisfying the groups of axioms II_1–II_2 and IV, respectively. Addition of vectors is defined by means of an equality which is fully analogous to axiom V_2. Then one proves the properties of vectors listed in the group of axioms I. Finally, it is required that the axioms II_3–II_4 and III be satisfied.

While this way of developing geometry has intuitive appeal [the relation $(ABCD)$ asserts that the points A, B, D, C (in this order!) are vertices of a parallelogram], it is somewhat arduous and therefore infrequently used.

In axiomatizing the **Galilean plane** and the **Minkowskian plane** we leave unchanged the groups of axioms I, II, III$^{(2)}$, and V. The axioms IV_1, IV_2, and IV_3 of the groups of axioms $\text{IV}^{(G)}$ ("G" for Galilean) and $\text{IV}^{(M)}$ ("M" for Minkowskian) are the same as the corresponding axioms of the group $\text{IV}^{(E)}$ in Euclidean geometry.

In the case of Galilean geometry, we keep axiom $\text{IV}_4^{(E)}$ (which we denote by $\text{IV}_4^{(G)}$) but replace axiom $\text{IV}_5^{(E)}$ with the axiom

$\text{IV}_5^{(G)}$. *There exists a vector* **a** *with* $\mathbf{a}^2 > 0$ *and a nonzero vector* **o** *with* $\mathbf{o}^2 = 0$.

One more axiom, which has no Euclidean analogue, will be added below.

To complete the axiomatization of Minkowskian geometry we need only replace the Euclidean axioms $\text{IV}_4^{(E)}$ and $\text{IV}_5^{(E)}$ with the single axiom

$\text{IV}_4^{(M)}$. *There exists a vector* **a** *with* $\mathbf{a}^2 > 0$ *and a vector* **b** *with* $\mathbf{b}^2 < 0$.

---

²One usually ignores the minor difference between the geometric structures determined by the groups of axioms I–V and I–IV, respectively. This difference amounts to the fact that in a "purely vectorial" space, where a point is defined as a vector, there exists a "special" point O. That is why the latter geometry is sometimes called *central Euclidean* (i.e., Eucidean with a special "center").

³It is easy to see that the relation $A, B \sim C, D$ on pairs of points is an *equivalence relation*. Axioms $\tilde{I}_2$ and $\tilde{I}_3$ say that the relation is symmetric and transitive. The reflexive nature of the relation follows easily from axioms $\tilde{I}_{2-4}$.

We repeat: the axioms of (plane) Minkowskian geometry consist of the groups I, II, III$^{(2)}$, and IV$^{(M)} = \{IV_1, IV_2, IV_3, IV_4^{(M)}\}$.

The coordinates of a vector **a** and the coordinates of a point A (defined as the coordinates of $\overline{OA}$, where O is the fixed origin) are introduced in Galilean and Minkowskian geometry in the same way as in Euclidean geometry. Differences arise when we define an orthonormal basis. In the case of Galilean geometry, axioms IV$_4^{(G)}$ and IV$_5^{(G)}$ readily imply that

$$\mathbf{oa} = 0$$

for every vector **a**. [Indeed, let **a** be any vector, and put $\mathbf{a}^2 = \alpha$. If $\mathbf{ao} = \beta \neq 0$, then for the vector $\mathbf{b} = \mathbf{a} + \lambda\mathbf{o}$ we have

$$\mathbf{b}^2 = (\mathbf{a} + \lambda\mathbf{o})^2 = \mathbf{a}^2 + 2\lambda(\mathbf{ao}) + \lambda^2\mathbf{o}^2 = \alpha + 2\lambda\beta.$$

But then $\mathbf{b}^2 < 0$ for suitable λ, which contradicts IV$_4^{(G)}$.] Now let **a** be a vector with $\mathbf{a}^2 = \alpha > 0$, and put $\mathbf{i} = (1/\sqrt{\alpha})\mathbf{a}$. Then

$$\mathbf{i}^2 = \frac{1}{\alpha}(\mathbf{a}^2) = \frac{1}{\alpha} \cdot \alpha = 1.$$

We define as *orthonormal* a basis of the Galilean plane formed by vectors **i**, **o** such that

$$\mathbf{i}^2 = 1, \quad \mathbf{o}^2 = 0, \quad \text{and} \quad \mathbf{io} = 0. \tag{19g}$$

In the case of Minkowskian geometry we choose vectors **a**, **b** with $\mathbf{a}^2 = \alpha > 0$ and $\mathbf{b}^2 = -\beta < 0$ (cf. IV$_4^{(M)}$). For the vector $\mathbf{i} = (1/\sqrt{\alpha})\mathbf{a}$ we have

$$\mathbf{i}^2 = \frac{1}{\alpha}(\mathbf{a}^2) = \frac{1}{\alpha} \cdot \alpha = 1.$$

Now let $\mathbf{ib} = \gamma$. Then for the vector $\mathbf{b}_1 = \mathbf{b} - \gamma\mathbf{i}$ we have

$$\mathbf{ib}_1 = \mathbf{i}(\mathbf{b} - \gamma\mathbf{i}) = \mathbf{ib} - \gamma(\mathbf{i}^2) = \gamma - \gamma = 0$$

and

$$(\mathbf{b}_1)^2 = (\mathbf{b} - \gamma\mathbf{i})^2 = \mathbf{b}^2 - 2\gamma\mathbf{bi} + \gamma^2(\mathbf{i}^2) = -\beta - 2\gamma^2 + \gamma^2 = -\beta - \gamma^2 < 0.$$

Putting $\mathbf{m} = (1/\sqrt{\beta + \gamma^2})\mathbf{b}$, we have

$$\mathbf{i}^2 = 1, \quad \mathbf{im} = 0, \quad \text{and} \quad \mathbf{m}^2 = -1. \tag{19m}$$

A basis of the Minkowskian plane satisfying (19m) is called *orthonormal*.

If $\{\mathbf{i}, \mathbf{o}\}$ is an orthonormal basis of the Galilean plane and **a**, **b** are vectors with coordinates (x, y) and (x_1, y_1) with respect to $\{\mathbf{i}, \mathbf{o}\}$, then the analogue of (20e) is

$$\mathbf{ab} = xx_1. \tag{20g}$$

Similarly, if $\{\mathbf{i}, \mathbf{m}\}$ is an orthonormal basis of the Minkowskian plane, then the analogue of (20e) is

$$\mathbf{ab} = xx_1 - yy_1. \tag{20m}$$

If, as before, we denote the *norm* $\mathbf{a}^2$ of a vector **a** by $\|\mathbf{a}\|$ and its *length*

$\sqrt{\mathbf{a}^2}$ by $|\mathbf{a}|$, then in place of (20″e) we obtain

$$\|\mathbf{a}\| = x^2, \qquad |\mathbf{a}| = x \qquad (20''\mathrm{g})$$

for Galilean geometry, and

$$\|\mathbf{a}\| = x^2 - y^2, \qquad |\mathbf{a}| = \sqrt{x^2 - y^2} \qquad (20''\mathrm{m})$$

for Minkowskian geometry. Thus in the Galilean plane we single out vectors $y\mathbf{o} = (0, y)$ with $|y\mathbf{o}| = 0$ ("special" vectors), while in the Minkowskian plane we distinguish between vectors for which $\|\mathbf{a}\| > 0$, $\|\mathbf{a}\| = 0$, and $\|\mathbf{a}\| < 0$ (vectors of positive, zero, and imaginary length, respectively). Vectors of zero norm are called *isotropic*, those of positive norm are called *spacelike*, and those of negative norm are called *timelike*; cf. Sections 12 and 13.

Lines in Galilean and Minkowskian geometry are introduced in the same way as in Euclidean geometry, i.e., by means of conditions (21)–(21a) or (22). All we need add is that in Galilean geometry we single out *special* lines with special direction vectors, and in the case of Minkowskian geometry we distinguish *spacelike*, *isotropic*, and *timelike* lines with appropriate direction vectors. Since the two types of Galilean lines are not comparable, it is natural to include in the class of lines of Galilean geometry only ordinary lines. Similarly, we include in the class of lines of Minkowskian geometry just spacelines or just timelines (since each of these classes is invariant under all the Lorentz transformations).

The *distance* d_{AA_1} between points $A(x,y)$ and $A_1(x_1,y_1)$ in the Galilean plane (the coordinates of A and A_1 are taken with respect to an orthonormal basis) is given by the formula

$$d_{AA_1} = x_1 - x \qquad (23\mathrm{g})$$

(and is therefore signed, i.e., positive, negative, or zero). The corresponding distance in the Minkowskian plane is given by

$$d_{AA_1} = \sqrt{(x_1 - x)^2 - (y_1 - y)^2} \qquad (23\mathrm{m})$$

(and is therefore positive real, zero, or imaginary). *Parallelism* and *perpendicularity* of lines with direction vectors $\mathbf{t}$ and $\mathbf{t}_1$ are defined in Galilean and Minkowskian geometry just as in Euclidean geometry, i.e., by means of the condition (24). We note that in Galilean geometry the condition (24) of *orthogonality of vectors* (and of perpendicularity of lines with these direction vectors) holds if and only if one of the two vectors is special (cf. the text on p. 43 and Fig. 37b).

It is somewhat more difficult to define the *angle* δ_{ll_1} between the lines l and l_1. In Minkowskian geometry it can be defined by an analogue of (25e):

$$\cosh^2 \delta_{ll_1} = \frac{(\mathbf{tt}_1)^2}{\|\mathbf{t}\| \|\mathbf{t}_1\|}, \qquad (25\mathrm{m})$$

where $\mathbf{t}$ and $\mathbf{t}_1$ are the direction vectors of l and l_1, respectively (cf. p. 187 ff). We require here that l and l_1 be both timelike or both spacelike.[4] In Galilean geometry, however, we run into a difficulty, since the right-hand side of (25e) reduces to 1.[5] The difficulty is fundamental: The axioms in groups I, II, III$^{(2)}$, V, and IV$^{(G)}$ = $\{$IV$_1$, IV$_2$, IV$_3$, IV$_4^{(G)} \equiv$ IV$_4^{(E)}$, IV$_5^{(G)}\}$ do not completely describe Galilean geometry. In fact, the class of transformations which preserve all the concepts in these axioms [in particular, the scalar product (20g)] consists of all transformations of the form

$$x' = x + a,$$
$$y' = vx + wy + b,$$

and is therefore larger than the class of motions (1) of the Galilean plane (see p. 33). One way of obtaining Galilean geometry is to supplement the above axioms with the axiom

IV$_6^{(G)}$. *The set of all vectors* $\mathbf{o}$ *such that* $\mathbf{oa} = 0$ *for all* $\mathbf{a}$ *is a Euclidean (vector) line.*

The axiom IV$_6^{(G)}$ states that in the set of "special" vectors $\mathbf{o}$ such that $|\mathbf{o}| = 0$ (and therefore $\mathbf{oa} = 0$ for all $\mathbf{a}$; cf. p. 414) we can define a ("special") scalar product $(\mathbf{pq})_1$ satisfying IV$_1$, IV$_2$, IV$_3$, IV$_4^{(E)}$, and IV$_5^{(E)}$. This scalar product enables us to define a "special" length $|\mathbf{p}|_1 = \sqrt{(\mathbf{p}^2)_1}$ for such vectors (in the coordinate system adopted above we can put $|\mathbf{p}(0,y)|_1 = y$). This special length allows us to introduce a measure on special lines which is closely related to the measure of angles between lines introduced in Sec. 1 of Chapter I.

We could approach the issue differently. In Euclidean geometry we can introduce, in addition to the scalar product (20e), the *cross product*

$$\mathbf{a} \times \mathbf{b} = xy_1 - yx_1 \qquad (26)$$

of vectors $\mathbf{a}(x,y)$ and $\mathbf{b}(x_1, y_1)$. This product has the following properties.

$\mathbf{a} \times \mathbf{b} = -(\mathbf{b} \times \mathbf{a})$ (anticommutativity);
$(\alpha \mathbf{a}) \times \mathbf{b} = \alpha(\mathbf{a} \times \mathbf{b})$ (associativity relative to multiplication of a vector by a scalar);
$(\mathbf{a} + \mathbf{b}) \times \mathbf{c} = \mathbf{a} \times \mathbf{c} + \mathbf{b} \times \mathbf{c}$ (distributivity over addition of vectors);
$\mathbf{i} \times \mathbf{j} = 1$ for an orthonormal basis $\{\mathbf{i}, \mathbf{j}\}$ which defines a "right-handed" coordinate system[6] (normalization condition).

---

[4] It is easily checked that $(\mathbf{tt}_1)^2 \geqslant \|\mathbf{t}\| \|\mathbf{t}_1\|$, and so the right side of (25m) is $\geqslant 1$. But then it is indeed of the form $\cosh^2 \theta$, where θ is real. Similar remarks apply in several places on the following pages. (Translator's note.)

[5] If we try to replace (25e) by its Galilean analogue,

$$\cosg \delta_{ll_1} = \frac{\mathbf{tt}_1}{|\mathbf{t}||\mathbf{t}_1|} \qquad (25g)$$

where $\cosg \alpha$ is the Galilean cosine of α (see Exercise 3, Sec. 3), then we run into the difficulty that $\cosg \alpha \equiv 1$ for all α.

[6] An orthonormal basis $\{\mathbf{i}, \mathbf{j}\}$ is right-handed if $\mathbf{i} \times \mathbf{j} = 1$ and left-handed if $\mathbf{i} \times \mathbf{j} = -1$. This corresponds to the distinction between right-handed and left-handed coordinate systems (and bases).

It is not difficult to show that these properties completely determine the product $\mathbf{a} \times \mathbf{b}$. We can use it to define the angle between lines l and l_1 with direction vectors $\mathbf{t}$ and $\mathbf{t}_1$ by means of the formula

$$\sin \delta_{l l_1} = \frac{\mathbf{t} \times \mathbf{t}_1}{|\mathbf{t}||\mathbf{t}_1|}. \tag{27e}$$

The definition (26) of a cross product (as well as its properties) is meaningful in Galilean and Minkowskian geometry.[7] We can use it to define the Galilean angle $\delta_{l l_1}$ between lines l and l_1 with direction vectors $\mathbf{t}$ and $\mathbf{t}_1$ by means of the formula

$$\delta_{l l_1} = \frac{\mathbf{t} \times \mathbf{t}_1}{|\mathbf{t}||\mathbf{t}_1|} \tag{27g}$$

(or the formula

$$\operatorname{sing} \delta_{l l_1} = \frac{\mathbf{t} \times \mathbf{t}_1}{|\mathbf{t}||\mathbf{t}_1|}, \tag{27'g}$$

where $\operatorname{sing} \alpha$ $(= \alpha)$ is the Galilean sine of the angle α; cf. Exercise 3, Sec. 3), and the Minkowskian angle by means of the formula

$$\sinh \delta_{l l_1} = \frac{\mathbf{t} \times \mathbf{t}_1}{|\mathbf{t}||\mathbf{t}_1|}. \tag{27m}$$

Thus far we have given complete characterizations of the geometries of Euclid, Galileo, and Minkowski. The basic concepts of **co-Euclidean** and **co-Minkowskian** geometry are a *doublet* (the analogue of a vector; see Sec. 6) and a *line*. Doublets (denoted by boldface capital letters) can be added and multiplied by numbers, and these operations are required to satisfy all the axioms of the groups I–III$^{(2)}$. Thus, $(\mathbf{A}+\mathbf{B})+\mathbf{C}=\mathbf{A}+(\mathbf{B}+\mathbf{C})$ for any three doublets $\mathbf{A},\mathbf{B},\mathbf{C}$; there exists a doublet $\mathbf{0}$ (the zero doublet) such that $\mathbf{A}+\mathbf{0}=\mathbf{A}$ for all $\mathbf{A}$; $\alpha(\mathbf{A}+\mathbf{B})=\alpha\mathbf{A}+\alpha\mathbf{B}$ for all $\mathbf{A}$, $\mathbf{B}$ and α; given three doublets $\mathbf{A},\mathbf{B},\mathbf{C}$ there exist numbers α,β,γ, not all zero, such that $\alpha\mathbf{A}+\beta\mathbf{B}+\gamma\mathbf{C}=\mathbf{0}$, and so on. In particular, these properties enable us to introduce coordinates (X,Y) of a doublet (relative to a basis of linearly independent doublets $\mathbf{E},\mathbf{F}$). The connection between doublets and lines consists in associating to each pair of lines a,b a unique doublet $\mathbf{A}=\overline{ab}$ (doublet with beginning a and end b). This, in turn, allows us to define the coordinates of an arbitrary line a as the coordinates of the doublet $\overline{oa}$, where o is the fixed "line origin." The association of a doublet to a pair of lines is required to satisfy axioms which are exact analogues of axioms V_1 and V_2. Finally, a *point* L is defined in our scheme as the set (pencil) of lines whose coordinates (X,Y) satisfy an equation of the form

$$\alpha X + \beta Y + \gamma = 0, \tag{22'}$$

where α,β,γ are given numbers [cf. (22)]. Such a pencil can also be specified by giving one of its lines, say a, and a doublet $\mathbf{T}$. The pencil then consists of all lines m

---

[7] It is easily seen that the geometric interpretation of the cross product $\mathbf{a} \times \mathbf{b}$ of (Euclidean) vectors $\mathbf{a}$ and $\mathbf{b}$ is the (signed) area of the parallelogram spanned by these vectors. Since the concept of area is meaningful in Galilean and Minkowskian geometry, the same is true of the concept of the cross product of vectors. We note that instead of supplementing IV_1–$IV_5^{(G)}$ with $IV_6^{(G)}$, we could have supplemented it with the requirement that there exist a cross product of vectors satisfying the conditions listed above. This approach would also have led us to the "usual" Galilean geometry whose study takes up most of this book.

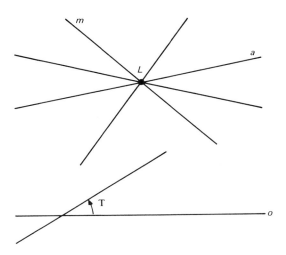

Figure 207

such that

$$\overline{am} = \lambda T \quad \text{or} \quad \overline{om} = \overline{oa} + \lambda T, \tag{21'}$$

where o is a preselected "doublet origin" (cf. Fig. 207). Thus we can speak of "the point on line a with doublet **T**." In order to measure *angles* between lines (and *distances* between points) we must still introduce the operation of scalar multiplication of doublets, which associates to each pair of doublets **A, B** a number **AB**, and satisfies analogues of the axioms IV$^{(E)}$ or IV$^{(M)}$. The choice of IV$^{(E)}$ leads to co-Euclidean geometry, and the choice of IV$^{(M)}$ to co-Minkowskian geometry.

In order to characterize by means of the above scheme **elliptic geometry, hyperbolic geometry, cohyperbolic geometry,** and **doubly hyperbolic geometry,** we must first introduce the concepts of a (*three-dimensional*) *Euclidean vector space* and that of a (*three-dimensional*) *Minkowskian vector space*. A Euclidean vector space is defined as a set of objects called vectors[8] satisfying the axioms in groups I, II, and IV$^{(E)}$. The axioms in group III are now replaced by the following.

III$^{(3)}$. Axioms of dimension

III$_1^{(3)}$. *Given four vectors* **a, b, c, d**, *there exist four numbers* $\alpha, \beta, \gamma, \delta$ *not all zero such that*

$$\alpha \mathbf{a} + \beta \mathbf{b} + \gamma \mathbf{c} + \delta \mathbf{d} = \mathbf{0}. \tag{16'}$$

III$_2^{(3)}$. *There exist three vectors* **a, b, c** *such that*

$$\alpha \mathbf{a} + \beta \mathbf{b} + \gamma \mathbf{c} = \mathbf{0} \quad \text{only if} \quad \alpha = \beta = \gamma = 0.$$

---

[8]In developing the usual (school) geometry we can identify "point" and "vector" and work with the single undefined term "vector." This approach, used in most university texts on linear algebra, is adopted, for example, in [8].

Supplement B. Axiomatic characterization of the nine plane geometries

An equivalent formulation of these axioms is, of course, the following.
III$_1^{(3)}$. *Any four vectors are linearly dependent.*
III$_2^{(3)}$. *There exist three linearly independent vectors.*

A *basis* is now a triple of linearly independent vectors $\mathbf{e}, \mathbf{f}, \mathbf{g}$, and the choice of a basis enables us to define the coordinates (x,y,z) of a vector $\mathbf{a}$ as the coefficients in the decomposition

$$\mathbf{a} = x\mathbf{e} + y\mathbf{f} + z\mathbf{g}. \tag{17'}$$

It is clear that the coordinates of $\alpha \mathbf{a}$ are obtained by multiplying the coordinates of $\mathbf{a}$ by α, and that the coordinates of $\mathbf{a} + \mathbf{b}$ are sums of the corresponding coordinates of $\mathbf{a}$ and $\mathbf{b}$. A (vector) line l with direction vector $\mathbf{t}$ can be defined as the set of all vectors $\mathbf{m}$ which are multiples of $\mathbf{t}$:

$$\mathbf{m} = \lambda \mathbf{t}. \tag{28$_1$}$$

A line can also be defined as the set of all vectors $\mathbf{p}(x,y,z)$ with coordinates proportional to given numbers α, β, γ (not all zero), i.e., by the condition

$$x : \alpha = y : \beta = z : \gamma. \tag{28$_2$}$$

Similarly, a (vector) plane π can be defined as the set of all linear combinations of two linearly independent vectors $\mathbf{u}, \mathbf{v}$, i.e., as the set of vectors $\mathbf{q}$ such that

$$\mathbf{q} = \lambda \mathbf{u} + \mu \mathbf{v}, \tag{29$_1$}$$

where λ and μ are arbitrary numbers. The plane π can also be defined as the set of vectors $\mathbf{q}(x,y,z)$ such that

$$\delta x + \varepsilon y + \zeta z = 0, \tag{29$_2$}$$

where $\delta, \varepsilon, \zeta$ are given numbers not all zero. The numbers α, β, γ in (28$_2$) and $\delta, \varepsilon, \zeta$ in (29$_2$) are determined only up to a nonzero multiple; for example, the number triples α, β, γ and $\lambda\alpha, \lambda\beta, \lambda\gamma$, $\lambda \neq 0$, determine the same line l.

The axioms IV$^{(E)}$ imply the existence of an *orthonormal basis* $\mathbf{i}, \mathbf{j}, \mathbf{k}$ such that

$$\mathbf{i}^2 = \mathbf{j}^2 = \mathbf{k}^2 = 1, \quad \mathbf{ij} = \mathbf{ik} = \mathbf{jk} = 0. \tag{19'e}$$

Indeed if $\{\mathbf{e}, \mathbf{f}, \mathbf{g}\}$ is a basis, then we can put $\mathbf{i} = (1/\sqrt{\alpha})\mathbf{e}$, where $\alpha = \mathbf{e}^2$. Then $\mathbf{i}^2 = 1$. Further, if $\mathbf{if} = \beta$, then the vector $\mathbf{f}_1 = \mathbf{f} - \beta \mathbf{i}$ is such that $\mathbf{if}_1 = 0$ and the vector $\mathbf{j} = (1/\sqrt{\gamma})\mathbf{f}_1$, where $\gamma = \mathbf{f}_1^2$, is such that $\mathbf{ij} = 0$ and $\mathbf{j}^2 = 1$. Finally, if $\mathbf{ig} = \delta$ and $\mathbf{jg} = \varepsilon$ then the vector $\mathbf{g}_1 = \mathbf{g} - \delta \mathbf{i} - \varepsilon \mathbf{j}$ satisfies the conditions $\mathbf{ig}_1 = \mathbf{jg}_1 = 0$, and the vector $\mathbf{k} = (1/\sqrt{\zeta})\mathbf{g}_1$, where $\zeta = \mathbf{g}_1^2$, satisfies the conditions $\mathbf{ik} = \mathbf{jk} = 0$, $\mathbf{k}^2 = 1$. Relative to an orthonormal basis $\{\mathbf{i}, \mathbf{j}, \mathbf{k}\}$, the scalar product of vectors $\mathbf{a}(x,y,z)$ and $\mathbf{b}(x_1,y_1,z_1)$ takes the form

$$\mathbf{ab} = xx_1 + yy_1 + zz_1, \tag{20'e}$$

so that if we define the norm $\|\mathbf{a}\|$ of $\mathbf{a}$ as $\|\mathbf{a}\| = \mathbf{a}^2$ and the length $|\mathbf{a}|$ of $\mathbf{a}$ as $|\mathbf{a}| = \sqrt{\mathbf{a}^2}$, then

$$\|\mathbf{a}\| = x^2 + y^2 + z^2, \quad |\mathbf{a}| = \sqrt{x^2 + y^2 + z^2}. \tag{20$_2'$e}$$

The angle φ between $\mathbf{a}$ and $\mathbf{b}$ is defined by the familiar formula $\cos\varphi = \mathbf{ab}/|\mathbf{a}||\mathbf{b}|$, so that the angle between lines l and l_1 with direction vectors $\mathbf{t}$ and $\mathbf{t}_1$ is computed by means of the formula (25e) used in two-dimensional space (cf. p. 413). Similarly, the plane with equation (29$_2$) is characterized by its normal vector $\mathbf{n}(\delta, \varepsilon, \zeta)$ (this vector is perpendicular to all vectors in π: if $\mathbf{q} \in \pi$, then $\mathbf{nq} = 0$), and the angle ψ between two planes π and π_1 with normals $\mathbf{n}$ and $\mathbf{n}_1$ is computed from the formula

$$\cos\psi = \frac{\mathbf{nn}_1}{|\mathbf{n}||\mathbf{n}_1|}. \tag{25'e}$$

(Three-dimensional) pseudo-Euclidean vector spaces are defined by the groups of axioms I, II, III$^{(3)}$, and two further axioms constituting the group IV$^{(M)}$. One is the familiar axiom IV$_4^{(M)}$ and the other is the axiom[9]

IV$_5^{(M)}$. *There is no nonzero vector* **o** *such that* **oa**=**o** *for all vectors* **a**.

The only difference between three-dimensional Euclidean and pseudo-Euclidean spaces occurs in connection with the definition of an orthonormal basis. We choose a basis $\{\mathbf{e},\mathbf{f},\mathbf{g}\}$ of pseudo-Euclidean space with $\mathbf{e}^2=\alpha>0$. It is obvious that this can be done. If, as before, we put $\mathbf{i}=(1/\sqrt{\alpha})\mathbf{e}$, then $\mathbf{i}^2=1$. Further, $\mathbf{f}_1$ is defined (as before) as the vector $\mathbf{f}-\beta\mathbf{i}$, where $\beta=\mathbf{if}$. Then $\mathbf{if}_1=0$. At this point we must consider the case $\mathbf{f}_1^2=\gamma>0$ and the case $\mathbf{f}_1^2=\gamma<0$. In the first case, putting $\mathbf{j}=(1/\sqrt{\gamma})\mathbf{f}_1$, we obtain the equalities $\mathbf{ij}=0$ and $\mathbf{j}^2=1$. In the second case, putting $\mathbf{m}=(1/\sqrt{|\gamma|})\mathbf{f}_1$, we obtain $\mathbf{im}=0$ and $\mathbf{m}^2=-1$. Finally, if we put $\mathbf{g}_1=\mathbf{g}-\delta\mathbf{i}-\varepsilon\mathbf{j}$ or $\mathbf{g}_1=\mathbf{g}-\delta\mathbf{i}-\varepsilon\mathbf{m}$, where $\delta=\mathbf{ig}$ and $\varepsilon=\mathbf{jg}$ or $\mathbf{mg}$, then we have $\mathbf{g}_1\mathbf{i}=\mathbf{g}_1\mathbf{j}=0$ or $\mathbf{g}_1\mathbf{i}=\mathbf{g}_1\mathbf{m}=0$. If $\mathbf{f}_1^2>0$, then necessarily $\mathbf{g}_1^2=\zeta<0$ (for otherwise IV$_4^{(M)}$ would not hold) and we put $\mathbf{m}=(1/\sqrt{|\zeta|})\mathbf{g}_1$. This yields the equalities

$$\mathbf{i}^2=\mathbf{j}^2=1, \quad \mathbf{m}^2=-1, \quad \text{and} \quad \mathbf{ij}=\mathbf{im}=\mathbf{jm}=0. \tag{19$_1'$m}$$

On the other hand, if $\mathbf{f}_1^2<0$, then we can have $\mathbf{g}_1^2=\zeta>0$ or $\mathbf{g}_1^2=\zeta<0$. In the first case, we put $\mathbf{j}=(1/\sqrt{\zeta})\mathbf{g}_1$ and in the second case we put $\mathbf{n}=(1/\sqrt{|\zeta|})\mathbf{g}_1$. In the first case we obtain the equalities

$$\mathbf{i}^2=\mathbf{j}^2=1, \quad \mathbf{m}^2=-1, \quad \text{and} \quad \mathbf{ij}=\mathbf{im}=\mathbf{jm}=0, \tag{19$_1'$m}$$

and in the second case, the equalities

$$\mathbf{i}^2=1, \quad \mathbf{m}^2=\mathbf{n}^2=-1, \quad \text{and} \quad \mathbf{im}=\mathbf{in}=\mathbf{mn}=0. \tag{19$_2'$m}$$

This shows that there exist two (quite similar) types of three-dimensional pseudo-Euclidean spaces in which the inner product of two vectors $\mathbf{a}(x,y,z)$ and $\mathbf{b}(x_1,y_1,z_1)$ is given, respectively, by

$$\mathbf{ab}=xx_1+yy_1-zz_1 \tag{20'm}$$

and

$$\mathbf{ab}=xx_1-yy_1-zz_1. \tag{20''m}$$

[We shall refer to the pseudo-Euclidean space with the scalar product (20'm) as *Minkowskian three-space*; it can be distinguished from the other pseudo-Euclidean space (with scalar product (20''m)) by the fact that it contains a *Euclidean* vector plane (29$_2$).] The norm $\|\mathbf{a}\|$ and length $|\mathbf{a}|$ of a vector $\mathbf{a}$ in Minkowskian three-space are given by the formulas

$$\|\mathbf{a}\|=x^2+y^2-z^2 \quad \text{and} \quad |\mathbf{a}|=\sqrt{x^2+y^2-z^2}\,; \tag{20$_2$m}$$

here it is necessary to distinguish *spacelike*, *isotropic*, and *timelike* vectors **a** with norms $\|\mathbf{a}\|>0$, $\|\mathbf{a}\|=0$ and $\|\mathbf{a}\|<0$, respectively. A line l of Minkowskian space defined by (28$_2$) is called *spacelike*, *isotropic*, or *timelike* according as

$$\|\mathbf{t}\|=\alpha^2+\beta^2-\gamma^2>0, \quad \|\mathbf{t}\|=0, \quad \text{or} \quad \|\mathbf{t}\|<0$$

(note that only the sign of $\|\mathbf{t}\|$ has geometric significance). It is clear that a spacelike line is a set of spacelike vectors, an isotropic line is a set of isotropic vectors, and a timelike line is a set of timelike vectors. The angle φ between two lines with direction vectors $\mathbf{t}$ and $\mathbf{t}_1$ is defined by the familiar formula (25m), and is

---

[9]IV$_2^{(M)}$ holds in the Minkowskian plane, but in this case it can be deduced from the other axioms and is thus a theorem rather than an axiom.

Supplement B. Axiomatic characterization of the nine plane geometries

meaningful only for two lines of the same kind (two spacelike or two timelike lines). The nature of a plane given by (29_2) and belonging to a Minkowskian (vector) space is determined by the nature of its normal vector $\mathbf{n}(\delta,\varepsilon,\zeta)$ ($\mathbf{n}$ is obviously orthogonal to all the vectors in π, that is, $\mathbf{nq}=0$ for all $\mathbf{q}\in\pi$). Depending on whether $\|\mathbf{n}\|>0$, $\|\mathbf{n}\|=0$, or $\|\mathbf{n}\|<0$ we call the plane *spacelike, isotropic,* or *timelike*.[10] It is easy to see that a timelike plane has the structure of a Euclidean plane and all its vectors are timelike; an isotropic plane contains a single isotropic line and represents a Galilean plane; and a timelike plane has the structure of a Minkowskian plane and contains timelike, isotropic, and spacelike vectors (why?). The angle ψ between two planes π and π_1 with normals $\mathbf{n}$ and $\mathbf{n}_1$ is defined by the formula

$$\cosh\psi = \frac{\mathbf{nn}_1}{|\mathbf{n}||\mathbf{n}_1|} ; \qquad (25'm)$$

it has meaning only if both planes are of the same kind (both spacelike or both timelike).

We can now define the *elliptic plane* as the set of all (vector) lines and planes of three-dimensional Euclidean space $\mathbb{E}^3$. The lines of $\mathbb{E}^3$ are called the *points* of the elliptic plane, and the planes of $\mathbb{E}^3$ are called the *lines* of the elliptic plane. The angle between two lines of $\mathbb{E}^3$ is called the *distance between the corresponding points* of the elliptic plane (represented by these lines), and the angle between two planes of $\mathbb{E}^3$ is called the *angle between the corresponding lines* of the elliptic plane. Similarly, we define the *hyperbolic plane* as the set of all timelike lines and timelike planes of three-dimensional Minkowskian space $\mathbb{M}^3$. The lines of $\mathbb{M}^3$ are called *points* of the hyperbolic plane and the angle between two lines is called the *distance between the corresponding points*. The planes of $\mathbb{M}^3$ are called *lines* of the hyperbolic plane (and the angle between two planes is called the *angle between the corresponding lines*). If we designate spacelike lines of $\mathbb{M}^3$ as points, and timelike planes of $\mathbb{M}^3$ as *lines*, then we arrive at *doubly hyperbolic geometry*. One way of constructing *cohyperbolic geometry* is to designate timelike planes of $\mathbb{M}^3$ as points, and timelike lines of $\mathbb{M}^3$ as lines (what are the other ways?).

---

[10]Clearly the quantity $\|\mathbf{n}\|$ depends on the choice of the normal vector $\mathbf{n}$ (and there are many such vectors). However, the sign of $\|\mathbf{n}\|$ is a geometric characteristic of the plane π (29_2).

Supplement C. Analytic models of the nine plane geometries

It is well known that the points of the Euclidean plane can be identified with the *complex numbers* by associating to the point with rectangular coordinates (x,y) or polar coordinates (r,φ) the complex number

$$z = x + iy = r(\cos\varphi + i\sin\varphi) \qquad (30)$$

(Fig. 208). The numbers x and y are called the real and imaginary part of the number z and are denoted by $\mathrm{Re}\,z$ and $\mathrm{Im}\,z$; r and φ are called the modulus and argument of z and are denoted by $|z|$ and $\arg z$. [The argument of $z \neq 0$ is defined to within an integral multiple of 2π—i.e., if φ is an argument of z, then so is $\varphi + 2k\pi$ for integral k; we assign no argument to $z=0$.] The modulus $|z|$ of z can be defined by the formulas

(a) $\quad |z|^2 = x^2 + y^2 \quad (=(\mathrm{Re}\,z)^2 + (\mathrm{Im}\,z)^2)$

or $\qquad\qquad\qquad\qquad\qquad\qquad\qquad\qquad\qquad\qquad\qquad (31)$

(b) $\quad |z|^2 = z\bar{z},$

where $\bar{z}$ is the complex conjugate of z:

(a) $\quad \mathrm{Re}\,\bar{z} = \mathrm{Re}\,z, \qquad \mathrm{Im}\,\bar{z} = -\mathrm{Im}\,z$

or $\qquad\qquad\qquad\qquad\qquad\qquad\qquad\qquad\qquad\qquad\qquad (32)$

(b) $\quad |\bar{z}| = |z|, \qquad \arg\bar{z} = -\arg z$

[i.e., if $z = x + iy = r(\cos\varphi + i\sin\varphi)$, then $\bar{z} = x - iy = r[\cos(-\varphi) + i\sin(-\varphi)] = r(\cos\varphi - i\sin\varphi)$]. The argument $\arg z$ of z is given by the relations

$$\frac{\mathrm{Re}\,z}{|z|} = \cos(\arg z), \qquad \frac{\mathrm{Im}\,z}{|z|} = \sin(\arg z), \qquad \frac{\mathrm{Im}\,z}{\mathrm{Re}\,z} = \tan(\arg z), \quad (33)$$

which most readers are used to seeing in the form

$$\frac{x}{r} = \cos\varphi, \qquad \frac{y}{r} = \sin\varphi, \qquad \frac{y}{x} = \tan\varphi. \qquad (33a)$$

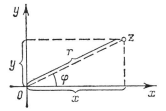

Figure 208

It is important to bear in mind that

$$\operatorname{Re}(z \pm z_1) = \operatorname{Re} z \pm \operatorname{Re} z_1, \qquad \operatorname{Im}(z \pm z_1) = \operatorname{Im} z \pm \operatorname{Im} z_1 \qquad (34)$$

[i.e., $(x + iy) \pm (x_1 + iy_1) = (x \pm x_1) + i(y \pm y_1)$],

$$|zz_1| = |z||z_1|, \qquad \arg(zz_1) = \arg z + \arg z_1 \qquad (34a)$$

(i.e., $r(\cos\varphi + i\sin\varphi) r_1(\cos\varphi_1 + i\sin\varphi_1) = rr_1[\cos(\varphi + \varphi_1) + i\sin(\varphi + \varphi_1)]$), and

$$\left|\frac{z}{z_1}\right| = \frac{|z|}{|z_1|}, \qquad \arg\left(\frac{z}{z_1}\right) = \arg z - \arg z_1 \qquad (34b)$$

(i.e., $r(\cos\varphi + i\sin\varphi)/r_1(\cos\varphi_1 + i\sin\varphi_1) = (r/r_1)[\cos(\varphi - \varphi_1) + i\sin(\varphi - \varphi_1)]$). The relations (32)–(34b) also imply that

$$\overline{z+z_1} = \bar{z} + \bar{z}_1, \quad \overline{z-z_1} = \bar{z} - \bar{z}_1, \quad \overline{zz_1} = \bar{z}\bar{z}_1 \quad \text{and} \quad \overline{\left(\frac{z}{z_1}\right)} = \frac{\bar{z}}{\bar{z}_1}, \quad (35)$$

and that the sum $z + \bar{z} = 2\operatorname{Re} z$ and the product $z\bar{z} = |z|^2$ are real, while the difference $z - \bar{z} = (2\operatorname{Im} z)i$ is pure imaginary (i.e., of the form ib, b real). The equality $z = \bar{z}$ holds only for real z, i.e., z such that $\operatorname{Im} z = 0$ and $\arg z = 0$ or π, and the equality $z = -\bar{z}$ holds only for pure imaginary z, i.e., z such that $\operatorname{Re} z = 0$ and $\arg z = \pm \pi/2$.

The *distance* d_{z,z_1} between two points of the Euclidean plane corresponding to the complex numbers z and z_1—or, as we shall say, between the complex numbers z and z_1—is defined by the formula

$$d_{z,z_1} = |z_1 - z| \quad \text{or} \quad d_{z,z_1}^2 = (z - z_1)(\bar{z} - \bar{z}_1) \qquad (36)$$

(Fig. 209a). The *angle* $\delta_{(z_0,z_1)(z_0,z_2)}$ between the lines joining z_1 and z_2 to z_0—or, as we shall say, between the lines (z_0, z_1) and (z_0, z_2)—is given by

$$\delta_{(z_0,z_1)(z_0,z_2)} = \arg(z_2, z_1; z_0) = \arg\frac{z_2 - z_0}{z_1 - z_0}, \qquad (37)$$

where $(z_2, z_1; z_0) = (z_2 - z_0)/(z_1 - z_0)$ is called the *simple ratio of three points* z_2, z_1, and z_0. Formula (37) follows from the fact that $\delta_{(z_0,z_1)(z_0,z_2)} = \varphi_2 - \varphi_1$, where φ_2 and φ_1 are the arguments of the complex numbers $z_2^0 = z_2 - z_0$ and $z_1^0 = z_1 - z_0$ (Fig. 209b). Here we must bear in mind that the angle $\delta_{(z_0,z_1)(z_0,z_2)}$ is the directed angle between the directed lines (z_0, z_1) and (z_0, z_2), i.e., the angle through which we must rotate the positive ray of the

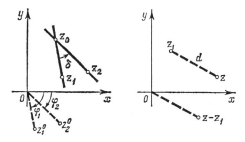

Figure 209a Figure 209b

directed line (z_0, z_1) (the ray from z_0 to z_1) to bring it into coincidence with the positive ray of the directed line (z_0, z_2).

It is clear that the (nondirected) *line* (z_1, z_2) is the set of points z such that

$$\arg(z, z_1; z_2) = \arg \frac{z - z_2}{z_1 - z_2} = 0 \quad \text{or } \pi \tag{38}$$

(Fig. 210a). It follows that a line (z_1, z_2) can be defined as the set of points z such that

$$\operatorname{Im}(z, z_1; z_2) = \operatorname{Im} \frac{z - z_2}{z_1 - z_2} = 0 \tag{38a}$$

i.e., that the number $(z, z_1; z_2)$ is real. But then

$$\frac{z - z_2}{z_1 - z_2} = \frac{\bar{z} - \bar{z}_2}{\bar{z}_1 - \bar{z}_2}. \tag{39}$$

In other words, *a line* (z_1, z_2) *is given by the equation* (39), which can be written as

$$(\bar{z}_1 - \bar{z}_2)z - (z_1 - z_2)\bar{z} + (z_1 \bar{z}_2 - \bar{z}_1 z_2) = 0,$$

or

$$Bz - \bar{B}\bar{z} + C = 0, \quad \operatorname{Re} C = 0, \tag{40}$$

where $B = \bar{z}_1 - \bar{z}_2$, and where $C = z_1 \bar{z}_2 - \bar{z}_1 z_2$ is indeed pure imaginary.

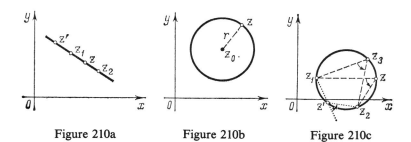

Figure 210a Figure 210b Figure 210c

Conversely, *an equation of the form* (40) *defines a line* passing through the points z_1, z_2 such that $\bar{z}_1 - \bar{z}_2 = B$, $z_1\bar{z}_2 - \bar{z}_1 z_2 = C$.

A *circle* with center z_0 and radius r is the set of points z such that

$$|z - z_0| = r \quad \text{or} \quad (z - z_0)(\bar{z} - \bar{z}_0) = r^2 \tag{41}$$

(Fig. 210b). Thus *the equation of a circle has the form*

$$z\bar{z} - \bar{z}_0 z - z_0 \bar{z} + (z_0 \bar{z}_0 - r^2) = 0,$$

or the form

$$az\bar{z} + bz + \bar{b}\bar{z} + c = 0, \quad \operatorname{Im} a = \operatorname{Im} c = 0. \tag{42}$$

Conversely, *an equation of the form* (42) *with* $a \neq 0$ *defines a circle*, namely the circle with center z_0 and radius r determined by the relations $\bar{z}_0 = -b/a$, $z_0 \bar{z}_0 - r^2 = c/a$. After multiplication by i, Eq. (40) becomes the special case of Eq. (42) with

$$a = 0. \tag{43}$$

We could also replace (42) by

$$Az\bar{z} + Bz - \bar{B}\bar{z} + C = 0, \quad \operatorname{Re} A = \operatorname{Re} C = 0, \tag{42a}$$

and view (40) as the special case of (42a) in which

$$A = 0. \tag{43a}$$

The circle passing through three points z_1, z_2, z_3 can be described as the set of points z such that

$$\delta_{(z_3, z_1)(z_3, z_2)} - \delta_{(z, z_1)(z, z_2)} = 0 \quad \text{or} \quad \pi$$

(here the angles and lines are directed; Fig. 210c) or such that

$$\arg \frac{z_1 - z_3}{z_2 - z_3} - \arg \frac{z_1 - z}{z_2 - z} = 0 \quad \text{or} \quad \pi. \tag{44}$$

In turn, (44) is equivalent to

$$\operatorname{Im}(z_1, z_2; z_3, z) = \operatorname{Im} \frac{(z_1, z_2; z_3)}{(z_1, z_2; z)} = \operatorname{Im} \frac{(z_1 - z_3)/(z_2 - z_3)}{(z_1 - z)/(z_2 - z)} = 0, \tag{44a}$$

where $(z_1, z_2; z_3, z)$ is called *the cross ratio of four points* z_1, z_2, z_3, and z. Thus *the condition for four points* z_1, z_2, z_3, z_4 *to lie on a circle (or line) is*

$$\operatorname{Im}(z_1, z_2; z_3, z_4) = \operatorname{Im} \frac{(z_1 - z_3)/(z_2 - z_3)}{(z_1 - z_4)/(z_2 - z_4)} = 0. \tag{45}$$

The equation (44a) of a circle passing through points z_1, z_2 and z_3 can also be written as

$$\frac{(z_1 - z_3)/(z_2 - z_3)}{(z_1 - z)/(z_2 - z)} = \frac{(\bar{z}_1 - \bar{z}_3)/(\bar{z}_2 - \bar{z}_3)}{(\bar{z}_1 - \bar{z})/(\bar{z}_2 - \bar{z})}, \tag{44b}$$

or as

$$Az\bar{z} + Bz - \bar{B}\bar{z} + C = 0, \quad \operatorname{Re} A = \operatorname{Re} C = 0 \tag{42a}$$

[cf. the earlier occurrence of (42a)], where

$$A = (z_1 - z_3)(\bar{z}_2 - \bar{z}_3) - (\bar{z}_1 - \bar{z}_3)(z_2 - z_3),$$
$$B = \bar{z}_2(\bar{z}_1 - \bar{z}_3)(z_2 - z_3) - \bar{z}_1(z_1 - z_3)(\bar{z}_2 - \bar{z}_3), \quad (42b)$$
$$C = \bar{z}_1 z_2(z_1 - z_3)(\bar{z}_2 - \bar{z}_3) - z_1 \bar{z}_2(\bar{z}_1 - \bar{z}_3)(z_2 - z_3).$$

The *motions* of the Euclidean plane are the transformations of the plane which take z to z', where[1]

(a) $\quad z' = pz + q,$

or

(b) $\quad z' = p\bar{z} + q \quad (p\bar{p} = 1).$ $\quad (46)$

Indeed, it is easy to show that if z and z_1 are sent to z' and z'_1, say, by (a) in (46), then

$$|z'_1 - z'|^2 = (z'_1 - z')(\overline{z'_1} - \overline{z'}) = [(pz_1 + q) - (pz + q)][(\bar{p}\bar{z}_1 + \bar{q}) - (\bar{p}\bar{z} + \bar{q})]$$
$$= p(z_1 - z)\bar{p}(\bar{z}_1 - \bar{z}) = p\bar{p}(z_1 - z)(\bar{z}_1 - \bar{z}) = (z_1 - z)(\bar{z}_1 - \bar{z}) = |z_1 - z|^2,$$

i.e., the transformation (a) in (46) preserves the distance between points:

$$d_{z'_1, z'} = d_{z_1, z}. \quad (47)$$

In particular, the very simple transformations

(a) $\quad z' = -z$

and

(b) $\quad z' = \bar{z}$ $\quad (48)$

represent, respectively, a half-turn about O (Fig. 211a) and a reflection in the line $\operatorname{Im} z = 0$ (Fig. 211b), and (a) in (46) represents a rotation about O through the angle $\arg p$ followed by the translation determined by the vector q with beginning O and end q (Fig. 212).

Finally, we note that the transformations

(a) $\quad z' = \dfrac{az + b}{cz + d}$

and

(b) $\quad z' = \dfrac{a\bar{z} + b}{c\bar{z} + d},$ $\quad (49)$

where $ad - bc \neq 0$, related to the motions (46) when $c = 0$, *either preserve the cross ratio* $(z_1, z_2; z_3, z_4)$ *of four points* z_1, z_2, z_3, z_4 [true of (a) in (49)] *or change it to its complex conjugate* [true of (b) in (49)]. Indeed, if the

---

[1] It is easy to see that (a) in (46) represents direct motions, and (b) in (46) represents opposite motions (cf. [19], [10]).

Supplement C. Analytic models of the nine plane geometries

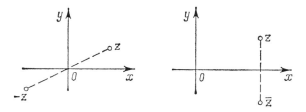

Figure 211a Figure 211b

quadruples z'_1, z'_2, z'_3, z'_4 and z_1, z_2, z_3, z_4 are related by (a) in (49), then

$$(z'_1, z'_2; z'_3, z'_4) = \frac{(z'_1 - z'_3)/(z'_2 - z'_3)}{(z'_1 - z'_4)/(z'_2 - z'_4)}$$

$$= \frac{\left(\dfrac{az_1+b}{cz_1+d} - \dfrac{az_3+b}{cz_3+d}\right) \Big/ \left(\dfrac{az_2+b}{cz_2+d} - \dfrac{az_3+b}{cz_3+d}\right)}{\left(\dfrac{az_1+b}{cz_1+d} - \dfrac{az_4+b}{cz_4+d}\right) \Big/ \left(\dfrac{az_2+b}{cz_2+d} - \dfrac{az_4+b}{cz_4+d}\right)}$$

$$= \frac{\dfrac{(ad-bc)(z_1-z_3)}{(cz_1+d)(cz_3+d)} \Big/ \dfrac{(ad-bc)(z_2-z_3)}{(cz_2+d)(cz_3+d)}}{\dfrac{(ad-bc)(z_1-z_4)}{(cz_1+d)(cz_4+d)} \Big/ \dfrac{(ad-bc)(z_2-z_4)}{(cz_2+d)(cz_4+d)}}$$

$$= \frac{[(z_1-z_3)/(z_2-z_3)] \cdot [(cz_2+d)/(cz_1+d)]}{[(z_1-z_4)/(z_2-z_4)] \cdot [(cz_2+d)/(cz_1+d)]}$$

$$= \frac{(z_1-z_3)/(z_2-z_3)}{(z_1-z_4)/(z_2-z_4)}$$

$$= (z_1, z_2; z_3, z_4).$$

It follows that if the cross ratio of four points z_1, z_2, z_3, and z_4 is real, then so is the cross ratio of their images z'_1, z'_2, z'_3, and z'_4 under (a) or (b) in (49);

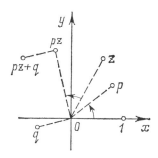

Figure 212

in other words, *a transformation* (49) *takes four points on a circle (or line) to four points on a circle (or line)*, and is thus a *circular transformation* (see p. 146). It is possible to show that *every circular transformation of the Euclidean plane is of the form (a) or (b) in* (49). In particular, the inversion with center O and coefficient 1 (cf. pp. 124–125) is given by

$$z' = \frac{1}{\bar{z}} \qquad (50)$$

[Fig. 213; (50) is a special case of (b) in (49) with $a=d=0$, $b=c=1$].

It is clear that a transformation (46) takes every point z of the Euclidean plane to some point z'. However, this is not the case for the transformations (49) with $c \neq 0$. Specifically, a transformation (a) in (49) assigns no image to the point $z = -d/c$ and a transformation (b) in (49) assigns no image to the point $z = -\bar{d}/\bar{c}$ (in each of these cases the denominator vanishes). Of course, this is connected with the fact that while the expressions $z_1 \pm z_2$ and $z_1 z_2$ always define a complex number, this is not the case for the quotient z_1/z_2 when $z_2 = 0$. To avoid this difficulty, we sometimes put

$$\frac{1}{0} = \infty$$

and

$$z + \infty = \infty, \quad z - \infty = \infty, \quad z \cdot \infty = \infty, \quad \frac{\infty}{z} = \infty, \quad \frac{z}{\infty} = 0$$

for an arbitrary complex number z (in the third equality we assume that $z \neq 0$ and in the others we assume that $z \neq \infty$). Thus, the transformations (a) and (b) in (49) are one-to-one mappings defined for all points of the complex plane, supplemented by the fictitious point at infinity ∞; in particular, a transformation (a) in (49) takes $z = -d/c$ to

$$z' = \frac{-a(d/c) + b}{0} = \frac{-(ad - bc)/c}{0} = \infty$$

and $z = \infty$ to

$$z' = \frac{a\infty + b}{c\infty + d} = \frac{a + (b/\infty)}{c + (d/\infty)} = \frac{a + 0}{c + 0} = \frac{a}{c}.$$

In slightly different terms, the natural domain of definition of the circular transformations (a) and (b) in (49) is not the Euclidean plane but rather the *inversive plane* obtained by adding to the Euclidean plane the point at infinity ∞.

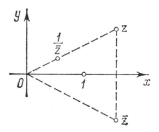

Figure 213

Supplement C. Analytic models of the nine plane geometries

It is possible to generalize the preceding constructions by considering, in addition to complex numbers, "dual numbers" and "double numbers." A *dual number* is an expression of the form $z = x + \varepsilon y$, where x and y are real and the "dual unit" ε satisfies the condition $\varepsilon^2 = 0$; like the "complex unit" i, the "dual unit" ε lies outside the domain of real numbers. A *double number* is an expression of the form $z = z + ey$, where x and y are real, and the "double unit" e (which again lies outside the reals) satisfies the condition $e^2 = +1$.

It seems that dual numbers were considered by the well-known German geometer E. STUDY (1862–1930) and are often referred to in the literature as the dual numbers of Study. Double numbers were apparently first introduced by the noted English mathematician W. K. CLIFFORD (1845–1879).[2]

The rules for addition and subtraction of dual and double numbers are similar to the rules for addition and subtraction of complex numbers:

$$(x + \varepsilon y) \pm (x_1 + \varepsilon y_1) = (x \pm x_1) + \varepsilon(y \pm y_1), \quad (51')$$

$$(x + ey) \pm (x_1 + ey_1) = (x \pm x_1) + e(y \pm y_1). \quad (51'')$$

The rules for multiplication are:

$$(x + \varepsilon y)(x_1 + \varepsilon y_1) = xx_1 + \varepsilon(xy_1) + \varepsilon(yx_1) + \varepsilon^2(yy_1)$$
$$= xx_1 + \varepsilon(xy_1 + yx_1), \quad (52')$$

$$(x + ey)(x_1 + ey_1) = xx_1 + e(xy_1) + e(yx_1) + e^2(yy_1)$$
$$= (xx_1 + yy_1) + e(xy_1 + yx_1) \quad (52'')$$

[compare with the equality $(x + iy)(x_1 + iy_1) = (xx_1 - yy_1) + i(xy_1 + yx_1)$]. It follows that if we put $x = \operatorname{Re} z, y = \operatorname{Im} z$ in either $z = x + \varepsilon y$ or $z = x + ey$, then, just as for complex numbers,

$$\operatorname{Re}(z \pm z_1) = \operatorname{Re} z \pm \operatorname{Re} z_1, \quad \operatorname{Im}(z \pm z_1) = \operatorname{Im} z \pm \operatorname{Im} z_1. \quad (34)$$

Also, for all three types of numbers (complex, dual, and double),

$$\operatorname{Im}(z \cdot z_1) = \operatorname{Re} z \cdot \operatorname{Im} z_1 + \operatorname{Im} z \cdot \operatorname{Re} z_1. \quad (53)$$

However, while for complex numbers

$$\operatorname{Re}(z \cdot z_1) = \operatorname{Re} z \cdot \operatorname{Re} z_1 - \operatorname{Im} z \cdot \operatorname{Im} z_1, \quad (54)$$

the corresponding expressions for dual and double numbers are

$$\operatorname{Re}(z \cdot z_1) = \operatorname{Re} z \cdot \operatorname{Re} z_1 \quad (54')$$

and

$$\operatorname{Re}(z \cdot z_1) = \operatorname{Re} z \cdot \operatorname{Re} z_1 + \operatorname{Im} z \cdot \operatorname{Im} z_1. \quad (54'')$$

The rules for division of dual and double numbers are closely linked to the rule of forming the conjugate $\bar{z}$ of z; the latter operation is defined in

---

[2] The term "complex numbers" is sometimes used for dual and double numbers as well as "ordinary" complex numbers. We shall not follow this usage.

the same way for all three systems of numbers:

If $z = x + Iy$, then $\bar{z} = x - Iy$, where $I = i, \varepsilon,$ or e \hfill (55)

(we continue to use the symbol I below). Hence for all three types of numbers

$$\operatorname{Re}\bar{z} = \operatorname{Re}z \quad \text{and} \quad \operatorname{Im}\bar{z} = -\operatorname{Im}z. \tag{32a}$$

In all three cases, the sum of two conjugate numbers is real, i.e., a number z such that $\operatorname{Im}z = 0$, and the difference of two conjugate numbers is pure imaginary, i.e., a number z such that $\operatorname{Re}z = 0$:

$$z + \bar{z} = 2\operatorname{Re}z, \quad z - \bar{z} = (2\operatorname{Im}z)I, \quad I = i, \varepsilon, e. \tag{56}$$

The condition $z = \bar{z}$ characterizes reals, and the condition $z = -\bar{z}$ characterizes pure imaginaries. What is more important is that in all three systems the product of conjugates is real:

$$(x+iy)(x-iy) = x^2 + y^2, \quad (x+\varepsilon y)(x-\varepsilon y) = x^2,$$
$$(x+ey)(x-ey) = x^2 - y^2. \tag{57}$$

It follows that in order to define the quotient z_1/z, it suffices to multiply its numerator and denominator by $\bar{z}$, for then the denominator $z\bar{z}$ of $(z_1\bar{z})/(z\bar{z})$ is real, and in order to divide $z_1\bar{z}$ by $z\bar{z}$ it suffices to divide $\operatorname{Re}(z_1\bar{z})$ and $\operatorname{Im}(z_1\bar{z})$, respectively, by $z\bar{z}$:

$$\frac{z_1}{z} = \begin{cases} \dfrac{x_1+iy_1}{x+iy} = \dfrac{(x_1+iy_1)(x-iy)}{(x+iy)(x-iy)} = \dfrac{(xx_1+yy_1)+i(xy_1-x_1y)}{x^2+y^2} \\ \qquad = \dfrac{xx_1+yy_1}{x^2+y^2} + i\dfrac{xy_1-x_1y}{x^2+y^2}; \hfill (58) \\[6pt] \dfrac{x_1+\varepsilon y_1}{x+\varepsilon y} = \dfrac{(x_1+\varepsilon y_1)(x-\varepsilon y)}{(x+\varepsilon y)(x-\varepsilon y)} = \dfrac{xx_1+\varepsilon(xy_1-x_1y)}{x^2} \\ \qquad = \dfrac{x_1}{x} + \varepsilon\dfrac{xy_1-x_1y}{x^2}; \hfill (58') \\[6pt] \dfrac{x_1+ey_1}{x+ey} = \dfrac{(x_1+ey_1)(x-ey)}{(x+ey)(x-ey)} = \dfrac{xx_1-yy_1+e(xy_1-x_1y)}{x^2-y^2} \\ \qquad = \dfrac{xx_1-yy_1}{x^2-y^2} + e\dfrac{xy_1-x_1y}{x^2-y^2}. \hfill (58'') \end{cases}$$

The formulas (58)–(58″) show that while in the domain of complex numbers the only inadmissible divisors are numbers $z = x + iy$ with $x^2 + y^2 = 0$, i.e., the number 0 ($=0+i0$), there are other inadmissible divisors in the domains of dual and double numbers. Specifically, in the domain of dual numbers we must not divide by numbers $z = x + \varepsilon y$ for which $x = 0$, i.e., by numbers of the form εy ($= 0 + \varepsilon y$); and in the domain of double numbers we must not divide by numbers $z = x + ey$ for which $x^2 - y^2 = 0$ [i.e., $(x+y)(x-y) = 0$]—in other words, numbers of the form $x + ex$ or $x - ex$.

Supplement C. Analytic models of the nine plane geometries

In defining the modulus $|z|$ of a complex number z we used the first of the equalities (57):
$$|z|^2 = z\bar{z} = x^2 + y^2, \quad \text{and} \quad |z| \geq 0$$
[cf. (31)]. Similarly, for dual and double numbers we define the *modulus* $|z|$ of z by the formulas
$$|z|^2 = z\bar{z} = x^2, \tag{31a}$$
$$|z|^2 = |z\bar{z}| = |x^2 - y^2|. \tag{31b}$$
More accurately, for a dual number $z = x + \varepsilon y$ we put
$$|z| = x \tag{31'}$$
(so that the modulus of a dual number may be positive, zero, or negative) and for a double number $z = x + ey$ we put
$$|z| = \begin{cases} \pm\sqrt{x^2 - y^2} & \text{for } |x| \geq |y|, \\ \pm\sqrt{y^2 - x^2} & \text{for } |x| \leq |y|, \end{cases} \tag{31''}$$
and make the convention that the sign of the modulus of a double number is the same as the sign of x if $|x| > |y|$ and the same as the sign of y if $|y| > |x|$ (so that the modulus of a double number may be positive, zero, or negative).

We can now say that in all three number systems (complex, dual, and double) *we cannot divide by numbers of zero modulus* (i.e., by numbers z with $|z| = 0$). Such numbers are called *divisors of zero*, since for each of them there exists a nonzero number z_1 such that $zz_1 = 0$:
$$0(a + ib) = 0, \quad (\varepsilon y)(\varepsilon b) = 0;$$
$$(x + ex)(a - ea) = (1 + e)(1 - e)(xa) = (1 - e^2)(xa) = 0.$$
[Note that, contrary to widespread usage, our definition make the number zero into a divisor of zero.]

Let z be a number with $|z| = r \neq 0$. We then have
$$z = x + \varepsilon y = r\left(\frac{x}{r} + \varepsilon\frac{y}{r}\right); \quad z = x + ey = r\left(\frac{x}{r} + e\frac{y}{r}\right).$$
For a dual number, $z = x + \varepsilon y$, where $|z| = r = x$. Hence
$$z = x + \varepsilon y = r\left(1 + \varepsilon\frac{y}{x}\right) \quad \text{if } x \neq 0.$$
The quotient $y/x = \varphi$ is called the *argument* of the dual number $z = x + \varepsilon y$, with $|z| \neq 0$, and is denoted by $\arg z$. Thus, *every dual number with nonzero modulus can be written as*[3]
$$z = x + \varepsilon y = r(1 + \varepsilon\varphi), \quad \text{with } r = |z| \quad \text{and} \quad \varphi = \arg z, \tag{30'}$$

---

[3]We can also write (30') as $z = r(\cos g\varphi + \varepsilon \sin g\varphi)$ (cf. Exercise 3, Sec. 3). Then $x = r\cos g\varphi$, $y = r\sin g\varphi$.

where $|z|$ and $\arg z$ are defined by the formulas
$$|z| = x = \operatorname{Re} z \tag{31'}$$
and
$$\arg z = \frac{y}{x} = \frac{y}{|z|} = \frac{\operatorname{Im} z}{|z|} = \frac{\operatorname{Im} z}{\operatorname{Re} z}. \tag{33'}$$

In the case of a double number $z = x + ey$ with $|z| \neq 0$, it is necessary to consider the cases $|x| > |y|$ and $|x| < |y|$ separately. If $|x| > |y|$, then we put

$$\text{(a)} \quad r = |z| = \pm \sqrt{x^2 - y^2}, \tag{31''}$$

where the sign of r is that of x. Hence

$$\left(\frac{x}{r}\right)^2 - \left(\frac{y}{r}\right)^2 = \frac{x^2 - y^2}{r^2} = \frac{x^2 - y^2}{x^2 - y^2} = 1.$$

It follows that there exists a number φ (which may be regarded as an angle in Minkowskian geometry; cf. pp 187–188) such that

$$\text{(a)} \quad \cosh\varphi = \frac{x}{r} = \frac{\operatorname{Re} z}{|z|}, \quad \sinh\varphi = \frac{y}{r} = \frac{\operatorname{Im} z}{|z|},$$

$$\tanh\varphi = \frac{\sinh\varphi}{\cosh\varphi} = \frac{y}{x} = \frac{\operatorname{Im} z}{\operatorname{Re} z}. \tag{33''}$$

If $|x| < |y|$, then we put

$$\text{(b)} \quad r = |z| = \pm \sqrt{y^2 - x^2}, \tag{31''}$$

where the sign of r is that of y. Hence

$$\left(\frac{y}{r}\right)^2 - \left(\frac{x}{r}\right)^2 = \frac{y^2 - x^2}{r^2} = \frac{y^2 - x^2}{y^2 - x^2} = 1.$$

In this case there exists a number φ such that

$$\text{(b)} \quad \sinh\varphi = \frac{x}{r} = \frac{\operatorname{Re} z}{|z|}, \quad \cosh\varphi = \frac{y}{r} = \frac{\operatorname{Im} z}{|z|},$$

$$\tanh\varphi = \frac{\sinh\varphi}{\cosh\varphi} = \frac{x}{y} = \frac{\operatorname{Re} z}{\operatorname{Im} z}. \tag{33''}$$

The number φ defined by the formulas (33'') (a) and (b) is called the *argument* of the double number z and is denoted by $\arg z$. Thus *every double number* $z = x + ey$ *with* $|z| \neq 0$ *can be put in one of the forms*

$$\text{(a)} \quad z = r(\cosh\varphi + e\sinh\varphi)$$

or

$$\text{(b)} \quad z = r(\sinh\varphi + e\cosh\varphi), \tag{30''}$$

where $r = |z|$ and $\varphi = \arg z$, with the modulus $|z|$ of $z = x + ey$ given by the formulas (31'') [in which the sign of $|z|$ is that of x for numbers of type (a) in (30'') and that of y for numbers of type (b) in (30'')], and with $\arg z$

Supplement C. Analytic models of the nine plane geometries

defined by the formulas (a) and (b) in (33″). We shall call the numbers (a) in (30″) (numbers $x + ey$ with $|x| > |y|$) *double numbers of the first kind*, and numbers (b) in (30″) (numbers $x + ey$ with $|x| < |y|$), *double numbers of the second kind*.

We note that if
$$z = x + ey = r(1 + ey),$$
then
$$\bar{z} = x - ey = r[1 + \varepsilon(-\varphi)],$$
so that the formulas
$$|\bar{z}| = |z|, \qquad \arg \bar{z} = -\arg z \tag{32b}$$
hold for dual (as well as for complex) numbers. These formulas also hold for double numbers of the first kind. In fact, if
$$z = x + ey = r(\cosh \varphi + e \sinh \varphi),$$
then
$$\bar{z} = x - ey = r\left[\cosh(-\varphi) + e \sinh(-\varphi)\right]$$
[see formulas (19a) and (19b) of Sec. 12]. However, for double numbers of the second kind these formulas change to
$$|\bar{z}| = -|z|, \qquad \arg \bar{z} = -\arg z. \tag{32c}$$
Indeed, if
$$z = x + ey = r(\sinh \varphi + e \cosh \varphi),$$
then
$$\bar{z} = x - ey = -r\left[\sinh(-\varphi) + e \cosh(-\varphi)\right]$$
[again see formulas (19a) and (19b) in Sec. 12].

We now form the product of the dual numbers $z = r(1 + \varepsilon \varphi)$ and $z_1 = r_1(1 + \varepsilon \varphi_1)$:
$$zz_1 = r(1 + \varepsilon \varphi) r_1 (1 + \varepsilon \varphi_1) = rr_1 \left[(1 + \varepsilon \varphi)(1 + \varepsilon \varphi_1)\right]$$
$$= rr_1 (1 + \varepsilon \varphi + \varepsilon \varphi_1 + \varepsilon^2 \varphi \varphi_1) = rr_1 \left[1 + \varepsilon(\varphi + \varphi_1)\right].$$
Thus, for dual numbers we have the formulas
$$|zz_1| = |z||z_1|, \qquad \arg(zz_1) = \arg z + \arg z_1. \tag{34a}$$
(These formulas are familiar under the name of de Moivre's formulas from the study of complex numbers and we shall refer to them by that name in the domain of dual numbers.) From (34a), we obtain the relations
$$\left|\frac{z}{z_1}\right| = \frac{|z|}{|z_1|}, \qquad \arg\left(\frac{z}{z_1}\right) = \arg z - \arg z_1 \tag{34b}$$
[i.e., $r(1 + \varepsilon \varphi)/r_1(1 + \varepsilon \varphi_1) = (r/r_1)[1 + \varepsilon(\varphi - \varphi_1)]$]. Similarly, if $z = r(\cosh \varphi +$

$e\sinh\varphi$) and $z_1 = r_1(\cosh\varphi_1 + e\sinh\varphi_1)$, then
$$zz_1 = r(\cosh\varphi + e\sinh\varphi)r_1(\cosh\varphi_1 + e\sinh\varphi_1)$$
$$= rr_1(\cosh\varphi + e\sinh\varphi)(\cosh\varphi_1 + e\sinh\varphi_1)$$
$$= rr_1[(\cosh\varphi\cosh\varphi_1 + \sinh\varphi\sinh\varphi_1) + e(\cosh\varphi\sinh\varphi_1 + \sinh\varphi\cosh\varphi_1)]$$
$$= rr_1[\cosh(\varphi + \varphi_1) + e\sinh(\varphi + \varphi_1)]$$

[cf. the formulas (21a) and (21b) of Section 12]; if $z = r(\sinh\varphi + e\cosh\varphi)$ and $z_1 = r_1(\sinh\varphi_1 + e\cosh\varphi_1)$, then
$$zz_1 = r(\sinh\varphi + e\cosh\varphi)r_1(\sinh\varphi_1 + e\cosh\varphi_1)$$
$$= rr_1[(\sinh\varphi + e\cosh\varphi)(\sinh\varphi_1 + e\cosh\varphi_1)]$$
$$= rr_1[(\sinh\varphi\sinh\varphi_1 + \cosh\varphi\cosh\varphi_1) + e(\sinh\varphi\cosh\varphi_1 + \cosh\varphi\sinh\varphi_1)]$$
$$= rr_1[\cosh(\varphi + \varphi_1) + e\sinh(\varphi + \varphi_1)];$$

and if $z = r(\cosh\varphi + e\sinh\varphi)$ and $z_1 = r_1(\sinh\varphi_1 + e\cosh\varphi_1)$, then
$$zz_1 = r(\cosh\varphi + e\sinh\varphi)r_1(\sinh\varphi_1 + e\cosh\varphi_1)$$
$$= rr_1[(\cosh\varphi + e\sinh\varphi)(\sinh\varphi_1 + e\cosh\varphi_1)]$$
$$= rr_1[(\cosh\varphi\sinh\varphi_1 + \sinh\varphi\cosh\varphi_1) + e(\cosh\varphi\cosh\varphi_1 + \sinh\varphi\sinh\varphi_1)]$$
$$= rr_1[\sinh(\varphi + \varphi_1) + e\cosh(\varphi + \varphi_1)]$$

[see formulas (21a) and (21b), Sec. 12]. Thus for double numbers we also have the de Moivre formulas

$$|zz_1| = |z||z_1|, \qquad \arg(zz_1) = \arg z + \arg z_1. \tag{34a}$$

Note, however, that *the product of two double numbers of the same* (first or second) *kind is a double number of the first kind and the product of two double numbers of different kinds is a double number of the second kind.* From the formulas (34a) it follows that

$$\left|\frac{z}{z_1}\right| = \frac{|z|}{|z_1|}, \qquad \arg\left(\frac{z}{z_1}\right) = \arg z - \arg z_1. \tag{34b}$$

Note, however, that *the quotient of double numbers of the same kind is a double number of the first kind and the quotient of double numbers of different kinds is a double number of the second kind:*

$$\frac{r(\cosh\varphi + e\sinh\varphi)}{r_1(\cosh\varphi_1 + e\sinh\varphi_1)} = \frac{r(\sinh\varphi + e\cosh\varphi)}{r_1(\sinh\varphi_1 + e\cosh\varphi_1)}$$
$$= \frac{r}{r_1}[\cosh(\varphi - \varphi_1) + e\sinh(\varphi - \varphi_1)]$$

and
$$\frac{r(\cosh\varphi + e\sinh\varphi)}{r_1(\sinh\varphi_1 + e\cosh\varphi_1)} = \frac{r(\sinh\varphi + e\cosh\varphi)}{r_1(\cosh\varphi_1 + e\sinh\varphi_1)}$$
$$= \frac{r}{r_1}[\sinh(\varphi - \varphi_1) + e\cosh(\varphi - \varphi_1)].$$

Supplement C. Analytic models of the nine plane geometries

For all three types of numbers we have the relations

$$\overline{z+z_1} = \bar{z}+\bar{z}_1, \quad \overline{z-z_1} = \bar{z}-\bar{z}_1, \quad \overline{zz_1} = \bar{z}\bar{z}_1, \quad \text{and } \overline{(z/z_1)} = \bar{z}/\bar{z}_1. \tag{35}$$

The first three of these relations follow from (34), (53), (54)–(54″), (58)–(58″), and the definition (55) of the conjugate of a number, and the last two follow from (34a,b) and (32a–c).

A final remark. In the domain of dual numbers, $\arg z = 0$ for real numbers and $\arg z$ is not defined for pure imaginary numbers, whereas in the domain of double numbers $\arg z = 0$ for real numbers (these are of the first kind) and well as for pure imaginary numbers (these are of the second kind).

After these preliminaries we can turn to geometry. Let M be a point in the *Galilean plane* (Secs. 3–10) with rectangular coordinates (x,y). Then the Galilean length $r = \overline{OM} = d_{OM} = x$ and the Galilean angle $y = \angle xOM = \delta_{Ox,OM}$ are called the *polar coordinates* of M. We associate to M the dual number

$$z = x + \varepsilon y = r(1 + \varepsilon y) \tag{30'}$$

(Fig. 214a). Now let M be a point in the *Minkowskian plane* (see Sec. 12) with rectangular coordinates (x,y). Define the *(signed) Minkowskian length* $\overline{OM}$ to be the quantity r given by the equations (31″) on p. 267. If OM is a line of the first kind (see p. 179) define y to be the Minkowskian angle $y = \angle xOM = \delta_{Ox,OM}$. If OM is a line of the second kind define y to be the Minkowskian angle $y = \angle yOM = \delta_{Oy,OM} = \delta_{Ox,OM'}$, where $OM \perp OM'$ in the sense of Minkowskian geometry (cf. footnote 20, Sec. 12). The numbers (r,φ) are called the *polar coordinates* of M. We associate to M the double number

$$z = x + ey = r(\cosh\varphi + e\sinh\varphi) \quad \text{or} \quad z = x + ey = r(\sinh\varphi + e\cosh\varphi), \tag{30''}$$

depending on whether $\varphi = \angle xOM$ or $\varphi = \angle yOM$ (Fig. 214b). Finally, in

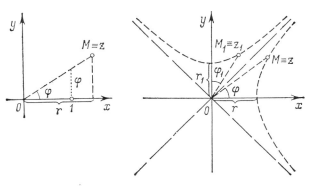

Figure 214a   Figure 214b

the case of the Galilean plane we associate to the points of the special line $x=0$ (cf. p. 34 and p. 39; we might also call this line a "null line") the divisors of zero in the domain of dual numbers, and in the case of the Minkowskian plane we associate to the points of the lines $x=y$ and $x=-y$ (cf. p. 177 and p. 179; we might also call these lines "null lines") the divisors of zero in the domain of double numbers. We have thus *identified the points of the Galilean plane with the dual numbers* (and will therefore feel free to speak of the "point z of the Galilean plane," meaning the point identified with the dual number z) and *the points of the Minkowskian plane with the double numbers* (and will likewise feel free to speak of the "point z").

The *(signed) distance* d_{z,z_1} between two points z and z_1 of the Galilean or Minkowskian plane (Figs. 215a and 215b) is defined by the expression

$$d_{z,z_1}=|z_1-z| \quad \text{so that} \quad d^2_{z,z_1}=(z_1-z)(\bar{z}_1-\bar{z}), \tag{36}$$

whose form is the same as that of the expressions obtained by identifying the points of the Euclidean plane with the complex numbers. This definition is a natural extension of definitions (31′) and (31″) on p. 267 for the modulus $|z|$ of a dual and double number respectively. For points in the Galilean plane (36) agrees with the distance formula (5) of Section 3. [The second of the formulas (36) also holds for points z and z_1 of the Minkowskian plane (Fig. 215a) if d_{zz_1} is defined by formula (15′a) in Section 12. However, the first formula of (36) does not hold in general. Indeed, the modulus of $|z_1-z|$ [as defined by (31″)] is always real, but may be either positive or negative. On the other hand, in Section 12, p. 180, we defined two different distances from z to z_1, neither of which behaves like $|z_1-z|$ in this respect. The distance defined by (15a) is always real and nonnegative, while the distance defined by (15′a) may not be real.]

The (directed) *angle* $\delta_{(z_0,z_1)(z_0,z_2)}$ between the (directed) lines (z_0,z_1) and (z_0,z_2) joining the points z_1 and z_2 to the point z_0 (whenever defined; cf.

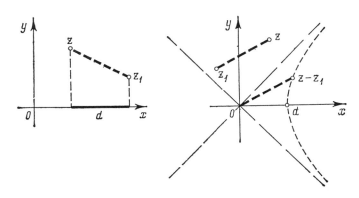

Figure 215a   Figure 215b

Supplement C. Analytic models of the nine plane geometries

pp. 40–41 and pp. 182–184) is given by the now familiar formula

$$\delta_{(z_0,z_1)(z_0,z_2)} = \arg(z_2,z_1;z_0) = \arg \frac{z_2 - z_0}{z_1 - z_0}, \tag{37}$$

where $(z_2,z_1;z_0)$ is again called *the simple ratio of three points* z_2, z_1, z_0 (in the Galilean or Minkowskian plane). The relations (37) follow from the fact that $\delta_{(z_0,z_1)(z_0,z_2)} = \varphi_2 - \varphi_1$, where

$$\varphi_1 = \arg z_1^0 = \arg(z_1 - z_0), \quad \varphi_2 = \arg z_2^0 = \arg(z_2 - z_0)$$

[compare, e.g., Fig. 216a for the case of the Galilean plane, and Figs. 216a–c for the case of the Minkowskian plane; note that in Fig. 216c we have $z_1'^0 = z_1' - z_0$, where $(z_0, z_1') \perp (z_0, z_1)$ in accordance with the definition (p. 186) of the angle between lines of different kinds in the Minkowskian plane]. We note that in the case of the Galilean and Minkowskian planes, Eq. (37) actually defines the signed magnitude of the directed angle between the lines (z_0, z_1) and (z_0, z_2) (cf. pp. 259–260).

Since a *line* (z_1, z_2) can be described as the set of points z such that

$$\text{Im}(z, z_1; z_2) = 0, \tag{38a}$$

where $(z, z_1; z_2) = (z - z_2)/(z_1 - z_2)$ is the simple ratio of the points z, z_1, z_2 (cf. Fig. 210a), it follows that *the equation of a line in the Galilean or Minkowskian plane has the* (by now familiar) *form*

$$\frac{z - z_2}{z_1 - z_2} = \frac{\bar{z} - \bar{z}_2}{\bar{z}_1 - \bar{z}_2} \tag{39}$$

or

$$Bz - \bar{B}\bar{z} + C = 0, \quad \text{Re}\, C = 0. \tag{40}$$

Conversely, just as in the case of the Euclidean plane, *every equation of the form* (40) *defines a line in the Galilean or Minkowskian plane* (according as z, B, and C are dual or double numbers); it joins points z_1 and z_2 such that $\bar{z}_1 - \bar{z}_2 = B$ and $z_1 \bar{z}_2 - \bar{z}_1 z_2 = C$.

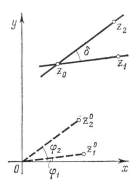
Figure 216a

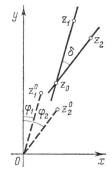
Figure 216b

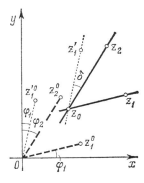
Figure 216c

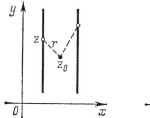 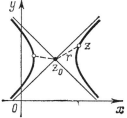

Figure 217a   Figure 217b

By a *circle* with center z_0 and radius of square ρ, where $\rho = r^2$ in Galilean geometry and $\rho = \pm r^2$ in Minkowskian geometry and $r > 0$ (see Figs. 217a and 217b, and p. 181), we mean the set of points z such that

$$(z - z_0)(\bar{z} - \bar{z}_0) = \rho; \qquad (41)$$

in Minkowskian geometry the (double) numbers $z - z_0$ must all be of the same kind (of the same form). It follows that *the equation of a circle has the form*

$$z\bar{z} - \bar{z}_0 z - z_0 \bar{z} + (z_0 \bar{z}_0 - \rho) = 0$$

or

$$az\bar{z} + bz + \bar{b}\bar{z} + c = 0, \qquad \operatorname{Im} a = \operatorname{Im} c = 0. \qquad (42)$$

Conversely, *every equation* (42) *defines a circle in the Galilean as well as the Minkowskian plane* with center z_0 and radius of square ρ given by the relations

$$\bar{z}_0 = -\frac{b}{a}, \qquad z_0 \bar{z}_0 - \rho = \frac{c}{a}.$$

A *cycle* of the Galilean plane passing through the points z_1, z_2, z_3 (Fig. 218a), as well as a circle of the Minkowskian plane, can each be defined as a set of points z such that

$$\delta_{(z_3, z_1),(z_3, z_2)} = \delta_{(z, z_1),(z, z_2)}$$

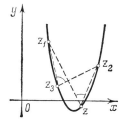 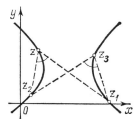

Figure 218a   Figure 218b

Supplement C. Analytic models of the nine plane geometries

(cf. p. 77 and pp. 186–187), i.e., such that

$$\text{Im}(z_1, z_2; z_3, z) = \text{Im} \frac{(z_1, z_2; z_3)}{(z_1, z_2; z)} = \text{Im} \frac{(z_1 - z_3)/(z_2 - z_3)}{(z_1 - z)/(z_2 - z)} = 0, \quad (44a)$$

where

$$(z_1, z_2; z_3, z) = \frac{(z_1 - z_3)/(z_2 - z_3)}{(z_1 - z)/(z_2 - z)}$$

is *the cross ratio of the four points* $(z_1, z_2; z_3, z)$ of the Galilean or Minkowskian plane. Thus *the equation of the cycle or the circle* determined by three points *is given by the equation*

$$\frac{(z_1 - z_3)/(z_2 - z_3)}{(z_1 - z)/(z_2 - z)} = \frac{(\bar{z}_1 - \bar{z}_3)/(\bar{z}_2 - \bar{z}_3)}{(\bar{z}_1 - \bar{z})/(\bar{z}_2 - \bar{z})}, \quad (44b)$$

which can be written in the form

$$Az\bar{z} + Bz - \bar{B}\bar{z} + C = 0, \quad \text{Re}\,A = \text{Re}\,C = 0, \quad (42a)$$

with coefficients A, B, C given by expressions identical with the expressions (42b) obtained in Euclidean geometry. However, while the equations (42) and (42a) are equivalent in Euclidean and in Minkowskian geometry (in the case of Minkowskian geometry we need only multiply one of these equations by the double unit e in order to obtain the other equation), this is not the case in Galilean geometry (since the dual unit ε is a divisor of zero, we must not divide the terms of an equation by ε). This corresponds to the fact that a Galilean circle (42) and a Galilean cycle (42a) are two essentially different curves, whereas in Euclidean and Minkowskian geometry, (42) and (42a) define the same curve.

In the case of Minkowskian geometry, we may view the equation (40) of a line as a special case of either of the equations (42) and (42a), from which (40) (or an equivalent equation) can be obtained by putting

$$a = 0 \quad (43)$$

and

$$A = 0, \quad (43a)$$

respectively. This is not so in Galilean geometry. Here we may view the line (40) as a special case[4] of the cycle (42a) [which reduces to (40) if (43a) holds] but not of the circle (42). [If (43) holds, then the equation (42) of a Galilean circle reduces to

$$bz + \bar{b}\bar{z} + c = 0, \quad \text{Im}\,c = 0;$$

putting $b = b_1 + \varepsilon b_2$, $z = x + \varepsilon y$, we obtain

$$(b_1 + \varepsilon b_2)(x + \varepsilon y) + (b_1 - \varepsilon b_2)(x - \varepsilon y) + c = 0$$

---

[4] More accurately, a limiting case (cf. p. 83).

or
$$2b_1 x + c = 0,$$
i.e.,
$$x = -\frac{c}{2b_1},$$

which is the equation of a special line of the Galilean plane.]

Four points z_1, z_2, z_3, z_4 are on the same cycle or line of the Galilean plane [on the same circle (or line) of the Minkowskian plane] if and only if

$$\operatorname{Im}(z_1, z_2; z_3, z_4) = \operatorname{Im} \frac{(z_1 - z_3)/(z_2 - z_3)}{(z_1 - z_4)/(z_2 - z_4)} = 0. \tag{45}$$

A *motion* of the Galilean or of the Minkowskian plane can be described as a mapping $z \to z'$ such that

$$\text{(a)} \quad z' = pz + q \quad \text{or} \quad \text{(b)} \quad z' = p\bar{z} + q \quad (p\bar{p} = 1). \tag{46}$$

In particular, just as in the case of the Euclidean plane, the maps

$$\text{(a)} \quad z' = -z \quad \text{and} \quad \text{(b)} \quad z' = \bar{z} \tag{48}$$

represent, respectively, a half-turn about the point O and a reflection in the line $\operatorname{Im} z = 0$ (cf. Figs. 211a and 211b). The general mapping (a) in (46) with $|p| = 1$, viewed as a mapping of the Galilean plane, represents a shear with coefficient $\arg p$ (cf. p. 25) followed by a translation by the vector $\mathbf{q}$ with beginning O and end q (Fig. 219a). Viewed as a mapping of the Minkowskian plane, (a) represents a (hyperbolic) rotation about O through the angle $\arg p$ followed by a translation by $\mathbf{q}$ (Fig. 219b).

It is easy to check that the mappings (a) and (b) in (46) are *square-distance-preserving*: $d^2_{z', z'_1} = d^2_{z, z_1}$, where z', z'_1 are the images of z, z_1 under a mapping (46) and d is interpreted as Galilean or Minkowskian distance (cf. p. 262). This, of course, is also true of the motions of the Galilean plane and of the Minkowskian plane.

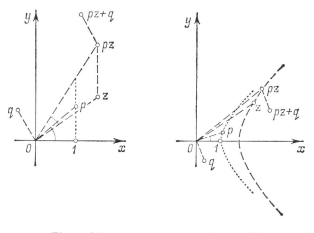

Figure 219a   Figure 219b

Supplement C. Analytic models of the nine plane geometries

Nevertheless, the mappings (46) do not coincide with the class of Galilean motions considered in Chap. I [cf. formulas (1), Sec. 3] or with the class of Minkowskian motions considered in Section 12 of the Conclusion [cf. formulas (11) or (23) in Sec. 12]. This is due to the fact that our exposition was limited to direct motions, while the mappings (46) include opposite motions as well (cf. footnote 12 of Sec. 1). Thus, for example, the coordinate representation of the Galilean motions in (46) is given by the formulas

$$x' = \pm x + a,$$
$$y' = vx \pm y + b.$$

We leave it to the reader to investigate the "generalized motions" (46) of the Galilean and the Minkowskian planes (see Exercise 4 in Sec. 2).

Finally, *the transformations*

$$\text{(a)} \quad z' = \frac{az+b}{cz+d} \quad \text{and} \quad \text{(b)} \quad z' = \frac{a\bar{z}+b}{c\bar{z}+d} \qquad (|ad-bc| \neq 0) \quad (49')$$

which are more general than the motions (46), *take four points in the Galilean plane that lie on a cycle (or line) to four points on a cycle (or line). The transformations* (49') *of the Minkowskian plane take four points on a circle (or line) to four points on a circle (or line).* These assertions, which are equivalent to the theorem that (49') are *circular transformations* of the Galilean as well as of the Minkowskian plane, follow from the criterion (45) for four points to lie on a cycle or circle. Indeed, if z'_1, z'_2, z'_3, and z'_4 are the images of z_1, z_2, z_3, and z_4 under the maps (a) in (49'), then $(z'_1, z'_2; z'_3, z'_4) = (z_1, z_2; z_3, z_4)$; and if they are the images of z_1, z_2, z_3, z_4 under the maps (b) in (49'), then $(z'_1, z'_2; z'_3, z'_4) = \overline{(z_1, z_2; z_3, z_4)}$. In particular, the mapping

$$z' = \frac{1}{\bar{z}}, \tag{50}$$

which is a special case of (b) in (49'), is an inversion (of the first kind) of the Galilean plane with center O and coefficient 1 (see pp. 128–129) when z and z' are dual numbers, and an inversion of the Minkowskian plane with center O and coefficient 1 when z and z' are double numbers (see pp. 199–197).

Consider the mappings (49') with $c \neq 0$. If z is such that $cz+d$ is a divisor of zero in the domain of dual or double numbers, then z has no image under (a) in (49'). Similarly, if $c\bar{z}+d$ is a divisor of zero, then z has no image under (b) in (49'). This makes it necessary to supplement the dual numbers by fictitious ("ideal") numbers

$$\lambda \omega = \frac{1}{(1/\lambda)\varepsilon} \quad \left(\text{and } \infty = \frac{1}{0}\right),$$

where λ ranges over the reals, and where we assume that

$$\bar{\omega} = -\omega, \qquad \overline{\infty} = \infty;$$

similarly, we supplement the double numbers by fictitious numbers

$$\lambda \omega_1 = \frac{1}{(1/\lambda)(1+e)}, \qquad \mu \omega_2 = \frac{1}{(1/\mu)(1-e)}$$

$$\left(\text{and} \quad \sigma_1 = \frac{1-e}{1+e}, \quad \sigma_2 = \frac{1+e}{1-e}, \quad \infty = \frac{1}{0}\right),$$

where λ and μ range over the reals, and where we assume that

$$\bar{\omega}_1 = \omega_2, \quad \bar{\omega}_2 = \omega_1, \quad \bar{\sigma}_1 = \sigma_2, \quad \bar{\sigma}_2 = \sigma_1, \quad \overline{\infty} = \infty.$$

Then we have, for example,

$$\frac{a\omega + b}{c\omega + d} = \frac{a(1/\varepsilon) + b}{c(1/\varepsilon) + d} = \frac{a + b\varepsilon}{c + d\varepsilon}$$

for dual numbers, and

$$\frac{a\omega_2 + b}{c\omega_2 + d} = \frac{a[1/(1-e)] + b}{c[1/(1-e)] + d} = \frac{a + b(1-e)}{c + d(1-e)} = \frac{(a+b) - be}{(c+d) - de},$$

or

$$\frac{a\sigma_1 + b}{c\sigma_1 + d} = \frac{a[(1-e)/(1+e)] + b}{c[(1-e)/(1+e)] + d} = \frac{a(1-e) + b(1+e)}{c(1-e) + d(1+e)} = \frac{(a+b) - (a-b)e}{(c+d) - (c-d)e}$$

for double numbers. Now each mapping (49′) (with $|ad - bc| \neq 0$, i.e., with $ad - bc$ not a divisor of zero) represents a one-to-one map of the extended dual (double) numbers onto the extended dual (double) numbers. In geometric terms, the natural domain of definition of the circular mappings (49′) is the inversive Galilean plane or the inversive Minkowskian plane. These planes are obtained by supplementing the "ordinary" Galilean and Minkowskian planes by "ideal" points ("points at infinity") corrsponding to the numbers $\lambda\omega$, ∞ and $\lambda\omega_1, \mu\omega_2, \sigma_1, \sigma_2, \infty$, respectively.

The similarities between complex, dual and double numbers often allow us to prove by means of a single computation (which is the same for complex, dual, and double numbers) theorems valid in Euclidean, Galilean, and Minkowskian geometry. For example, *let S_1, S_2, S_3, S_4 be four circles in the Euclidean or Minkowskian plane (four cycles in the Galilean plane). Let z_1 and w_1, z_2 and w_2, z_3 and w_3, and z_4 and w_4 be the points of intersection of the pairs S_1 and S_2, S_2 and S_3, S_3 and S_4, and S_4 and S_1, respectively. We shall show that if z_1, z_2, z_3, z_4 lie on a circle or line (on a cycle or line), then the corresponding statement is true of w_1, w_2, w_3, w_4* (see Figs. 220a and 220b which refer, respectively, to Euclidean and to Galilean geometry).

We use the fact that z_1, z_2, w_1, w_2 are on S_2; z_2, z_3, w_2, w_3 are on S_3; z_3, z_4, w_3, w_4 are on S_4; and z_4, z_1, w_4, w_1 are on S_1. This implies that the four cross ratios

$$(z_1, w_2; z_2, w_1) = \frac{(z_1 - z_2)/(w_2 - z_2)}{(z_1 - w_1)/(w_2 - w_1)},$$

$$(z_2, w_3; z_3, w_2) = \frac{(z_2 - z_3)/(w_3 - z_3)}{(z_2 - w_2)/(w_3 - w_2)},$$

$$(z_3, w_4; z_4, w_3) = \frac{(z_3 - z_4)/(w_4 - z_4)}{(z_3 - w_3)/(w_4 - w_3)},$$

and

$$(z_4, w_1; z_1, w_4) = \frac{(z_4 - z_1)/(w_1 - z_1)}{(z_4 - w_4)/(w_1 - w_4)}$$

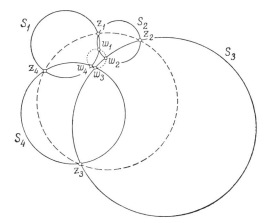

Figure 220a

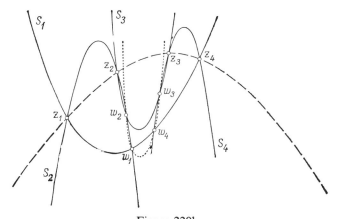

Figure 220b

are real [cf. (45)]. But then the expression

$$\frac{(z_1, w_2; z_2, w_1)}{(z_2, w_3; z_3, w_2)} \cdot \frac{(z_3, w_4; z_4, w_3)}{(z_4, w_1; z_1, w_4)}$$

$$= \left\{ \frac{(z_1 - z_2)/(z_3 - z_2)}{(z_1 - z_4)/(z_3 - z_4)} \right\} \cdot \left\{ \frac{(w_1 - w_2)/(w_3 - w_2)}{(w_1 - w_4)/(w_3 - w_4)} \right\}$$

$$= (z_1, z_3; z_2, z_4) \cdot (w_1, w_3; w_2, w_4)$$

is also real. Since the cross ratio $(z_1, z_2; z_3, z_4)$ is real, we conclude that the cross ratio $(w_1, w_2; w_3, w_4)$ is also real. But then the quadruple w_1, w_2, w_3, w_4 lies on a circle (cycle), as asserted. In other words, either both quadruples lie on a circle (cycle), or neither quadruple lies on a circle (cycle).

Supplement C. Analytic models of the nine plane geometries

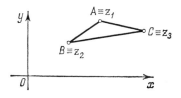

Figure 221

What follows is a more important example of such a "universal" computation. Let ABC be a (Euclidean, Galilean, or Minkowskian) triangle. As usual, we denote the lengths of its sides by a, b, c and the magnitudes of its angles by A, B, C. Let the numbers (complex, dual, or double) corresponding to the vertices be z_1, z_2, z_3 (Fig. 221). It is clear that if we put

$$w_1 = z_3 - z_2, \qquad w_2 = z_1 - z_3, \qquad w_3 = z_2 - z_1,$$

then[5]

$$|w_1| = a, \qquad |w_2| = b, \qquad |w_3| = c. \qquad (59)$$

Putting

$$\arg w_1 = \varphi_1, \qquad \arg w_2 = \varphi_2, \qquad \arg w_3 = \varphi_3$$

[so that

$$\arg \bar{w}_1 = -\varphi_1, \qquad \arg \bar{w}_2 = -\varphi_2, \qquad \arg \bar{w}_3 = -\varphi_3;$$

see formula (b) in (32) on p. 258 and formulas (32b) and (32c) on p. 269), and assuming, for the sake of simplicity, that in the case of Minkowskian geometry the numbers w_1, w_2, w_3 are dual numbers of the first kind, we have

$$w_1 = a(\text{Cos}\,\varphi_1 + I\,\text{Sin}\,\varphi_1), \qquad w_2 = b(\text{Cos}\,\varphi_2 + I\,\text{Sin}\,\varphi_2), \qquad w_3 = c(\text{Cos}\,\varphi_3 + I\,\text{Sin}\,\varphi_3),$$

where $I = i, \varepsilon$ or e, and $\text{Cos}\,\varphi$ and $\text{Sin}\,\varphi$ stand for the ordinary ("Euclidean") trigonometric functions $\cos\varphi$ and $\sin\varphi$, the "Galilean trigonometric functions" $\cos_g \varphi = 1$, $\sin_g \varphi = \varphi$ [cf. Exercise 3, Sec. 3 and formula (30′) in this Supplement], and the "Minkowskian trigonometric functions" $\cosh\varphi$ and $\sinh\varphi$, respectively. Finally, we note that

$$\varphi_2 - \varphi_3 = A, \qquad \varphi_3 - \varphi_1 = B, \qquad \varphi_1 - \varphi_2 = C \qquad (60)$$

(see Fig. 221),[6] so that in view of the formulas (34a) of this supplement, we have, respectively,

$$w_2 \bar{w}_3 = bc(\text{Cos}\,A + I\,\text{Sin}\,A),$$
$$w_3 \bar{w}_1 = ca(\text{Cos}\,B + I\,\text{Sin}\,B),$$
$$w_1 \bar{w}_2 = ab(\text{Cos}\,C + I\,\text{Sin}\,C),$$

and

$$\bar{w}_2 w_3 = bc(\text{Cos}\,A - I\,\text{Sin}\,A),$$
$$\bar{w}_3 w_1 = ca(\text{Cos}\,B - I\,\text{Sin}\,B),$$
$$\bar{w}_1 w_2 = ab(\text{Cos}\,C - I\,\text{Sin}\,C).$$

---

[5]Thus in Galilean and Minkowskian geometry a, b, c are signed lengths of the sides of $\triangle ABC$. (For an earlier use of signed lengths of the sides of a triangle see, for example, p. 51.)

[6]Thus A, B, C are directed angles of $\triangle ABC$ (cf. p. 51).

Now
$$w_1 + w_2 + w_3 = 0 \quad [=(z_3 - z_2) + (z_1 - z_3) + (z_2 - z_1)],$$
i.e.,
$$-w_1 = w_2 + w_3,$$
which implies
$$-\bar{w}_1 = \bar{w}_2 + \bar{w}_3.$$
Hence
$$a^2 = |w_1|^2 = w_1\bar{w}_1 = (w_2 + w_3)(\bar{w}_2 + \bar{w}_3) = w_2\bar{w}_2 + w_3\bar{w}_3 + w_2\bar{w}_3 + \bar{w}_2 w_3$$
$$= |w_2|^2 + |w_3|^2 + bc(\operatorname{Cos} A + I\operatorname{Sin} A) + bc(\operatorname{Cos} A - I\operatorname{Sin} A)$$
$$= b^2 + c^2 + 2bc\operatorname{Cos} A. \tag{61}$$
This is just the law of cosines of Euclidean, Galilean, and Minkowskian geometry. In Euclidean geometry, it takes the form
$$a^2 = b^2 + c^2 + 2bc\cos A \tag{61e}$$
[note that using the notation in Fig. 221, which we now interpret in Euclidean terms, $A = 180° - \angle BAC$, so that (61) is equivalent to the relation $a^2 = b^2 + c^2 - 2bc\cos\angle BAC$]. In Galilean geometry, the law takes the form
$$a^2 = b^2 + c^2 + 2bc \cdot 1 = (b+c)^2 \tag{61g}$$
[cf. formula (11), Sec. 4], and in Minkowskian geometry it takes the form
$$a^2 = b^2 + c^2 + 2bc\cosh A. \tag{61m}$$
Form the expressions
$$w_2\bar{w}_3 - \bar{w}_2 w_3 = bc(\operatorname{Cos} A + I\operatorname{Sin} A) - bc(\operatorname{Cos} A - I\operatorname{Sin} A) = (2bc\operatorname{Sin} A)I,$$
$$w_3\bar{w}_1 - \bar{w}_3 w_1 = ca(\operatorname{Cos} B + I\operatorname{Sin} B) - ca(\operatorname{Cos} B - I\operatorname{Sin} B) = (2ca\operatorname{Sin} B)I,$$
$$w_1\bar{w}_2 - \bar{w}_1 w_2 = ab(\operatorname{Cos} C + I\operatorname{Sin} C) - ab(\operatorname{Cos} C - I\operatorname{Sin} C) = (2ab\operatorname{Sin} C)I.$$
Now
$$w_2\bar{w}_3 - \bar{w}_2 w_3 = (z_1 - z_3)(\bar{z}_2 - \bar{z}_1) - (\bar{z}_1 - \bar{z}_3)(z_2 - z_1)$$
$$= z_1\bar{z}_2 + z_2\bar{z}_3 + z_3\bar{z}_1 - \bar{z}_1 z_2 - \bar{z}_2 z_3 - \bar{z}_3 z_1,$$
and as can be readily shown,
$$w_3\bar{w}_1 - \bar{w}_3 w_1 = w_1\bar{w}_2 - \bar{w}_1 w_2 = z_1\bar{z}_2 + z_2\bar{z}_3 + z_3\bar{z}_1 - \bar{z}_1 z_2 - \bar{z}_2 z_3 - \bar{z}_3 z_1.$$
[Note that the expression on the right-hand side is symmetric in z_1, z_2, z_3.] Therefore, $(2bc\operatorname{Sin} A)I = (2ca\operatorname{Sin} B)I = (2ab\operatorname{Sin} C)I$, or
$$bc\operatorname{Sin} A = ca\operatorname{Sin} B = ab\operatorname{Sin} C,$$
from which we readily obtain
$$\frac{a}{\operatorname{Sin} A} = \frac{b}{\operatorname{Sin} B} = \frac{c}{\operatorname{Sin} C}, \tag{62}$$
the law of sines of Euclidean, Galilean, and Minkowskian geometry:
$$\frac{a}{\sin A} = \frac{b}{\sin B} = \frac{c}{\sin C} \tag{62e}$$
(Euclidean geometry);
$$\frac{a}{A} = \frac{b}{B} = \frac{c}{C} \tag{62g}$$

(Galilean geometry; cf. p. 49) and

$$\frac{a}{\sinh A} = \frac{b}{\sinh B} = \frac{c}{\sinh C} \tag{62m}$$

(Minkowskian geometry, cf. p. 191).

It is well known that the points of the hyperbolic plane can be represented by means of complex numbers (compare, e.g., [80] or [81]). As the points of our model of the hyperbolic plane, we take the points of the unit disk K defined by $z\bar{z} < 1$. The hyperbolic distance d_{z,z_1} of points z, z_1 in K is defined as

$$\tanh^2 \frac{d_{z,z_1}}{2} = \frac{(z_1 - z)(\bar{z}_1 - \bar{z})}{(1 - z\bar{z}_1)(1 - \bar{z}z_1)}. \tag{63}$$

The hyperbolic motions are given by those linear fractional transformations (49) which map K to itself, i.e., the transformations

(a) $\quad z' = \dfrac{pz + q}{\bar{q}z + \bar{p}}$

and

(b) $\quad z' = \dfrac{p\bar{z} + q}{\bar{q}\bar{z} + \bar{p}} \qquad (p\bar{p} - q\bar{q} \neq 0). \tag{64}$

Particularly simple examples of the transformations (64) are

$$\text{(a)} \quad z' = -z \quad \text{and} \quad \text{(b)} \quad z' = \bar{z}, \tag{48}$$

which represent a half-turn about the point O of the hyperbolic plane (cf. Fig. 211a) and a reflection in the line $\operatorname{Im} z = 0$ (cf. Figs. 211b and 222), respectively.

The arc MM' in Figure 222 represents a segment of a hyperbolic line perpendicular to the hyperbolic line IJ, the segment of $\operatorname{Im} z = 0$ in K, which is also a diameter of K. The hyperbolic distance (63) between two points of IJ coincides with the hyperbolic distance (1b) in Supplement A. All the remaining hyperbolic lines can be obtained by applying a motion (64) to the line

$$z - \bar{z} = 0. \tag{65}$$

This implies that *the equation of a hyperbolic line is given by*

$$az\bar{z} + bz + \bar{b}\bar{z} + a = 0, \quad \operatorname{Im} a = 0, \tag{66}$$

or, equivalently, by

$$Az\bar{z} + Bz - \bar{B}\bar{z} + A = 0, \quad \operatorname{Re} A = 0, \tag{66a}$$

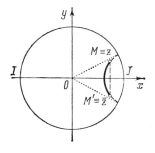

Figure 222

Supplement C. Analytic models of the nine plane geometries

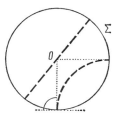

Figure 223

where $z \in K$. The hyperbolic lines are given geometrically by the diameters of K and arcs in K which belong to circles perpendicular to the boundary circle Σ of K (Fig. 223; an angle between two circles is defined as the angle between the tangents to the circles at a point of intersection of the circles).

The circles

$$az\bar{z} + bz + \bar{b}\bar{z} + c = 0, \quad \mathrm{Im}\, a = \mathrm{Im}\, c = 0 \tag{42}$$

or

$$Az\bar{z} + Bz + \bar{B}\bar{z} + C = 0, \quad \mathrm{Re}\, A = \mathrm{Re}\, C = 0 \tag{42a}$$

(more accurately, their parts in K) represent the hyperbolic cycles [i.e., curves which have the same "structure" at each point; curves S such that if P and Q are any given points of S, there is a motion mapping S onto itself and sending P to Q]. We know by now that a cycle (42) or (42a) is a hyperbolic line if

$$a = c \tag{67}$$

or

$$A = C \tag{67a}$$

[cf. Eqs. (66) and (66a) above]. A Euclidean circle (42) or (42a) represents a hyperbolic circle [the set of points at fixed (hyperbolic) distance from a given point —its center; cf. p. 233], a horocycle (p. 233) or an equidistant curve [the locus of points at fixed (hyperbolic) distance from a line—the base of the equidistant curve; cf. p. 233], according as it is contained in K, is internally tangent to K or cuts K in two points M and N. The three curves in question are shown in Figures 224a–c.

In the case of an equidistant curve, the above description must be amended. Instead of saying that an equidistant curve is the arc (in K) of a circle (42) [or

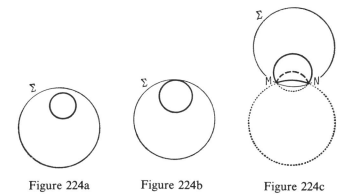

Figure 224a   Figure 224b   Figure 224c

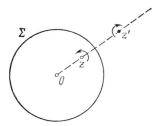

Figure 225

(42a)] intersecting Σ in two points M and N, we should say that it consists of the points on two such arcs (see Fig. 224c) lying on either side of the base of the equidistant curve (represented in Fig. 224c by the dotted arc MN) at a fixed distance from the base.

The last remark underlines the principal defect of our model of the hyperbolic plane represented by the disk K. None of the simple equations (66), (66a), (42), (42a), and not even the extremely simple equation (65), represents a line or cycle of hyperbolic geometry, for in each case we must restrict ourselves to the complex numbers z with $z\bar{z} < 1$. Consequently, it is very difficult to find analytic expressions for circular transformations of the hyperbolic plane (cycle-preserving transformations). In our model, these transformations would be required to permute arcs of the circles (42) [or (42a)]. But the task of obtaining sufficiently simple analytical expressions for such arcs (rather than for the circles of which they are a part) is not easy.[7] And, surely, the representation of an equidistant curve (Fig. 224c) as a lens consisting of two circular arcs with common ends is not likely to strike anyone as simple.

The careful reader may well have guessed how to surmount these difficulties. We shall view each point of the hyperbolic plane as oriented, i.e., supplied with an indication of which direction of rotation about it is to be taken as positive; in our drawings each point is oriented by means of a circular arrow (Fig. 225). We shall assign opposite orientations to distinct points represented by complex numbers z and z' such that

$$z' = \frac{1}{\bar{z}}, \tag{50}$$

i.e., to distinct points which correspond under inversion in Σ (inversion with center O and coefficient 1; cf. Fig. 225). Finally, we shall include among the points of the hyperbolic plane the points of Σ (given by the equation $z\bar{z} = 1$); we shall call them "points at infinity" of the hyperbolic plane and assign to them no orientation. We can think of a hyperbolic point at infinity as a pencil of hyperbolic parallels (which "converge at that point") just as a point at infinity in the Euclidean plane is often thought of as a pencil of Euclidean parallels.

Thus the set of all points of the hyperbolic plane (including the oriented points as well as the points at infinity) is represented by the set of all complex numbers (including the "number" ∞ associated to the "negatively oriented" point 0). Each hyperbolic cycle will also be oriented (i.e., supplied with an arrow indicating the positive direction of motion on the cycle). An oriented point A is to belong to an oriented cycle S if A is a point of S and if the orientations of A and S are related as in the schematic diagram in Figure 226a; there A belongs to S but B does not. By an (oriented) equidistant curve with base l we shall mean the set of points at a

---

[7] We suggest that the reader consider the problem of producing a simple equation describing the segment IJ in Figure 222.

Supplement C. Analytic models of the nine plane geometries   285

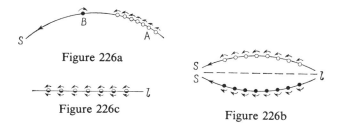

Figure 226a

Figure 226c

Figure 226b

fixed distance h from l, with points on opposite sides of l assigned opposite orientations (Fig. 226b); we thus have two equidistant curves with given base l and width $h>0$, and we could agree to say that their widths are h and $-h$, respectively). It is convenient not to orient lines[8] and to regard points of a hyperbolic line as "doubled" (Fig. 226c), i.e., as having both orientations.[9] Finally, the (non-oriented) "circle at infinity" represented by Σ (the circle $z\bar{z}=1$) is also regarded as a cycle in our model. Apart from Σ, *the set of all (oriented) cycles and lines of the hyperbolic plane is represented by the set of circles and lines of the complex plane* (see, in particular, Fig. 227 which shows a hyperbolic line l and an equidistant curve S

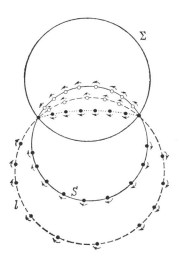

Figure 227

[8]We recall, by way of comparison, that in co-Euclidean geometry (cf. pp. 233–237) we found it convenient to regard oriented lines of the Euclidean plane as the fundamental elements of the geometry, and that while we assigned orientation to circles ("cycles") we assigned none to points.

[9]An orientation of a cycle induces a natural orientation of its points: The orientation of the curved circular arrow associated to a point of a cycle and drawn on the convex side of the cycle must agree with the orientation of the cycle (Fig. 226a). In the case of a line, there is no "convex side" and the scheme which works for points on cycles fails. The simplest way out of the dilemma is to assign to a point on a line both possible orientations (Fig. 226c). (If we regard a line as "an equidistant curve of zero width," both of whose branches coincide with the base, then it is again reasonable to assign to its points both possible orientations; cf. Figs. 226b and 226c.)

with base l). Now a line in the hyperbolic plane is given by an equation (66) [or (66a)], and a cycle by an equation (42) [or (42a)].

Four (oriented) points z_1, z_2, z_3, z_4 lie on a circle or line if and only if

$$\operatorname{Im}(z_1, z_2; z_3, z_4) = 0. \tag{45}$$

It follows that the transformations

$$\text{(a)} \quad z' = \frac{az+b}{cz+d} \quad \text{and} \quad \text{(b)} \quad z' = \frac{a\bar{z}+b}{c\bar{z}+d}, \tag{49}$$

where $ad - bc \neq 0$, take four points on a line or cycle to four points with the same property, i.e., (46) are the *circular transformations* of the hyperbolic plane. This fact enables us to give a geometric description of all possible circular transformations of the hyperbolic plane.[10]

We saw that the points of the Euclidean, Galilean, and Minkowskian planes can be represented by various "numbers": complex numbers $z = x + iy$ with $i^2 = -1$, dual numbers $z = x + \varepsilon y$ with $\varepsilon^2 = 0$ and double numbers $z = x + ey$ with $e^2 = +1$. On the other hand, the (oriented) points of the hyperbolic plane can also be represented by complex numbers. This way of looking at plane Cayley–Klein geometries is very fruitful: The (directed) points of the elliptic plane can be represented by the complex numbers; the (directed) points of the co-Euclidean and co-Minkowskian plane can be represented by the dual numbers $z = x + \varepsilon y$; and the (directed) points of the doubly hyperbolic and cohyperbolic plane can be represented by the double numbers $z = x + ey$. Thus, *the points of the three Cayley–Klein planes with elliptic metric of angles*—i.e., those of the elliptic, Euclidean, and hyperbolic planes—*can be represented by the complex numbers*; *the points of the three planes with parabolic metric of angles*—i.e., those of the co-Euclidean, Galilean, and co-Minkowskian planes—*can be represented by the dual numbers*; and *the points of the three planes with hyperbolic metric of angles*—i.e., those of the cohyperbolic, Minkowskian, and doubly hyperbolic plane—*can be represented by the double numbers*. Also, *the distance d_{z,z_1} between points z and z_1 of a plane with elliptic metric of distance* (i.e., the elliptic, co-Euclidean, and cohyperbolic planes) is defined by the formula

$$\tan^2 \frac{d_{z,z_1}}{2} = \frac{(z-z_1)(\bar{z}-\bar{z}_1)}{(1+z\bar{z}_1)(1+\bar{z}z_1)}; \tag{63'}$$

in a plane with parabolic metric of distance (Euclidean, Galilean, and Minkowskian planes) *it is defined by the formula*

$$d_{z,z_1}^2 = (z-z_1)(\bar{z}-\bar{z}_1); \tag{36}$$

and *in a plane with hyperbolic metric of distance* (hyperbolic, co-Minkowskian and doubly hyperbolic planes) *it is defined by the* (familiar) *formula*

$$\tanh^2 \frac{d_{z,z_1}}{2} = \frac{(z_1-z)(\bar{z}_1-\bar{z})}{(1-z\bar{z}_1)(1-\bar{z}z_1)}. \tag{63}$$

In Cayley–Klein planes with distance (63'), the motions are given by

$$z' = \frac{pz+q}{\bar{q}z - \bar{p}} \quad \text{or} \quad z' = \frac{p\bar{z}+q}{\bar{q}\bar{z} - \bar{p}} \quad (|p\bar{p}+q\bar{q}| \neq 0); \tag{64'}$$

in planes with distance (36), the motions are given by

$$z' = pz + q \quad \text{or} \quad z' = p\bar{z} + q \quad (p\bar{p} = 1); \tag{46}$$

---

[10]See, for example, [80].

Supplement C. Analytic models of the nine plane geometries

and in planes with distance (63), *the motions are given by*

$$z' = \frac{pz+q}{\bar{q}z+\bar{p}} \quad \text{or} \quad z' = \frac{p\bar{z}+q}{\bar{q}\bar{z}+\bar{p}} \quad (|p\bar{p}-q\bar{q}|\neq 0). \tag{64a}$$

The equations

$$Az\bar{z} + Bz - \bar{B}\bar{z} + A = 0, \quad \operatorname{Re} A = 0, \tag{66a}$$

$$Bz - \bar{B}\bar{z} + C = 0, \quad \operatorname{Re} C = 0, \tag{40}$$

and

$$Az\bar{z} + Bz - \bar{B}\bar{z} - A = 0, \quad \operatorname{Re} A = 0, \tag{66b}$$

describe the lines in planes with distance (63), (36), *and* (63'), *respectively.*[11] In all three cases, however, *the equation of a cycle is given by*

$$Az\bar{z} + Bz - \bar{B}\bar{z} + C = 0, \quad \operatorname{Re} A = \operatorname{Re} C = 0 \tag{42a}$$

(see footnote 11). It follows that, corresponding to the distance formulas (63), (36), and (63'), the conditions for a cycle (42a) to reduce to a line are, respectively,

$$A - C = 0, \tag{67a}$$

$$A = 0, \tag{43a}$$

and

$$A + C = 0 \tag{67b}$$

(cf. Sec. 13).

Using the equation (42a) of a cycle or the condition

$$\operatorname{Im}(z_1, z_2; z_3, z_4) = 0 \tag{45}$$

for four points to lie on a cycle, we can show that in all nine Cayley–Klein planes the *circular* (i.e., cycle-preserving) *transformations are given by*

$$z' = \frac{az+b}{cz+d} \quad \text{or} \quad z' = \frac{a\bar{z}+b}{c\bar{z}+d}, \quad (|ad-bc|\neq 0). \tag{49}$$

Our findings are summarized in Tables V–VII.

We shall not apply our machinery to prove specific theorems, but we wish to point out one example which hints at the wide range of its applicability. Specifically, the computations on pp. 280–281 used to prove the result stated on p. 278 for three plane Cayley–Klein geometries prove it for all nine of these geometries.

Table V Numbers representing the points of Cayley–Klein planes

Metric of angles	Type of numbers	
E	$z = x + iy$,	$i^2 = -1$
P	$z = x + \varepsilon y$,	$\varepsilon^2 = 0$
H	$z = x + ey$,	$e^2 = +1$

---

[11] If the numbers in the formulas (66b), (40), (66a), and (42a) are complex or dual, then these formulas can be replaced by (42) and by suitable variants of (66b), (40), and (66a) [see, for example, (66)]. However, if the coefficients A, B, and C and the variable z stand for dual numbers, then (42) and (42a) are no longer equivalent.

Table VI   The distance d_{z,z_1} between points z and z_1 of a Cayley–Klein plane

Metric of distances		
E	$\tan^2 \dfrac{d_{z,z_1}}{2} =$	$\dfrac{(z-z_1)(\bar{z}-\bar{z}_1)}{(1+z\bar{z}_1)(1+\bar{z}z_1)}$
P	$d^2_{z,z_1} =$	$(z-z_1)(\bar{z}-\bar{z}_1)$
H	$\tanh^2 \dfrac{d_{z,z_1}}{2} =$	$\dfrac{(z-z_1)(\bar{z}-\bar{z}_1)}{(1-z\bar{z}_1)(1-\bar{z}z_1)}$

Table VII   The motions of Cayley–Klein planes

Metric of distances	Motions	
E	$z' = \dfrac{pz+q}{\bar{q}z-\bar{p}},$	$\|p\bar{p}+q\bar{q}\| \neq 0$
P	$z' = pz+q,$	$p\bar{p}=1$
H	$z' = \dfrac{pz+q}{\bar{q}z+\bar{p}},$	$\|p\bar{p}-q\bar{q}\| \neq 0$

Table VIII   The equations of lines in Cayley–Klein planes

Metric of distances	Equation of line	
E	$Az\bar{z} + Bz - \bar{B}\bar{z} - A = 0,$	$\operatorname{Re} A = 0$
P	$Bz - \bar{B}\bar{z} + C = 0,$	$\operatorname{Re} C = 0$
H	$Az\bar{z} + Bz - \bar{B}\bar{z} + A = 0,$	$\operatorname{Re} A = 0$

We leave it to the reader to find other relevant examples.

We note that many of the constructions developed in this Supplement can be carried over to algebraic systems other than our three systems of complex numbers. Examples of such systems are the *quaternions* (five systems; cf. [80], pp. 24–25) and *octaves* (seven systems; cf. [80], pp. 21 and 221–223), as well as numbers of the form

$$Z = a_0 + a_1 E + a_2 E^2 + \cdots + a_{n-1} E^{n-1},$$

where $E^p E^q = E^{p+q}$ and $E^n = -1, 1,$ or 0 (*cocyclic, cyclic,* or *plural* numbers). The associated geometries are sometimes of the Cayley–Klein type, but more often are not. We refer the interested reader to [83].

Bibliography

A Textbooks on Euclidean Geometry[1]

1–3 Heath, T. L., *The Thirteen Books of Euclid's Elements*, Vols. I–III. Dover, New York, 1956.

4 Hilbert, D., *The Foundations of Geometry, 10th ed.* Open Court, LaSalle, IL, 1971.

5 Moise, E. E. and Downs, F. L., *Geometry* . Addison Wesley, Reading, MA, 1971.

5a Moise, E. E., *Elementary Geometry From an Advanced Standpoint*. Addison Wesley, Reading, MA, 1974.

5b Boltyanski, V. G. Equivalent and Equidecomposable Figures, Heath, Boston, 1963.

6 Jacobs, H. R., *Geometry*. Freeman, San Francisco, 1974.

7 Brumfiel, C. E., Eicholz, R. E., and Shanks, M. E., *Geometry, 3rd ed.* Addison Wesley, Reading, MA, 1975.

8 Dieudonné, J., *Linear Algebra and Geometry*. Houghton Mifflin, Boston, 1969.

B Section 1 of the Introduction[2]

9 Klein, F., *Vergleichende Betrachtungen über neure geometrische Forschungen.* Gesammelte mathematische Abhandlungen, Vol. I, 1921, pp. 460–497. (English version is found in Sommerville, D. M. Y., *Bibliography of Non-Euclidean Geometry, 2nd ed.*, Chelsea, New York, 1970.)

10–13 Yaglom, I. M. The items [10]–[12] are *Geometric Transformations* I–III. New Mathematics Library; Volumes 8 (1962), 21 (1968), and 24 (1973), Random House, New York. (The NML Series is now published and distributed by the Mathematical Association of America.) Item [13] is the untranslated part of Yaglom's two-volume Russian work *Geometric Transformations*. GITTL, Moscow, 1955–1956 (Chap. II, Vol. II, pp. 169–354, 483–605).

---

[1] See also Yaglom [10]–[13], Coxeter [19], and Wylie [20].
[2] See also Coxeter [19], Pedoe [31], Ewald [32], and Yaglom-Ashkinuze [57].

14 Jaglom (Yaglom), I. M. and Atanasjan, L. S., *Geometrische Transformationen*, Enzyklopädie der Elementarmathematik (EEM), Vol. IV (Geometrie). Deutscher Verlag der Wissenschaften, Berlin (DDR), 1969, pp. 43–151 [German]

C Section 2 of the Introduction[3]

15 Galilei Galileo, *Dialogue Concerning the Two Chief World Systems*. University of California Press, Berkeley, 1970.
16 Newton, I., *Principia*. University of California Press, Berkeley, 1970.
17 Haber-Schaim, U., Cross, J. B., Dodge, J. H., Walter, J. A., *PSSC Physics*, D. C. Heath, Lexington, MA, 1971.
18 Haikin, S. E., *The Physical Foundations of Mechanics*. Fizmatgiz, Moscow, 1962 [Russian].

D What is 4-dimensional Space?

19 Coxeter, H. S. M., *Introduction to Geometry*, 2nd ed. Wiley, New York, 1969.
20 Wylie, C. R., *Foundations of Geometry*. McGraw-Hill, New York, 1964.
20a Manning, H. P., *Geometry of Four Dimensions*. Dover, New York, 1955.
21 Rosenfeld, B. A., Jaglom, I. M., *Mehrdimensionale Räume*, EEM, Vol. V (Geometrie). Deutscher Verlag der Wissenschaften, Berlin, 1971, pp. 337–383 [German].

E Systems of Forces in Statics and Sliding Vectors

22 de la Vallée-Poussin, Ch., *Leçons de mécanique analytique*, Vol. I. Paris, 1932 [French].
23 Synge, J. L. and Griffith, B. A., *Principles of Mechanics*, 2nd ed. McGraw-Hill, New York, 1949.
24 Boltjanski, W. G. and Jaglom, I. M., *Vektoren und ihre Anwendungen in der Geometrie*. EEM, Vol. IV, pp. 295–390 [German].

F Galilean Geometry[4]

25 Rosenfeld, B. A. Jaglom, I. M., *Nichteuklidische Geometrie*. EEM, Vol. V, 1971, pp. 385–469.
26 Kuiper, N., *On a certain plane geometry*. Simon Stevin 30, 1954, pp. 94–105 [Dutch].
27 Strubecker, K., "Geometrie in einer isotropen Ebene I–III," Der mathematischer und naturwissenschaftlicher Unterricht **15**: 297–306, 343–351, 385–394, 1962-1963 [German]
28 Makarova, N. M., *Two-dimensional Noneuclidean Geometry with Parabolic Angle and Distance Metric*. Dissertation, Leningrad, 1962 [Russian].

---

[3]See also Schwartz [45].
[4]See also Klein [56].

G Chapter II[5]

29 Jaglom, I. M., *Geometrie der Kreise*. EEM, Vol. IV, pp. 457–526 [German].
30 Coolidge, J. L., *A Treatise on the Circle and the Sphere*. Oxford, 1916.
31 Pedoe, D., *A Course of Geometry for Colleges and Universities*. Cambridge University Press, Cambridge, 1970.
32 Ewald, G., *Geometry: An Introduction*. Wadsworth, Belmont, CA, 1971.
33 Johnson, R. A., *Advanced Euclidean Geometry*. Dover, New York, 1960.
34 Pedoe, D. *Circles*. Pergamon, New York, 1957.
35 Makarova, N. M., On the theory of cycles in plane parabolic geometry, Siberian Journal of Mathematics 2 (1): 68–81, 1961 [Russian].

H Properties of Plane Curves[6]

36 Struik, D. J., *Lectures on Classical Differential Geometry*. Addison Wesley, Reading, MA, 1961.
37 Boltjanski, W. G. and Jaglom, I. M., *Geometrische Extremwertaufgaben*. EEM, Vol. V, pp. 259–335 [German].

I Section 11 of the Conclusion

38 Einstein, A. (translated by R. W. Lawson), *Relativity, the Special and General Theory*. Crown, New York, 1961.
39 Weyl, H., *Space, Time, Matter*, Dover, New York, 1950.
40 Born, M., *Einstein's Theory of Relativity*. Dover, New York, 1962.
41 Taylor, E. F. and Wheeler, J. A., *Spacetime Physics*. Freeman, San Francisco, 1966.
42 Bergmann, P. G. *Introduction to the Theory of Relativity*. Prentice–Hall, Englewood Cliffs, NJ, 1942.
43 Bohm, D., *The Special Theory of Relativity*. Benjamin, New York, 1965.
44 Durell, C. V., *Readable Relativity*. Harper & Row, New York, 1960.
45 Schwartz, J., *Relativity in Illustrations*. New York University Press, New York, 1965.
46 Gardner, M., *Relativity for the Millions*. Macmillan, New York, 1962.
47 Eddington, A. S., *Space, Time and Gravitation*. Harper and Row, New York, 1959.
48 Landau, L. D. and Rumer, Yu. B., *What is Relativity?* Fawcett World Library, 1972.
49 Nevanlinna, R., *Raum, Zeit und Relativität*, Birkhäuser, Basel, 1964 [German].

---

[5]See also Dieudonné [8], Yaglom [13], Yaglom-Atanasjan [14], Coxeter [19], Kuiper [25], Strubecker [27], Makarova [28], Yaglom [80], Schwerdfeger [81], and Deaux [82].
[6]See also Coxeter [19].

50 Lanczos, C., *Albert Einstein and the Cosmic World Order*. Interscience, New York, 1965.
51 Bergmann, P. G., *The Riddle of Gravitation*. Scribners, New York, 1968.
52 Kourganoff, V., *Initiation á la Théorie de la Relativité*, Presses Universitaires de France, Paris, 1964 [French].
53 Bondi, H., *Relativity and Common Sense*. Doubleday, Garden City NY, 1964.
54 Melcher, H., *Relativitätstheorie in elementarer Darstellung*. Deutscher Verlag der Wissenschaften, Berlin, 1971 [German].
55 Synge, J. L., *Talking about Relativity*. North-Holland, Amsterdam–London, 1970.
55a Rindler, W., *Essential Relativity*. Springer-Verlag, New York, 1977.

J Section 12 of the Conclusion[7]

56 Klein, F., *Vorlesungen über nicht-Euklidische Geometrie*. Springer, Berlin, 1928 [German].
57 Yaglom, I. M. and Ashkinuze, V. G., *The Ideas and Methods of Affine and Projective Geometry, Part I*. Ucpedgiz, Moscow, 1963 [Russian].
58 Shervatov, V. G., *Hyperbolic Functions*. Heath, Boston, 1963.

K Hyperbolic Geometry[8]

59–61 Bonola, R., *Non-Euclidean Geometry*. Dover, New York, 1955. Single Volume containing three separate works: [59] Bonola, R., *Non-Euclidean Geometry; A Critical and Historical Study of its Development;* [60] Lobachevski, N., *The Theory of Parallels;* [61] Bolyai, J., *The Science of Absolute Space*.
62 Gauss, C. F., *Werke, Vol. VIII*. 1900, pp. 157–268; see also, Stäckel, P., *Gauss als Geometer*, in Gauss C. F., *Werke, Vol. X, Part II*. 1923, pp. 3–46 [German].
63 Kagan, V. F., *Lobatshevski*, Mir, Moscow, 1974 [French].
64 Coxeter, H. S. M., *Non-Euclidean Geometry*. University of Toronto Press, Toronto, 1968.
65 Norden, A. P., *Elementare Einführung in die Lobatschewskische Geometrie*. Deutscher Verlag der Wissenschaften, Berlin, 1958 [German].
66 Sommerville, D. M. Y., *The Elements of Non-Euclidean Geometry*. Dover, New York, 1958.
67 Busemann, H. and Kelly, P. J., *Projective Geometry and Projective Metrics*. Academic, New York, 1953.
68 Liebmann, H., *Nichteuklidische Geometrie*. Berlin, 3d ed., 1923 [German].
69 Baldus, R. and Lobell, F., *Nichteuklidische Geometrie*, 4th ed. Sammlung Göschen, Berlin, 1964 [German].
70 Carslaw, H. S., *The Elements of Non-Euclidean Plane Geometry and Trigonometry*. Longmans, London, 1916.

---

[7]See also Born [40], Norden [65], and Rosenfeld/Yaglom [25].

[8]See also Moise [5a], Yaglom [12], [13], Coxeter [19], Wylie [20], Rosenfeld/Yaglom [25], Pedoe [31], Ewald [32], Klein [56], Klein [73], Yaglom [80], Schwerdtfeger [81], Rosenfeld [78].

71 Karzel, H. and Ellers, E., *Die klassische euklidische und hyperbolische Geometrie*. Grundzüge der Mathematik, Vol. II, Part A (Grundlagen der Geometrie, Elementargeometrie, herausgegeben von H. Behnke, F. Bachmann, K. Fladt), Vandenhoeck und Ruprecht, Göttingen, 1967, pp. 187–213 [German].

72 Kagan, V. F., *Foundations of Geometry, Vol. I*. GITTL, Moscow, 1949 [Russian].

72a Nöbeling, G. *Einführung in die nichteuklidischen Geometrien der Ebene*. Walter de Gruyter, Berlin–New York, 1976 [German].

72b Caratheodory, C. *Theory of Functions of a Complex Variable, Vol. 1*. Chelsea, New York, 1954, Part 1, pp. 1–86.

L Supplements[9]

73 Klein, F., "Über die sogenannte Nicht-Euklidische Geometrie," *Gesammelte Math Abh* **I**: 254–305, 311–343, 344–350, 353–383, 1921 [German].

74 Riemann, B., *Über die Hypothesen, welche der Geometrie zu Grunde liegen*. Springer, Berlin, 1923 [German].

75 Sommerville, D. M. Y., "*Classification of geometries with projective metrics*," Proc Edinburgh Math Soc., **28**: 25–41, 1910–1911.

76 Rosenfeld, B. A., *Die Grundbergriffe der sphärischen Geometrie und Trigonometrie*. EEM, Vol. IV, pp. 527–567 [German].

77 Yaglom, I. M., Rosenfeld, B. A., and Yasinkaya, E. U., *Projective Metrics*. Russian Mathematical Surveys, Vol. 19, No. 5, 1964, pp. 49–107.

78 Rozenfeld, B. A., *Noneuclidean Spaces*. Nauka, Moscow, 1969 [Russian].

79 Klein, F., *Vorlesungen über die Entwicklung der Mathematik im 19. Jahrhundert*. Springer, Berlin, 1926 [German].

80 Yaglom, I. M., *Complex Numbers in Geometry*. Academic, New York, 1968.

81 Schwerdtfeger, H., *Geometry of Complex Numbers*. University of Toronto Press, Toronto, 1962.

82 Deaux, R., *Introduction to the Geometry of Complex Numbers*. Ungar, New York, 1956.

83 Rozenfeld, B. A. and Yaglom, I. M., "On the geometries of the simplest algebras," Mat. Sb. **28**: 205–216, 1951 [Russian].

---

[9]See also Klein [56], Rosenfeld/Jaglom [25].

Answers and Hints to Problems and Exercises

Introduction

2   (a) For opposite motions formulas (6) are replaced by
$$x' = x\cos\alpha + y\sin\alpha + a, \quad y' = \pm x\sin\alpha \pm -y\cos\alpha + b \tag{6'}$$

3   (a) It suffices to prove that it is always possible to map one of the two congruent segments to the other by means of a rotation or a translation (direct motions), or a glide reflection (in particular, a reflection in a line). For more details see [10] or [19].

I   Every direct motion in three-dimensional space is a *screw displacement* (i.e., the product of a rotation and a translation along the axis of the rotation; special cases of screw displacements are rotations and translations). Every opposite motion is a *glide reflection* (i.e., the product of a reflection in a plane and a translation along a line in that plane; a special case of a glide reflection is a reflection in a plane) or a *rotatory reflection* (i.e., the product of a rotation and a reflection in a plane perpendicular to the axis of the rotation; a special case of a rotatory reflection is a half turn about a point).

II   (a) The concepts of triangle, quadrilateral, trapezoid, parallelism, median, area. (b) The theorems about the midline of a triangle and a trapezoid; the theorem about the concurrence of the medians of a triangle (including the assertion that the point of intersection of the medians divides each of them in the ratio 2:1); the theorem which asserts that triangles with equal and *parallel* bases and equal altitudes have equal areas (in our geometry, the condition of equality of the altitudes of the triangles can be stated in terms of congruence of the "strips" containing the triangles, where each strip is formed by a base and a line parallel to that base and passing through the vertex opposite it). The theorem which asserts that triangles polygons with equal areas are equidecomposable does not hold in our geometry; for conditions of equidecomposability of polygons with equal areas, see, for example, [5b].

IV   (b) The theorem about the equidecomposability of polygons with equal areas holds in our geometry (compare, e.g., the book by Boltyanski cited above).

6 (b) "Opposite motions" are defined in our geometry by the formulas

$$x' = \pm x + \lambda y + \mu z + a,$$
$$y' = \pm y + \nu z + b,$$
$$z' = \pm \gamma z + c,$$

where at least one of the equations contains a minus sign, while similitudes are defined by the formulas

$$x' = \alpha x + \lambda y + \mu z + a,$$
$$y' = \beta y + \nu z + b,$$
$$z' = \gamma z + c,$$

where $\alpha\beta\gamma \neq 0$.

VI Since every system of forces $\mathbf{f}_1, \mathbf{f}_2, \cdots, \mathbf{f}_n$ applied at points $A_1, A_2, \cdots, A_n$ in space can be reduced to a single vector $\mathbf{F} = \mathbf{f}_1 + \mathbf{f}_2 + \cdots + \mathbf{f}_n$ [the *principal vector* of *total moment* $\mathbf{U} = OA_1 \times \mathbf{f}_1 + OA_2 \times \mathbf{f}_2 + \cdots + OA_n \times \mathbf{f}_n$ (a sum of cross products) of the system, which is also characterized by its three coordinates (u, v, w)], it follows that the domain of definition of three-dimensional Poinceau geometry is a six-dimensional space with coordinates (x, y, z, u, v, w). Investigate the effect on the vectors $\mathbf{F}$ and $\mathbf{U}$ of a translation of the origin and of a rotation of the axes [thus, for example, translation of the origin by a vector $\mathbf{c}$ has no effect on $\mathbf{F}$ and changes $\mathbf{U}$ to $\mathbf{U}' = \mathbf{U} - (\mathbf{c} \times \mathbf{F})$].

VII If a plane in three-dimensional central Galilean geometry does not pass through the origin, then it can be described by an equation of the form $Ax + By + Cz = 1$. Explain how the coefficients A, B, C (the "coordinates" of the plane) change under a "central Galilean motion" (12'a).

VIII (a) A direct similitude of the plane which is not a motion is a *spiral similarity* (i.e., the product of a dilatation with center O and a rotation about O). An opposite similitude of the plane which is not a motion is a *dilatative reflection* (i.e., the product of a dilatation with center O and a reflection in a line through O). (b) A similitude of three-space which is not a motion is a *spiral similarity* (i.e., the product of a dilatation with center O and a rotation with axis through O). The coefficient of the dilatation is positive or negative according as the similitude is direct or opposite.

IX (a) A proper motion (13a) of the Galilean plane is either a *translation* (14b) or a *shear* (14a) (where the role of the y-axis is played by an appropriately selected special line, the *axis* of the shear), or a *cyclic rotation* [see Sec. 8, Chap. II; in particular, see formulas (12) of Sec. 8]. (b) An "opposite motion of the first kind" (characterized by the choice of $-x$ and $+y$ in (13'a) in Sec. 1; such motions reverse the signs of distances but preserve the signs of angles] is a *glide reflection* (in particular, a reflection) *relative to a special line*. An "opposite motion of the second kind" (characterized by the choice of $+x$ and $-y$ in (13'a); such motions reverse the signs of angles but preserve the signs of distances) is a *glide reflection* (in particular, a reflection) *relative to an (ordinary) line*. An "opposite motion of the third kind" (characterized by the choice of $-x$ and $-y$ in (13'a); such motions reverse the signs of angles and distances) is a *glide reflection relative to a point* (a particular case of which is a half-turn), i.e., a half-turn about a point O followed by a shear whose axis passes through O. (c) In classifying the similitudes (13"a) in Section 2 it is convenient to distinguish "similitudes of the first kind" which alter distances and preserve the magnitude of angles [these are transformations (13"a) with "second similitude coefficient" $\kappa = |\beta/\alpha| = 1$; cf. p. 53 and Fig. 47a]; "similitudes of the second kind" [transformations (13"a) with "first similitude coefficient" $k = |\alpha| = 1$; cf. p. 53 and Fig. 47b); and "general" similitudes (with $k \neq 1$ and $\kappa \neq 1$). Also, depending on the signs of α and β in (13"a), there is one variety of "direct similitudes" and there are three varieties of "opposite similitudes."

Chapter I

I It is natural to define the distance d_{AA_1} between points $A(x,y,z)$ and $A_1(x_1,y_1,z_1)$ by means of the formula $d_{AA_1} = z_1 - z$; if $d_{AA_1} = 0$, then we introduce the "special distance" $\delta_{AA_1} = \sqrt{(x_1-x)^2 + (y_1-y)^2}$ between these points. It is also helpful to bear in mind that a plane of our space not parallel to the z-axis may be regarded as a Euclidean plane, and a plane parallel to that axis may be regarded as a Galilean plane. This suggests reasonable definitions of angles between lines and analogues of circles.

II If $A(x,y,z)$ and $A_1(x_1,y_1,z_1)$ are two points of our geometry, then the (first or "basic") distance between them is defined as $d_{AA_1} = z_1 - z$. If $d_{AA_1} = 0$, then we introduce "the second distance" $\delta_{AA_1} = y_1 - y$. Finally, if $\delta_{AA_1} = 0$, then we introduce "the third distance" $\Delta_{AA_1} = x_1 - x$. There are analogous definitions of angles between lines and angles between planes. These angles may also be of different types. It must be remembered that the lines of our geometry are of three different (noncongruent) kinds: (*i*) "general lines," (*ii*) lines parallel to the yz-plane but not parallel to the z-axis, and (*iii*) lines parallel to the z-axis. All the planes have the structure of Galilean planes (which suggests reasonable definitions of angles between lines, as well as analogues of circles). Nevertheless there are three types of planes. In particular, each coordinate plane is of a different type [provided that the coordinate system is chosen in such a way that the motions are given by (12'), Sec. 2].

III (a) If $A(x,y,u)$ and $A_1(x_1,y_1,u_1)$ are two points in Poinceau space, then it is natural to take as their distance the quantity $d_{AA_1} = \sqrt{(x_1-x)^2 + (y_1-y)^2}$; if $d_{AA_1} = 0$, then we introduce the "special distance" $\delta_{AA_1} = u_1 - u$. (b) See Problem **VI** in the Introduction. In particular, it is clear that the length of a vector **F** and the scalar product of vectors **F** and $\mathbf{F}_1$ corresponding to two systems of forces are invariants of the Poinceau geometry of three-dimensional statics.

VI Note that all "affine" properties of a tetrahedron, such as the theorem which asserts that its *four medians (i.e., segments joining each vertex to the centroid of the opposite face) intersect in a point which divides each median (beginning with the vertex) in the ratio* 3:1, remain valid in three-dimensional semi-Galilean geometry (as well as in three-dimensional Galilean geometry!). Application of the principle of duality (see Problem **VIII**, Chap. I) to these results yields new results. Also, note that the properties of trihedral angles in our geometry are closely related to the properties of triangles in a Galilean plane. To relate the two concepts, it suffices to cut the trihedral angle by means of a plane parallel to the yz-plane and at a distance 1 from the vertex of the angle. Another example: if $OABC$ is a ("general") tetrahedron in our space with volume V (volume is a valid concept in our geometry), and if $OA = a$, $OB = b$, $OC = c$, $\angle(OA, OB) = \alpha$, and $\angle(OC, \text{plane } OAB) = \varphi$ (with distances and angles measured in terms of our geometry), then $V = (1/6)\, abc\alpha\varphi$.

9 The angle bisectors of $\triangle ABC$ form a triangle whose sides are equal to the corresponding sides of $\triangle ABC$ and whose angles are half of the corresponding angles of $\triangle ABC$. This triangle is obtained from $\triangle ABC$ by the compression with axis m, Figure 57b, and coefficient $-\frac{1}{2}$ (cf. Fig. 47b).

11 (b) If the lines u, v, w pass through the vertices A, B, C of $\triangle ABC$, then u, v, w are concurrent if and only if

$$\frac{\delta_{aw}}{\delta_{bw}} \cdot \frac{\delta_{bu}}{\delta_{cu}} \cdot \frac{\delta_{cv}}{\delta_{av}} = 1$$

(illustrate with diagram).

12 If BD is a special line, then *the diagonal AC of the cotrapezoid $ABCD$, the line p joining the points of intersection of its opposite sides AB and DC, AD and BC, and the bisectors m and n of the angles B and D are concurrent.*

18 (a) The set of points M is a cycle (see Chap. II) and the set of lines m is the set of tangents to a cycle.

IX Compare, e.g., Sec. 4 of [12], or [14].

X In axiomatic terms, three-dimensional semi-Galilean geometry can be described as a three-dimensional point-vector space (cf. Supplement B) with three "scalar products." Specifically, we have the "basic" scalar product **ab**, the "special" scalar product $(\mathbf{pq})_1$ defined on the set of vectors $\mathbf{p}, \mathbf{q}, \ldots$ such that $|\mathbf{p}| = |\mathbf{q}| = \cdots = 0$, and the "second special" scalar product $(\mathbf{uv})_2$ defined on the set of (special) vectors $\mathbf{u}, \mathbf{v}, \ldots$ for which $(\mathbf{u}^2)_1 = (\mathbf{v}^2)_1 = \cdots = 0$ (cf. hint to Problem II, Chap. II).

Chapter II

1 The relation $\mathbf{r}' = d\mathbf{r}/ds = \mathbf{t}$, where $\mathbf{t}^2 = 1$, $|\mathbf{t}| = 1$ (cf. p. 251, Supplement B), implies that if $\mathbf{r} = (x, y)$, then $s = x$. Hence, in the present case, a curve $\mathbf{r} = \mathbf{r}(s)$ can be given as $\mathbf{r} = (x, f(x))$ or $y = f(x)$. The "moving bihedron" of the curve C at a point M of C consists of vectors $\mathbf{t}$ and $\mathbf{n}$, where $\mathbf{n} \perp \mathbf{t}$ in the sense of Galilean geometry, i.e., $\mathbf{n}$ is the special vector $(0,1)$ of special length 1 (cf. pp. 251 and 252, Supplement B). Further, $\mathbf{t}' = \rho \mathbf{n}$, where it is natural to call $\rho = \rho(x)$ the *curvature* of the curve. (Note that $\mathbf{n}' = \mathbf{0}$.) It is again clear that $\rho = f''(x)$. This immediately implies the theorem on the natural equation of a curve and the fact that all curves of constant curvature are *cycles*.

2 Here the role of the "natural parameter" (or "arc length") is played by the variable angle formed by the line m and the fixed "angle origin" (the fixed line o), and the role of the curvature ρ is played by the radius of curvature $r = 1/\rho$.

3 In this case, these concepts are not of great significance. Thus, for example, the involutes of a curve are the special lines.

I In Galilean three-dimensional space, it is natural to define the scalar product **ab** of vectors $\mathbf{a}(x,y,z)$ and $\mathbf{b}(x_1,y_1,z_1)$ by the formula $\mathbf{ab} = zz_1$ [assuming that the coordinate system is chosen so that the motions of the geometry are given by the formulas (12′) of Sec. 2]; here the condition $\mathbf{ap} = 0$ or $\mathbf{a} \perp \mathbf{p}$ (in the sense of Galilean geometry) implies, for $\mathbf{a}^2 \neq 0$, that $\mathbf{p} = \mathbf{p}(x,y,0)$ is a special vector [i.e., $|\mathbf{p}| = 0$, where the length $|\mathbf{a}|$ of a vector is defined (up to sign) by the equality $|\mathbf{a}|^2 = \mathbf{a}^2$]. In the set of special vectors $\mathbf{p}, \mathbf{q}, \ldots$ there is defined a "special" scalar product $(\mathbf{pq})_1$. Specifically, if $\mathbf{p} = \mathbf{p}(x,y,0)$ and $\mathbf{q} = \mathbf{q}(x_1,y_1,0)$, then $(\mathbf{pq})_1 = xx_1 + yy_1$. The special scalar product enables us to define in the set of special vectors the length $|\mathbf{p}|_1$ of a vector $\mathbf{p}$ [by means of the equality $(|\mathbf{p}|_1)^2 = (\mathbf{p}^2)_1$] and the orthogonality of vectors $[\mathbf{p} \perp_1 \mathbf{q}$ means that $(\mathbf{pq})_1 = 0]$.

Now let $\mathbf{r} = \mathbf{r}(u) = (x(u), y(u), z(u))$ be the radius vector of a curve in our space. Then the condition $\mathbf{r}' = \mathbf{t}$, $|\mathbf{t}| = 1$, is equivalent to the condition $u = z$. Thus the natural equation of our curve is $\mathbf{r} = \mathbf{r}(f(z), g(z), z)$. It follows that it is reasonable to give a curve in Galilean space by means of a pair of equations $x = f(z)$, $y = g(z)$. The moving trihedron at a point M of the curve consists of three vectors $\mathbf{t}, \mathbf{n}, \mathbf{b}$, where $\mathbf{t}$ is the unit tangent vector, $\mathbf{n} \perp \mathbf{t}$ is the unit special vector in the osculating plane (i.e., $|\mathbf{n}| = 0$, $|\mathbf{n}|_1 = 1$), and $\mathbf{b}$ is the special unit vector perpendicular to $\mathbf{n}$ [i.e., $(\mathbf{nb})_1 = 0$]. The Frenet formulas for the curve take the form $\mathbf{t}' = \rho\mathbf{n}$, $\mathbf{n}' = \tau\mathbf{b}$, $\mathbf{b}' = -\tau\mathbf{n}$, where the scalar functions $\rho = \rho(z)$, $\tau = \tau(z)$ play the role of *curvature* and *torsion*, respectively. These formulas enable us to compute the curvature ρ and torsion τ; they imply the theorem on the natural equations of the curve [the functions $\rho = \rho(z)$ and $\tau = \tau(z)$ determine the curve up to position in space]; and they give the structure of curves of constant curvature and torsion.

II In semi-Galilean space, it is natural to define the scalar product **ab** of vectors $\mathbf{a}(x,y,z)$ and $\mathbf{b}(x_1,y_1,z_1)$ by the formula $\mathbf{ab} = zz_1$ [assuming that the coordinate system is chosen so that the motions of the geometry are given by the formulas (12″) of Sec. 2]. Now we can define the length $|\mathbf{a}|$ ($=z$) of a vector $\mathbf{a}$ (so that $|\mathbf{a}|^2 = \mathbf{a}^2$), as well as the relation of orthogonality $\mathbf{a} \perp \mathbf{p}$, of vectors: for $|\mathbf{a}| \neq 0$, $\mathbf{a} \perp \mathbf{p}$ means that $\mathbf{p}(x,y,0)$ is a special vector ($|\mathbf{p}| = 0$). In the set of special vectors $\mathbf{p}(x,y,0), \mathbf{q}(x_1,y_1,0), \ldots$ we can introduce a special scalar product $(\mathbf{pq})_1 = yy_1$. This scalar product enables us to define a ("special") length $|\mathbf{p}|_1 = y$, as well as a relation of orthogonality $\mathbf{p} \perp_1 \mathbf{u}$ which means that $(\mathbf{pu})_1 = 0$. If $\mathbf{p}_1 \neq 0$, then $(\mathbf{pu})_1 = 0$ means that $\mathbf{u} = \mathbf{u}(x,0,0)$. Finally, in the set of vectors $\mathbf{u}(x,0,0), \mathbf{v}(x_1,0,0), \ldots$ we define a new scalar product by means of the formula $(\mathbf{uv})_2 = xx_1$ and a new length $|\mathbf{u}|_2 = x$.

If $\mathbf{r} = \mathbf{r}(u) = (x(u), y(u), z(u))$ is a curve in our space, then $\mathbf{r}' = \mathbf{t}$, $|\mathbf{t}| = 1$, implies that $u = z$, and the natural equation of the curve is $\mathbf{r} = \mathbf{r}(f(z), g(z), z)$ or $x = f(z)$, $y = g(z)$. The moving trihedron at a point M of the curve is formed by vectors $\mathbf{t}, \mathbf{n}, \mathbf{b}$, where $\mathbf{t}$ is a unit tangent vector; $\mathbf{n}$ is a vector in the osculating plane with $|\mathbf{n}| = 0$, $|\mathbf{n}|_1 = 1$; and $|\mathbf{b}| = |\mathbf{b}|_1 = 0$, $|\mathbf{b}|_2 = 1$. The Frenet formulas take the form $\mathbf{t}' = \rho\mathbf{n}$, $\mathbf{n}' = \tau\mathbf{b}$, $\mathbf{b}' = 0$, where the scalar functions $\rho = \rho(z)$, $\tau = \tau(z)$ play the role of *curvature* and *torsion*, respectively. These formulas permit us to compute the curvature and torsion [namely, $\rho(z) =$

$g''(z)$, $\tau(z)=(f''(z)/g''(z))']$; they imply the theorem on the natural equations; and they give the structure of curves of constant curvature and torsion [namely, $\rho=a$ and $\tau=b$ imply $y=(a/2)z^2+cz+d$, $x=(ab/6)z^3+ez^2+fz+g$], as well as the structure of curves of constant curvature and arbitrary torsion.

IV–V If the coordinate systems are chosen so that the motions in the two spaces are given by formulas (12′) and (12″) (of Sec. 2), respectively, then the surfaces in question are given by equations of the form

$$az^2+2bx+2cy+2dz+e=0$$

(for $b=c=0$ they reduce to spheres in the sense of loci of points at a given distance from a given point), and so are, from an affine or Euclidean point of view, parabolic cylinders. These "cyclic surfaces" are the only ones which to some extent share with Euclidean spheres the property of being able to "glide along themselves" (cf. Sec. 8, in particular, Problems **VII** and **VIII**).

4 Let AB and CD be two parallel chords of a cycle Z, and let PA and PB, QC and QD be tangents to Z at the endpoints of the chords. To prove that the midpoints of the chords are on the same special line (which implies the desired result), use the equality

$$0=d_{CQ}-d_{QD}=(d_{CA}+d_{AP}+d_{PQ})-(d_{QP}+d_{PB}+d_{BD})=2d_{PQ}.$$

Justify the latter.

8 The bisectors of the angles formed by the pairs of tangents to a given cycle Z from the points of a given line m are tangent to a cycle Z_1.

11 (a) Let AB be tangent to z at F and let t be tangent to Z at F'. If Q is a point of FF' such that $QF=4QF'$, then the dilatation with center Q and coefficient 4 maps Z to z. If C_1 is the midpoint of AB, then the fact that $F'C_1$ is a special line implies that QM is also a special line (use the fact that $FQ=(4/3)FF'$, $FF'=FC_1$, $FC_1=C_1C$, and $C_1M=\frac{1}{3}C_1C$). (b) The dilatation γ_2 is the product of the dilatations γ and γ_1. (c) To show that L is a point of the cycle Z_1 (a dilatation with center at a point of a cycle maps it to a cycle tangent to the original cycle at the point in question), it suffices to show that, say, $\delta_{LC_1,LR}=\delta_{B_1C_1,B_1R}$, where B_1 is the midpoint of the side AB and R is the projection of C to AB (i.e., R is the point of AB such that the line CR is special). Now $\delta_{B_1C_1,B_1R}=B-A$. It remains to compute $\delta_{LC_1,LR}$. Note that if $H=\delta_{F'C_1}$ and $h=\delta_{CR}$ are altitudes of $\triangle ABF'$ and $\triangle ABC$, and N is the projection of Q (and M) to AB, then $QN=(4/3)H$ and $MN=\frac{1}{3}h$. It follows that $QM=(4/3)H-\frac{1}{3}H$ and $NL=ML-MN=(4/9)(H-h)$. But then, with $H=(c/2)\cdot(C/2)$ and $h=bA=aB$ (here it suffices to remember the definition of a Galilean angle), we conclude that $NL=(b-a)(B-A)/9$. Now it follows readily that $\delta_{C_1N,C_1L}=\frac{2}{3}(B-A)$ and $\delta_{RN,RL}=\frac{1}{3}(B-A)$, so that $\delta_{LC_1,LR}=B-A$ $(=\delta_{B_1C_1,R_1R})$.

13 Apply an inversion (of the first kind) with center A. *Answer*: A circle (cycle).

16 The theorem (of Galilean geometry) in Exercise **14** goes over into the following theorem: *Let a_1 and b_1, a_2 and b_2, a_3 and b_3, a_4 and b_4 be common tangents of the pairs of cycles Z_1 and Z_2, Z_2 and Z_3, Z_3 and Z_4, Z_4 and Z_1. If a_1,a_2,a_3,a_4 are tangent to a cycle, then b_1,b_2,b_3,b_4 are also tangent to a cycle.*

17 **(b)** Compare the proofs of the fundamental theorem of the theory of circular transformations of the Euclidean plane presented, for example, in the books [13] and [19]; use the fact that *all (point) transformations of the Galilean plane which map lines to lines and cycles to cycles are similitudes* (cf. Exercise **4** in the Introduction).

XIII **(b)** The "general" equation of a cycle (which includes lines and circles) is of the form $ax^2 + 2bx + 2cy + d = 0$. The equation of the "dual cycle" (the set of tangents of a cycle; cf. pp. 100–103) in "rectangular line coordinates" (ξ, η) (see pp. 73–74 and, in particular, footnote 23 in Sec. 6 and Fig. 65) can be written in the form $A\xi^2 + 2B\xi + 2C\eta + D = 0$ (this equation includes points, pencils of parallel lines, and "dual circles"—pairs of pencils of parallel lines). If the coordinate systems $\{x,y\}$ and $\{\xi,\eta\}$ are adjusted in a natural way, then the coefficients of the equations of a cycle Z and of its set of tangents are related as follows:

$$A = c \quad \text{and} \quad a = C, \quad B = b, \quad D = \frac{b^2 - ad}{c} \quad \text{and} \quad d = \frac{B^2 - AD}{C}.$$

Consequently, the homogeneous coordinates $u_0 : u_1 : u_2 : u_3 : u_4 = a : b : c : d : (b^2 - ad)/c = C : B : A : [(B^2 - AD)/C] : D$, where $u_1^2 - u_0 u_3 - u_2 u_4 = 0$, describe the most general cycles, including points and lines. In these coordinates, the condition of contact of cycles takes a particularly simple form, which makes it very easy to give an analytical description of a transformation of the set of general cycles which preserves contact of cycles. These constructs can be interpreted geometrically, with the concept of *the power of a cycle with respect to a cycle* playing an important role (cf. pp. 238–239 of [80]).

XIV–XV Let M be a point and Z a "cyclic surface" in our space (cf. Problems **IV–V** and the hints for their solution). If A and B are the points at which a line l passing through M intersects Z, then the product $d_{MA} \cdot d_{MB}$ (d is the distance between points in each geometry; cf. Problems **I–II** of Chap. I and the corresponding hints) is independent of the choice of l. It is natural to call this product *the power of M with respect to Z*. The set of points which have the same power with respect to two (three; four) cyclic surfaces Z_1 and Z_2 (Z_1, Z_2 and Z_3; Z_1, Z_2, Z_3 and Z_4) is a plane (line; point) called *the radical plane* of Z_1 and Z_2 (*the radical axis* of Z_1, Z_2 and Z_3; *the radical center* of Z_1, Z_2, Z_3, and Z_4). These results suggest the way in which one could develop in our geometries the study of pencils and bundles of circles, and define an *inversion of the first kind* (*reflection in a sphere*) which maps cyclic surfaces to cyclic surfaces. The concept of a *reflection in a cyclic surface* (an *inversion of the second kind*) is defined differently in our two geometries (since it involves the special distances in our geometries; cf. the hint relating to Problems **I–II** of Chap. I).

XVI In the Euclidean plane, in addition to degenerate quadrics (i.e., quadrics that are "empty," consist of single points, or of one or two lines), there are just three types of nondegenerate quadrics: an ellipse $x^2/a^2 + y^2/b^2 - 1 = 0$, a hyperbola $x^2/a^2 - y^2/b^2 - 1 = 0$, and a parabola $y - ax^2 = 0$. In the Galilean plane, in addition to 10 types of degenerate quadrics, there are six types of nondegenerate quadrics: an ellipse $x^2/a^2 + y^2/b^2 - 1 = 0$, hyperbolas of the first and second kind $x^2/a^2 - y^2/b^2 \pm 1 = 0$, a special hyperbola $xy = k$, a parabola $x - by^2 = 0$, and a cycle $y - ax^2 = 0$. Give a geometric description of these quadrics. Find the Galilean invariants of a general quadratic equation in two variables as well as the algebraic characteristics of the equations associated with each of the 16 types of curves.

Conclusion

1   48°

2   Its volume contracts by the same factor by which a rod, moving in the direction of the solid, contracts. For proof, think of the solid as a collection of small cubes one of whose edges coincides with the direction of the motion.

6   See Exercise 5.

4   59 years.

13   (b) Try to carry over to Minkowskian geometry the proofs of the corresponding theorem of Euclidean and Galilean geometry, presented on pp. 110–116 and 138–141.

Index of Names

BOLYAI, Johann, 1802–1860, vi

CAYLEY, Arthur, 1821–1891, vii
CLIFFORD, William Kingdon, 1845–1879, 265

DESCARTES, René, 1596–1650, 4

EINSTEIN, Albert, 1879–1955, 167
EUCLID of Alexandria, about 300 BC, 24

FERMAT, Pierre, 1601–1665, 4
FITZGERALD, George Francis, 1851–1901, 166

GALILEI, Galileo, 1564–1642, x, 17
GAUSS, Carl Friedrich, 1777–1855, vi
GERGONNE, Joseph Diez, 1771–1859, 67

HILBERT, David, 1862–1943, 243

KLEIN, Felix, 1849–1925, vii

LOBACHEVSKY, Nicholas Ivanovich, 1793–1856, vi
LORENTZ, Hendrik Antoon, 1853–1928, 162

MAXWELL, James Clerk, 1831–1879, 173
MICHELSON, Albert Abraham, 1852–1931, 161
MINKOWSKI, Hermann, 1864–1909, 175
MÖBIUS, August Ferdinand, 1790–1868, 67

NEWTON, Isaac, 1642–1727, 158

POINCARÉ, Henri, 1857–1912, 173
POINCEAU, Louis, 1777–1859, 30
PONCELET, Jean Victor, 1787–1867, 67

RIEMANN, Bernard, 1826–1866, 13

STUDY, Eduard, 1862–1930, 265

WEYL, Hermann, 1885–1955, vii

Index of Subjects

Acceleration 17
Arc length
  Galilean 81
  Minkowskian 183
Angle
  between two curves 128
  elliptic measure of 217
  Galilean 41
  hyperbolic measure of 217
  Minkowskian 182
  parabolic measure of 217

Cayley–Klein geometries
  list of 218
  plane
    analytic models of 258
    axiomatic characterization of 242
    models of, in 3-spaces 229
  three-dimensional 240
Circle
  alternate definition of Euclidean 77
  Galilean 39
  Minkowskian 181
  six-point (nine-point) 109, 194
Congruence
  criteria for Galilean triangles 52–53
  of figures 2
  of Minkowskian segments 180

Coparallelogram 58
Cotrapezoid 58
Cross product
  of doublets 75
  of vectors 252
Cross ratio 104
Curvature
  of circle 82
  of cycle 83
  of plane Euclidean curve 85
  of plane Galilean curve 86
  radius of, of plane Euclidean curve 85
  radius of, of plane Galilean curve 86
Cycle 77
  circumcycle 50
  incycle 50
  radius of 81

Distance
  elliptic measure of 215
  Euclidean, between parallel lines 38
  Euclidean, between points 2, 37, 272
  Galilean
    between parallel lines 42
    between points 38, 272
    from point to line 42
  hyperbolic measure of 215

Minkowskian
  between parallel lines 186
  between points 180, 272
  from point to line 186
  parabolic measure of 214
  special Galilean distance, points 39
Doublet 71
Duality, principle of (in Galilean geometry) 56

Erlanger Programm, vii

Fitzgerald contraction 166

Geometric property 3, 6
Geometry
  central Euclidean 14 (Problem III)
  Euclidean 24
  four-dimensional Galilean 26
  Klein's definition of 3
  Minkowskian 174
  of parallelism 15
  of plane statics (Poinceau geometry) 28
  of translations 14
  plane Galilean 24
  three-dimensional central Galilean 31 (Problem VII)
  see also Cayley–Klein geometries

Inertial
  force 21
  reference frame 21
Inversion
  in a Euclidean circle 124
  in a Galilean circle 128
  in a Galilean cycle 130
  in a Minkowskian circle 196
Inversive
  Euclidean plane 142
  Galilean plane 149
  Minkowskian plane 198
  see also Stereographic projection

Line
  Minkowskian
    of the first kind 179
    of the second kind 179
  special Galilean 34
  special (isotropic) Minkowskian 177, 179

Motion
  in geometry 3
  in mechanics 15
  of Euclidean plane 8
  of Galilean plane 25
  of Galilean space 20
  of Minkowskian plane 190
  plane-parallel 16
  rectilinear 16
  see also Rotation; Shear; Similitude; Translation; Transformation

Numbers
  complex 258
  double 265
  dual 265

Parallel
  curves 97
  points 56
Perpendicular
  Galilean, to a line 43
  Minkowskian, to a line 185
Power of a point
  with respect to a Euclidean circle 117
  with respect to a Galilean circle 119
  with respect to a Galilean cycle 120
  with respect to a Minkowskian circle 195

Relativity
  Einsteinian principle of 162
  Galilean principle of 18

Index of Subjects

Rotation
  cyclic 93
  Euclidean 12
  hyperbolic 177

Scalar product 245
Similitude
  Galilean, 30 (Exercise 4)
  of the first kind 53
  of the second kind 53
Shear 25
Stereographic projection
  of inversive Euclidean plane to sphere 142
  of inversive Galilean plane to cylinder 149
  of inversive Minkowskian plane to hyperboloid 198

Transformation
  circular
    of inversive Euclidean plane 147
    of inversive Minkowskian plane 198
  cyclic, of inversive Galilean plane 152
  Lorentz 162, 167
  see also Motion
Translation 12
Trigonometry
  Galilean 47
  Minkowskian 187

Velocity 17
  relative 22

Encounter with Mathematics

By **Lars Gårding**

1977. ix, 270p. 82 illus. cloth

The purpose of this text is to provide an historical, scientific, and cultural frame for the basic parts of mathematics encountered in college.

Nine chapters cover the topics of Number Theory, Geometry and Linear Algebra, Limiting Processes of Analysis and Topology, Differentiation and Integration, Series and Probability, and Applications. Each of these chapters moves from an historical introduction to a basic factual account, and finally into a presentation of the present state of the subject, including, wherever possible, most recent research. Most end with passages from historical mathematical papers, as well as references to additional literature. Three remaining chapters deal with models and reality, the sociology, psychology, and teaching of mathematics, and the mathematics of the seventeenth century, providing a fuller historical background to infinitesimal calculus. Intended for beginning undergraduates, the text assumes background in high school or some college mathematics.

Essential Relativity

Special, General, and Cosmological

Second Edition

By **W. Rindler**

1977. xv, 284p. 44 illus. cloth
(Texts and Monographs in Physics)

Essential Relativity has been completely revised and updated, largely rewritten, and expanded by more than a third for this new edition. It contains new sections on Kruskal space, gravitational waves and the linear approximation, new appendices on curvature components and Maxwell's theory, as well as many new problems.

The New Cosmos

Second Revised and Enlarged Edition

By **Albrecht Unsöld**

Translated from the German by R.C. Smith

1977. xii, 451p. 166 illus. paper

(Heidelberg Science Library)

This revised and enlarged second edition of **The New Cosmos** provides a comprehensive and straightforward introduction to present day astronomy and astrophysics. The text thoroughly covers the subject, from apparent motions on the celestial sphere to studies of the solar system, stellar atmospheres and evolution, radio astronomy, high energy astrophysics and cosmology, and concluding with considerations regarding the origin of life on earth.

From Reviews of the First Edition

"With the excellent photographs and diagrams... the book can be read with more excitement and pleasure than one would think possible for such a concise and comprehensive handbook. This is due to the fact that the author not only knows which developments are really significant, but never fumbles for words to make his meaning clear." *The Ohio Journal of Science*

"...penetrates particularly in the later chapters to the basic, more philosophical questions which are the fundamental stimulus for the rest of the book." *Physics in Canada*